Tourism at the Grassroots

In two regions where tourism is of considerable economic importance, eastern Asia and the Pacific, there have been remarkably few studies of the impacts of tourism in rural areas. Moreover, the shift towards ecotourism, touted as a more environmentally benign form of tourism than mass tourism, has extended the reach of tourism into more remote and fragile environments. This shift has drawn more local people in rural and remote areas into a partly tourism economy, involving them as participants in the tourist industry. Yet little is known about who have been the beneficiaries of these developments.

This new collection focuses on both the interactions between tourists and villagers and the impacts of tourism at the local level, considering economic, social, cultural and environmental changes. It traces changes in structures of vulnerability as tourism becomes more prominent, studies the role of tourism in community development (or localised tension) and examines issues of governance, the role of tour operators as intermediaries, cultural change and other local impacts. In short, it will examine the changing role of tourism in local development (or its absence).

It includes case studies drawn from a broad geographical area across eastern Asia and the island Pacific. This book will be useful to those researching and studying Tourism, Geography and Development Studies.

John Connell is Professor of Human Geography at the University of Sydney, Australia. John's publications include (with Chris Gibson) *Music and Tourism: On the Road Again* (Channel View, 2005), papers on tourism in Bali, the Caribbean and elsewhere, and books on Papua New Guinea, New Caledonia and *The Last Colonies*.

Barbara Rugendyke is an Associate Professor in Geography at the University of New England, Australia. Barbara's publications have focused on the impacts of nature-based tourism at the local level in Vietnam, development assistance in the Asia-Pacific region, community development planning and the role of development NGOs, including *NGOs as Advocates for Development in a Globalising World* (Routledge, 2007).

Tourism at the Grassroots

Villagers and visitors in the Asia-Pacific

Edited by John Connell and Barbara Rugendyke

LONDON AND NEW YORK

First published 2008
by Routledge
2 Park Square, Milton Park, Abingdon, Oxfordshire OX14 4RN

Simultaneously published in the USA and Canada
by Routledge
711 Third Avenue, New York, NY 10017

First issued in paperback 2014

Routledge is an imprint of the Taylor & Francis Group, an informa business

Typeset in Times New Roman by Prepress Projects Ltd, Perth, UK

British Library Cataloguing in Publication Data
A catalogue record for this book is available from the British Library

Library of Congress Cataloging in Publication Data
Tourism at the grassroots: villagers and visitors in the Asia Pacific/edited by John Connell and Barbara Rugendyke.
        p. cm.
    Include bibliographical references and index.
    1. Tourism—Social aspects—Pacific Area I. Connell, John, 1946– II. Rugendyke, Barbara.
    G155.P25T64 2007
    306.4'819095—dc22
                    2007032657

ISBN 13: 978-0-415-40555-3 (hbk)
ISBN 13: 978-1-138-01051-2 (pbk)

Contents

List of figures	xii
List of maps	xiii
List of tables	xiv
List of contributors	xv
Preface and acknowledgements	xvii
Map	xix

1 Tourism and local people in the Asia-Pacific region 1
JOHN CONNELL AND BARBARA RUGENDYKE

The tourist area life cycle 3
On the edge of the tourist industry 4
Tourism as transformation 8
The complexities of culture 25
New directions 31

**2 Another (unintended) legacy of Captain Cook?: the evolution
of Rapanui (Easter Island) tourism** 41
GRANT MCCALL

*Contact and colonial history as tourism narrative: visitors
   and souvenirs 42*
Rapanui curbed and curtailed: the island of innocent inmates 44
Turning disadvantage to advantage 46
Towards the modern era 49
Autonomous futures in modernity 52

**3 Moderate expectations and benign exploitation: tourism on
the Sepik River, Papua New Guinea** 58
ERIC KLINE SILVERMAN

Introduction: the obvious 58

11 **Communities on edge: conflicts over community tourism
     in Thailand**   198
     TIM WONG

     The business of parks and the role of local people 198
     *Park ideology in Thailand: territorialisation and excluding
        local people 199*
     *A history of Khao Yai – tourism and growing state control over
        local affairs 202*
     Livelihoods, transition and tourism at the local level 204
     Analysing the politics at the local level 208
     New forms of resource conflicts 210
     Conclusion 211

12 **Community-based ecotourism in Thailand**   214
     ANUCHA LEKSAKUNDILOK AND PHILIP HIRSCH

     Ecotourism as an alternative and the move toward CBET 215
     *From mainstream to community-based ecotourism
        in Thailand 216*
     Four case studies of community-based ecotourism 220
     *Environment, society and economy in
        community-based ecotourism 227*
     Conclusion: inventing tradition and claiming resources 231

13 **Ecotourism and indigenous communities: the Lower
     Kinabatangan experience**   236
     RAJANATHAN RAJARATNAM, CAROLINE PANG AND
     ISABELLE LACKMAN-ANCRENAZ

     The sanctuary and its environs 239
     The people of the Kinabatangan 240
     Nature tourism in the Lower Kinabatangan 240
     The sanctuary: friend or foe 242
     *Red Ape Encounters and Adventures: a community-based
        orang-utan ecotourism project 243*
     Conclusion 253

14 **Adventures, picnics and nature conservation: ecotourism
     in Malaysian national parks**   256
     NORMAN BACKHAUS

     Tourism in protected areas 256
     National parks and impacts of tourism in Malaysia 258

Forms of tourism in Malaysian national parks 260
A World Heritage site: Gunung Mulu National Park 263
Conclusion 269

15 Marginal people and marginal places? 274
BARBARA RUGENDYKE AND JOHN CONNELL

With the tourists 276
Local participation 277
Onwards and outwards 278

Index 283

Figures

2.1	Lithograph of the La Pérouse visit of 1786	43
4.1	Advertising *kastom*	81
4.2	Yakel village	84
4.3	Yeniar village	85
5.1	Mass tourists dressed up in 'Tibetan' clothing	99
5.2	Official minority village special entrance	107
6.1	Isles of Smiles	115
6.2	Promotional image of Fijian man in grass skirt	121
6.3	Promotional poster featuring Fijian waiter and tourists	122
7.1	The coastal landscape dotted with beach *fale*	142
10.1	Tourists' impact on the environment at Cuc Phuong: soil compaction and litter at the 'Big Tree'	183
10.2	Villagers' observations of use of forest resources, Cuc Phuong National Park	191
10.3	Ban Khanh	193
12.1	The transformation from tourism in the community to CBET	219
13.1	The orang-utan	243
13.2	Village benefit-sharing, community-based ecotourism project, Kampung Sukau, Sabah, Malaysia	248
13.3	*Warisan* cultural troupe	249
13.4	Reforestation in action	251

Maps

	The Asia-Pacific region	xix
4.1	Vanuatu	78
7.1	*Fale* in Samoa	136
9.1	Lombok regencies and tourism zones	166
10.1	Study villages at Cuc Phuong National Park	188
12.1	Case study villages, Thailand	221
13.1	Lower Kinabatangan Wildlife Sanctuary, Sabah, Malaysian Borneo	238
14.1	Gunung Mulu National Park: location, visitors and climate	264

Tables

7.1	Visitor numbers for Heavenly Beach *Fale*, 2002	140
9.1	Number of tourists to NTB for selected years from 1990 to 2003	167
9.2	Origin of hotel staff (contract and casual) from selected hotels at Senggigi and Kute, 1999 and 2000, and at Senggigi, 2005	170
9.3	Percentage of positions held by Sasaks in selected hotels, 2000 and 2005	171
10.1	Occupation by age of male villagers living near Cuc Phuong National Park	189
10.2	Occupation by age of female villagers living near Cuc Phuong National Park	190
12.1	Tourism resources in the case study areas, Thailand	225
12.2	Tourism services and activities in the case study areas, Thailand	226
12.3	Ratio of income from tourism to total income of some involved families, Thailand	232
14.1	Tourism potential of Mulu	266
14.2	Visits to and evaluation of Mulu's major attractions	267

Contributors

Norman Backhaus, Department of Geography, University of Zurich, Switzerland.

John Connell, School of Geosciences, University of Sydney, Sydney, Australia.

Kelly Dombroski, Department of Human Geography, Research School of Asian and Pacific Studies, Australian National University, Canberra, Australia.

Fleur Fallon, University of Nottingham Ningbo, Ningbo, Zhejiang, P.R. China.

Philip Hirsch, School of Geosciences, University of Sydney, Sydney, Australia.

Yoko Kanemasu, School of Sociology, University of the South Pacific, Suva, Fiji.

Isabelle Lackman-Ancrenaz, Kinabatangan Orang-utan Conservation Project, Sandakan, Sabah, Malaysia.

Anucha Leksakundilok, Thailand Institute for Science and Technology Research, Bangkok, Thailand.

Grant McCall, School of Sociology, University of New South Wales, Sydney, Australia.

Nguyen Thi Son, Faculty of Geography, Hanoi University of Education, Hanoi, Vietnam.

Caroline Pang, Environmental Marketing Consultant, Singapore.

Rajanathan Rajaratnam, Geography and Planning, University of New England, Armidale, Australia.

Prue Robinson, Mediterranean Association to Save the Sea Turtles (MEDASSET), Athens, Greece.

Barbara Rugendyke, Geography and Planning, University of New England, Armidale, Australia.

Regina Scheyvens, School of People, Environment and Planning, Massey University, Palmerston North, New Zealand.

Eric Kline Silverman, School of American Studies and Human Development, Wheelock College, Boston, Massachusetts.

Susan Tarplee, School of Geosciences, University of Sydney, Sydney, Australia.

Tim Wong, Safeguarding Biodiversity for Poverty Reduction Project, Ban Wat Chan, Vientiane, Laos.

Preface and acknowledgements

Books begin in various ways. This one stemmed from the experiences of its editors. Two recent experiences in Fiji fascinated one. The first involved a three-day visit with 24 students to the small village on the Coral Coast, which involved payments of money to 'the village' for lodging and food. Near the end of that three-day period a chance conversation with a villager from the distant end of the village revealed that he had no idea who the visitors were, why they were there and what was involved. He was certainly gaining no monetary benefits from their presence. The outsiders had in fact been hosted by only one of the five landowning groups (*mataqali*) in the village and they were the only financial beneficiaries of their presence. A second experience involved an ecotourism tour of about ten people to a village on the island of Ovalau. Here the only financial beneficiaries were the extended family of the guide, though the visitors had travelled over the land of other *mataqali* and there was some resentment of their presence in the village. Although in both cases villagers were benefiting financially from these particular examples of small-scale tourism they were certainly not benefiting equally. Tourism was contributing to inequality and potentially enhancing divisions within the villages.

Wandering onto a beach near Hua Hin in Thailand, the other editor was confronted by guards. A subsequent conversation with a beach-side massage worker revealed their presence was to ensure the beach was frequented only by paying guests from an adjacent resort complex; overseas tourists, who could afford to holiday in luxury, were unwittingly responsible for the exclusion of local people from local beaches. Gracious, silk-clad, smiling Thais waited on those same guests. A chance turn on an evening walk through grounds alongside the resort revealed the accommodation provided for the young Thai workers – poorly ventilated corrugated iron sheds. The contrast between their living conditions and those in which the tourists luxuriated could not have been greater. Moreover, while tourists wallowed in large swimming pools, indulged in lengthy showers to remove beach sand, enjoyed extensive watered gardens and played on water-hungry golf courses, nearby town residents experienced water shortages. Many local people no doubt benefited economically from the tourists, but the costs were obvious.

These and various other experiences emphasised that local communities were diverse and differentiated, and that the impacts of tourism were more complex than many studies suggested. Reflecting on such experiences revealed relatively few detailed attempts to examine tourism at the local level, but also that what is true in parts of Fiji and Thailand did not necessarily parallel what occurred elsewhere. This, then, was the rationale for this volume – to present substantial empirical research about the complex impacts of visitors on villagers, and about the interactions between villagers and visitors, in a range of locations in the Asia-Pacific region. The authors all share a commitment to understanding the multiple impacts of tourism, which might assist in attempts to ensure that those who bear the costs of tourism are not excluded from reaping its rewards.

We are grateful to various individuals for permission to use the illustrative material included in this volume. Most photographs were taken by the authors of individual chapters and are credited to them. Several copyright holders kindly provided images and granted permission for their use: the Linda Hall Library for the use of the lithograph in Figure 2.1; Jamil Sinyor for Figures 13.2 and 13.4 and Suhaile Kahar for Figure 13.5. Figure 7.1 was prepared by Tim Nolan, and the general map and Figures 4.1 and 10.2 by Michael Roach. We are grateful to them all. Figures 6.1, 6.2 and 6.3 are reproduced from materials published by the Fijian Visitors Bureau. Every effort was made to contact copyright holders and obtain permission to reproduce material. In the event that proper acknowledgement has not been made, copyright holders are invited to contact the editors about this oversight.

The manuscript would not have been completed without Colin Hearfield's meticulous editing of most chapters, Deb Vale's editorial assistance in the early stages of the project and Suzanne Peake's superb proofeading. We are indebted to Jennifer Page at Routledge for being ever patient and cheerful, despite disappearing deadlines. Above all, countless villagers in the Asia-Pacific region, and those visiting them, contributed to the stories about tourism and its impacts which appear here. We are all grateful to them and to other stakeholders in the tourism industry who have willingly shared their time, thoughts and recollections. They not only allowed this project to eventuate, but helped to build greater understanding of the ways in which tourism has shaped their worlds.

John Connell and Barbara Rugendyke

July 2007

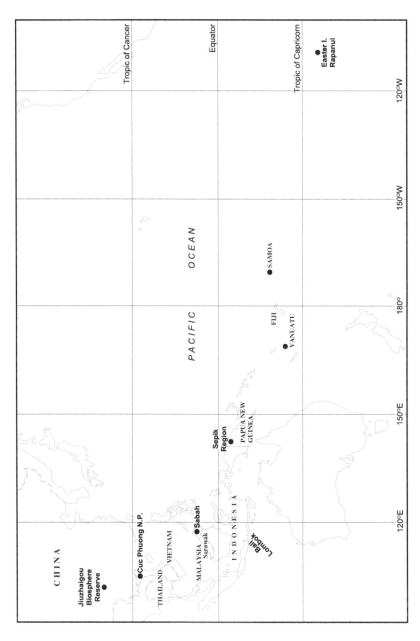

Map The Asia-Pacific region

1 Tourism and local people in the Asia-Pacific region

John Connell and Barbara Rugendyke

Even in the wake of 9/11 in 2001 tourism has experienced a global boom, a boom evident in the Asia-Pacific region (here east and southeast Asia and the island Pacific), as the fastest growing service sector in the world continues to expand. After the 1997 Asian economic crisis, especially as rural areas were places of economic stagnation, governments throughout the region placed new emphasis on tourism, which became 'the priority tool of rural planning orthodoxy' (Liu 2006: 878; Briedenhann and Wickens 2004). Tourism is a valuable and growing source of foreign exchange, and some parts of the region, but certainly not all, have been seen as safe places, distanced from global troubles, and thus ideal destinations.

Part of that boom has involved the rise of domestic tourism, at least in Asia. In 2004 Chinese domestic travellers apparently made more than 800 million trips, twice the number of a decade earlier, made possible by rising incomes, unprecedented freedom to travel and more leisure time. Similar but less dramatic trends have become evident in other parts of the Asia-Pacific region (Erb 2003; Halvaksz 2006), while countries such as Laos, Cambodia and Brunei experience new tourist flows as they open up to tourism and tourists seek new experiences and places (Cochrane 2006a; Winter 2007). After some initial reluctance on cultural grounds, Pacific island states similarly embraced tourism, usually less successfully, because of costs and distance from markets. Nonetheless, in most countries in the region, tourism has grown relative to other economic sectors, as they become increasingly involved in tourism, in a variety of different and complex ways.

Expansion of tourism has occurred not just in accessible areas but in more remote areas, such as Yunnan (China) and in northern Vietnam, with even claims of 'first contact' tourism in West Papua (Indonesia), where tourists are taken to see tribes who are said to have never been hitherto contacted. Getting off the beaten track continues to claim new spaces (Azarya 2004). Tourism is both putting pressure on local communities and their resources, whether economic, environmental or cultural, while simultaneously offering new possibilities to diversify incomes. National parks and other conservation areas have been set aside, and a range of distinctive forms of tourism have emerged, including nature-based tourism, wildlife tourism and ecotourism. Ubiquitously tourism has changed and is changing economic, social and cultural systems. This book seeks to examine tourism at the grassroots – the rural communities and regions where it is playing an increasing role in people's lives, but sometimes in a manner significantly dictated elsewhere.

Conventional wisdom has long been that tourism offers economic benefits alongside social costs. Yet in most contexts, at the very least, such generalisations take little note of which tourists and what local people (sometimes known by the unfortunate word 'tourees', or even as 'hosts', which, like its supposed converse 'guests', are all words which beg important questions). A second element of that conventional wisdom is that, always loosely defined, 'ecotourism' operates at a scale that involves local gains, values local cultures and avoids substantial change ('take only photographs, leave only footprints'), so that it is the most appropriate form of tourism in rural and remote areas. The third, if declining, element of that 'wisdom' remains an urban perception that 'lack of an adequate capacity and an understanding of tourism culture by the peasantry is a common problem' (Liu 2006: 878), alongside lack of capital and cultural constraints, hence tourism must be directed from a distance.

Despite such pervasive notions, surprisingly few detailed studies provide valuable insights into the economic and social effects of tourism and the role of local people in the vast Asia-Pacific region, hence such conclusions are hard to substantiate. This book seeks to consider these issues and add to existing knowledge, as tourism continues to evolve. It centres on the 'people who both service tourists' needs and are the object of their attention' (Boissevain 1996: 1), and examines the uneven relationship between places, local people, transient tourists and intermediaries. What exactly are 'local' people and places varies enormously in the multiple studies of tourism; here the impact of tourism in villages and small towns is considered, alongside the people who live or who have moved there, and who are active or passive participants in the tourist industry.

This book emerged from the need to capture local experiences and examine the impact of tourism at the local level in a variety of contexts and circumstances. Necessarily, in a region that spans southern China through to Easter Island (Rapanui) – with vast social, cultural, economic and physical differences – and with equally wide variations in tourist numbers and types, the effects vary enormously, and are a function of many obvious factors – the sights and facilities, the origins of the tourists, and their length of stay, pattern of expenditure and interests. Some countries have pursued more elite resort-based tourism whereas others, like China, have officially promoted tourism as a poverty alleviation strategy for poorer regions (Luk 2005), even though outcomes may not be so different. Not only tourism influences change; a multitude of factors, loosely subsumed as globalisation (including education, migration, media and trade), affect local lives even if tourism may sometimes be the most direct contact between different cultures.

Tourism also has environmental consequences, especially where much tourist activity is in fragile coastal locations (like Halong Bay or the Coral Coast of Fiji) or national parks. Any construction activity has environmental costs but tourism may have less severe consequences than most economic activities, if only because tourists expect and prefer relatively pristine environments. Although the relationship between tourism and sustainable development is an important one, it is given relatively less attention here, compared with the social and economic issues that influence it. This chapter reviews contemporary literature that examines these themes in the Asia-Pacific.

The tourist area life cycle

Centred on Butler's now famous tourist area life cycle model, large numbers of studies have suggested that initial impacts of tourism are largely benign. Little initially changes, tourists respect the values of local people and there are economic gains without significant cultural change. Over time tourist numbers and their impacts increase, both local people and (at least some) tourists are more likely to become alienated by the experience, some part of the economy may pass out of local hands and eventually a combination of over-familiarity and the search for the new, inherent in the 'business cycle', and pressure on the environment, within the 'ecological cycle', degrade place and discourage tourism, so that destinations may even experience decline (Butler 1980, 1991). The gains from small-scale tourism become outweighed by the economic, environmental and social costs of what increasingly becomes mass tourism.

Of course the impacts of tourism are never straightforward; as tourist numbers increase, destinations evolve and generate new 'attractions'. Though few longitudinal studies of tourism demonstrate clear relationships, there are obvious 'ethnographic time traps' in seeking to deduce the rationale and impact of change from studies at particular, brief time periods, when other factors besides tourism influence change (Wilson 1993: 36). Kuta, in Bali, in the 1960s a small village destination for backpackers and surfers on the Asian overland trail, evolved to become all things to everyone, by constantly linking into new tourist niches (from raves and honeymoons to whitewater rafting and camel safaris) that accommodated the needs and whims of all kinds of tourists: it became 'whatever you want it to be' (Connell 1993). Through these changes local people played a declining economic role, as more distant Javanese and international interests constructed hotels and other facilities, and secured employment. Land was absorbed into the tourist industry as Balinese were displaced, yet most retained crucial elements of local culture, though visual culture took on a variety of transformations, simultaneously being diminished, transformed, reinvented and globalised for the tourist gaze (Howe 2005). As Kuta changed, other parts of Bali responded quite differently; some specialised in particular tourist artifacts, as tourists followed conventional paths through the island, and other remote villages became the hangouts of travellers who felt displaced from Kuta. Nusa Dua became an enclave of expensive, elegant hotels and in the village of Pengosekan, near Ubud, a group of workshops manufactured what had hitherto been Australian didjeridus for a European market (Gibson and Connell 2005: 110–11). As one of the most important destinations for regional and global tourism, it is unsurprising that parts of the island have changed in complex ways and that there are continuous plans for new forms of tourism in Bali, such as a Grand Prix motor racing circuit (Howe 2005). Thus across a relatively small island the form and impact of tourism varied enormously as people adopted mixed responses to tourism and tourists, according to their perceptions, needs, abilities, location and resources, and the desires of governments, tourists and tourist companies. While Kuta and Bali went through approximations to the tourist cycle, they showed that the impact of tourism is more complex than any simple model can address, and that the agencies of tourists and local people, and of governments and distant entrepreneurs, play a crucial role in outcomes. Sadly, more recently, they also demonstrate the vulnerability of tourism to violence and political disruption (see Tarplee in this volume; Darma Putra and Hitchcock 2006; Baker and Coulter 2007). Most places in the Asia-Pacific region have changed more slowly and less dramatically, and remain on the periphery of the tourist industry.

On the edge of the tourist industry

Some countries, such as East Timor and Papua New Guinea, and most towns and villages in the Asian-Pacific region have little to do with tourism. They may be distant from conventional tourist circuits, have no particular attractions – whether cultural or natural – be perceived as unsafe or expensive and, if they benefit at all from the industry, experience some trickledown from national economic growth or gain from the remittances of people who have migrated to work in the industry.

However in some remote places tourists may be virtually the only source of cash income. In Makira (Solomon Islands) even a small, infrequent ecotourism venture brought in substantial income. One ecotour alone generated 40 per cent of the cash incomes of highland villages, where alternative incomes mainly came from long trips with few goods to distant markets (Russell and Stabile 2003: 50). Yet these are the exception that proves the rule: such remote villages are rarely beneficiaries of any kind of tourism. Moreover, ironically, the Makira venture ended with the onset of sustained violence in the Solomon Islands at the end of the 1990s. In the smaller of the Cook Islands tourists were so few that they were seen as no different from other visitors such as researchers, teachers or volunteer workers, and thus as friends who would be invited to participate in local events and of whom reciprocity was expected (Berno 1996) rather than as contributors to the local economy.

In the Fijian village of Korotogo, where tourists are rare, since the village has never sought to encourage tourism, independent travellers are unusual and there are no distinctive attractions, the direct impact of tourism on the village is trivial. However Korotogo is a kilometre from one of the largest hotels on the Fijian Coral Coast, hence two-thirds of men and a third of the women work in the tourist industry. Both groups earn incomes four times those of Fijian villagers who are too distant for employment in the tourist industry (hence distant Fijians have mobilised kinship ties to move to Korotogo and take advantage of tourism employment), and no villagers are involuntarily unemployed. All have ready access to employment since the hotel (and, in a broader sense, the tourist industry) prefers local Fijians. Tourist incomes have enabled all villagers to have permanent homes, with electricity, toilets and piped water, and to send their children through high school. They have retained certain dances, with many performing the *meke* (actually partly designed for tourists) almost nightly, though new sources of extra-village income have reduced chiefly authority. Agriculture declined but pig husbandry increased, because of the ease of access to food left over in the hotels, but villagers retained all their land, and hence could revert to the old order. Here therefore, in a village where tourists are absent, there have been significant economic gains, people have made the social changes they would like (better infrastructure, more modern clothes and education) and land tenure remains secure. Similarly, when a major resort unexpectedly closed near a village 30 kilometres away, which had experienced changes like those of Korotogo, villagers were able to either find work in the tourist industry elsewhere (because of their experience) or return to the village agricultural economy (Sofer 1990).

At Piliura, one of four villages on the tiny Vanuatu island of Pele, small groups of tourists (between five and forty) visit two or three days a week, stay for an hour or so while they have a village tour, do some snorkeling and have lunch, which they bring themselves. The village receives 1000 vatu (about US$10) for each tourist, and individual villagers may sell small amounts of handicrafts (including shells and baskets) while the Nguna-Pele Marine Protection Agency, based in the village, mounts an exhibition and sells tee-shirts. Tourists stay for such a short time that their environmental and cultural impact is limited. Income from hosting tourists, distributed by the Village Tourism Committee, supports such aims as village water supply projects. The bulk of tourist payments (the day tour of which the Pele visit is part costs 7500 vatu) remain with a large foreign owned company. Although it may seem that the village is benefiting only marginally from tourism as a distant company reaps the main rewards, it has no other good source of income and, through the company having an employee from Pele, has secured a seemingly sustainable and beneficial outcome.

In the earliest phases of tourism in Nias (Indonesia) returns from dance performances were distributed throughout the key village, and between the dancers, while villagers supplemented this with handicraft sales where cash incomes were otherwise negligible (Mauer and Zeigler 1988: 85). In all these places tourism, tangential or in its infancy, has largely benefited villagers through projects that support the whole village, or enabled universal access to incomes, especially where other incomes are limited. However, ongoing employment is limited and villagers have no real control over the industry. Nonetheless their relative success emphasises the differences between them and more distant villages, where income generation and employment possibilities are fewer, hence local migration to 'tourist' villages has sometimes followed (Ulack 1993).

Even in regions dominated by tourism, village benefits may be few. In Nusa Dua (Bali), and Denarau and the Coral Coast of Fiji, where tourist developments are increasingly likely to be within enclaves and tourists have limited contact with the world beyond the perimeter, incomes from employment must be the principal gain from tourism. Ironically the most exclusive resorts in the world are on some of the most isolated islands, such as Wakaya and Vatulele in Fiji, the former with prices starting at US$1900 per night (2006) and the latter advertising on its website that 'in keeping with the owner's philosophy there are no money transactions'. All obvious needs are satisfied within the resort, almost all payments are made outside the country and local people gain only from some lease payments and employment. Retention of tourist expenditure is minimised. However such resorts, notably Turtle Island (Fiji), have a high ratio of employees to guests and hence make a substantial contribution to employment, while choosing to operate through environmental principles (Weaver 2002: 131).

Many villages are simply bypassed by tourism. At the Komodo National Park on Komodo island (Indonesia), intent on seeing large, rare 'Komodo dragons', tourists stay in nearby urban centres on other islands and hire boats to visit Komodo. None stay overnight in the lone village (though about 5000 tourists a year pass through) and few spend money there, other than on carved wooden dragons, since the village has no store. A handful of carvers and boat crew are the only village beneficiaries; the village remains dependent on fisheries, despite the Park's 20-year presence. Villagers were further disadvantaged by being unable to expand their agricultural systems because of the Park, becoming increasingly dependent on expensive imported food, though they could sell goats as bait for the dragons. Local people are disadvantaged by their lack of relevant skills and knowledge, the absence of training opportunities and vested interests elsewhere (Hitchcock 1993; Walpole and Goodwin 2000).

On the Sumatran island of Siberut tourists are brought by guides from the mainland to see and photograph 'primitive tribes'. Islanders cannot speak Indonesian, let alone English, there are no handicrafts to buy and local houses are unsuitable for tourists, hence the only tourist income comes from occasional portering, paid in tobacco rather than money:

> there are no means by which the communities can earn much through tourism, and they have no control over how many tourists come or where they go – yet the tourists bring cultural dislocation in their wake, inadvertently causing the breakdown of traditional barter systems and hospitality . . . A clear case of exploitation of indigenous peoples by more educated, entrepreneurial outsiders.
>
> (Cochrane 2006a: 251)

Such extremes of exploitation are now rare. Indeed in somewhat similar contexts elsewhere in Sumatra, villagers resent the presence of such tourists and steal from them (Cochrane 2006a: 252).

Even whole regions may not benefit significantly from tourism. In the first year of tourism in the remote Himalayan kingdom of Mustang (Nepal) no incomes reached Mustang as trekkers hired guides and porters from elsewhere, brought their own accommodation and food and found no artifacts to purchase (Shackley 1994). Here too local people initially lacked the resources and knowledge to participate effectively. In other words, though both ecotourism and trekking have been widely seen as providing direct local benefits, the specific local conditions made Mustang quite different from Makira and elsewhere. Mustang, Siberut and Komodo villagers, as in other remote places (Nyaupane *et al.* 2006), make at best trivial gains from tourism, though environmental costs are local, while the real economic gains are elsewhere.

Despite such problems, and conscious of the gains that others have made, local people have usually been anxious to develop tourism, especially since the combination of the tourism industry and formal sector employment are favoured as reliable, clean, relatively pleasant and easy sources of income compared with many local alternatives (see below). However, many people lacked the access, resources and skills to participate. In Papua New Guinea local development initiatives, such as guesthouses in Tufi (Oro province), failed because of overall declining tourist numbers, lack of knowledge of their existence, inaccessibility, costs and local conflict over returns, whilst proprietors lacked experience and encountered disputes over land tenure and trouble maintaining properties, organising food supplies and even preventing malaria (Ranck 1987). The few successful ventures tended to be European owned or with very specialised functions, such as hunting and fishing, and thus with a definite local market (Tapari 1988). A small ecotourism project in Vanua Levu, Fiji, seemingly a 'model' project but two hours by road from a small urban centre, collapsed after a very short time as too few tourists came (Sinha and Bushell 2002). In much of Vanuatu, bungalow owners, who receive few if any tourists in any single year, struggle to get their presence acknowledged by wholesalers and complain bitterly of having constructed bungalows on government advice only to see them disintegrate (Slatter 2006: 6–7). Many small projects, especially those centred around ecotours and village guesthouses, have failed, had chequered careers, or been dormant for long periods, because of problems of management, disputes over land and incomes or inadequate access to markets: they are simply unable to get onto the 'tourist map'. In Thailand there are considerable regional variations in success, yet in Samoa small projects have flourished (see Leksakundilok and Hirsch, and Scheyvens, in this volume).

In several Asian contexts, tourism is primarily an urban phenomenon, at least for domestic tourism, and few tourists visit villages or see the countryside in other than regimented and sanitised form, as at Jiuzhaigou, China (see Dombroski in this volume). Thus in highlands Vietnam, at the small town of Sapa, the new Vietnamese 'affluent population is not that keen for prolonged contact with nature, and enjoying the town's amenities proves far more attractive than visiting unclean highlander villages', hence even rapidly growing tourist numbers have limited effect on rural people other than those few who visit the town to sell artifacts or put on cultural performances there (Michaud and Turner 2006: 793). Even the economic impact of overseas tourists on local villages is limited, since it is officially illegal to stay there overnight; only villages within walking distance of Sapa can be visited, and there are linguistic constraints to social contact (ibid.: 797). Obvious political, social and geographical factors constrain the impact and extent of tourism.

A relatively small number of remote villages are significant destinations, yet some have seemingly benefited little from tourism, often despite years of participation in the industry. As in Makira and Pele, many small and remote communities in Thailand gained little from tourism, but even that income – despite being a fraction of that spent by tourists – is rather more than might be obtained from other local economic pursuits, in which they are just as likely to be exploited by others. Some Thai hilltribes, like the Akha, are largely passive participants in tourism, involved in little more than hawking cheap trinkets, such as pipes and bracelets, while others without such goods resort to begging (Cohen 1996: 250). Ban Suay, a Hmong hilltribe village, first experienced trekking tourism in the 1980s but, after 12 years of increasing tourist presence, the economic impact was 'astonishingly negligible'; economic gains most affected 'Hmong marginals' for whom agriculture had become inaccessible because of opium addiction. Trekkers joined package tours in Thai towns, were transported by Thai vehicles, drivers and guides, and ate Thai food brought with them. Only 2.3 per cent of the money spent by tourists (about US$10 per person in the 1990s) actually reached the village, the rest mostly going to Thai middlemen. A lowland Thai trader then opened a store in the village, which absorbed additional tourist expenditure and much of the income the Hmong made from tourism. Even the limited economic gains of the 'marginals' were threatened as their new pattern of consumption eroded the 'authenticity' tourists sought (Michaud 1997). This is a frequent (if localised) example of tourists consuming imported goods and services (with minimal local expenditure), but within a particularly extreme form of exploitation.

Other studies of trekkers in hilltribe villages suggest that local people have usually been more successful in generating income and using it effectively, in circumstances where there would otherwise be poverty and inadequate food security. However incomes are concentrated in the hands of relatively few households in each village and amongst already relatively well off households (Dearden 1996: 214). Even then, in every case, the principal beneficiaries were still the intermediary guides rather than local people (Toyota 1996: 236–7).

At Ban Suay and in other hilltribe contexts, as also in Siberut, Mustang and elsewhere, particular groups who are ethnically different from their neighbours or from the majority population are spatially, socially and linguistically disadvantaged in terms of substantial participation in the tourist industry (and other economic sectors) and are dependent on intermediaries. Some, like the indigenous Orang Asli of Kedah (Malaysia), after periods of peripheral participation in the tourist industry, eventually rejected the idea of being a 'tourist attraction' and simply stopped participating (Liu 2006). The economic rewards were inadequate. Thus, at the margins of the tourism industry incomes are small and uneven, and exploitation more evident, such that people may withdraw their participation even though this is a significant source of income where alternatives are few.

Tourism as transformation

In cases like those described above the impacts of tourism are slight, even with a significant flow of tourists, whereas in more accessible places tourism has had a significant influence on economic relationships, social change is considerable and cultures have been transformed. The following sections examine development and change in these contexts.

Employment

In developed countries tourism employment tends to be regarded as a low-status menial activity, often performed by migrants, whereas in most developing countries tourism employment is relatively popular. Regular, secure and relatively high wages, away from the dirt, drudgery and uncertainty of agricultural labour, are attractive, and have even led to migration away from skilled employment. Success may so transform particular destinations that 'old' jobs in agriculture and other arduous activities are abandoned. Tourism employment is usually clean, if repetitive, and requires no high education levels (but, conversely, little training exists to enable career advancement, and limited skills mean low wages), though it may be seasonal.

In some contexts informal sector workers, generally those without regular hours and wages who are often elsewhere marginalised, have higher status because of their need for special skills, such as English language ability. More than half the informal traders of Bali could speak more than three languages, despite being young and with limited formal education (Cukier 1996). Tourist guides are often accorded particularly high status, even seen in Bali as 'cultural emissaries', where most are well-educated younger men (Cukier 1998) and in Nepal where guiding is almost exclusively a male activity. Guides in remote Fijian villages had high status and acquired superior English language skills and reasonable incomes, hence ecotourism projects ensured that each social group provided a guide to ensure equity of status (Bricker 2001). Some employment is particularly prestigious. Even informal tourist employment becomes a means of acquiring new skills that may enable better paid, regular employment elsewhere.

Frequently there is competition for good jobs. Especially where alternative income-earning opportunities are few, as on Beqa Island (Fiji), those who have tourism employment are envied by others. Hostility to the lone resort's presence on the island also existed because of its disruption of more limited but more equitable structures of development and its access to the most modern boat on the island and to telephones that were denied to local islanders (Burns 2003). In such circumstances hotel managements may seek to encourage equity and thus loyalty and wider support. One small elite hotel off the larger island of Vanua Levu (Fiji) hired two or three teams of neighbouring villagers, rotated on a fortnightly basis, enabling most villagers in this remote place to gain cash incomes (Tuinabua 2000: 189). Managements though usually prefer the continuity of a more stable workforce.

Many tourism projects hire local workers, who may have claims to land ownership, or are familiar with local geography, culture and management practices, but also because it is less costly. Even in large hotels kin may be employed to secure loyalty (Fairbairn-Dunlop 1994). In Phuket (southern Thailand) ecotourism companies hire local people and may give preference to the relatively poor, hiring only skilled elephant handlers from a distance (Kontogeorgopoulos 2004). Similarly on Beqa the only non-local people hired by a resort had high-level management or diving skills (Burns 2003) and at the Bukit Lawan orang-utan station in Sumatra technical guides were also outsiders (Cochrane 2006b). In Fiji indigenous Fijians, rather than Indo-Fijians, are employed in positions that involve dealing with tourists since they meet the cultural perceptions of the industry (see Kanemasu in this volume).

Sometimes, however, distant workers are preferred in the belief that they work harder than local people and are less involved in local social and domestic activities that disrupt continuous, formal employment. In Saipan migrant Filipinos are preferred to local Marianas residents (Mansperger 1995) and at Koh Samui local people were regarded as unreliable, lazy and dishonest and likely to take much time off work, whereas migrant workers would work longer hours and take less time off (Williamson and Hirsch 1996). However, when tourist operators hire more distant workers conflicts can arise, as at Anuha (Solomon Islands), which contributed to the eventual closure of the resort (Sofield 1996). Local employment too may create tensions, as in such Muslim islands as Lombok (Indonesia), where cultural requirements resulted in workers withdrawing from the tourist industry on a short- or long-term basis, so that the tourist industry shifted towards more compliant workers who would put up with harsher conditions. For similar reasons local people may also be employed in more menial positions, though in the larger hotels of Koh Samui it was argued by the managers that these were 'training' positions and, if local workers stayed, they could graduate to superior positions in the chain's other hotels elsewhere (Williamson and Hirsch 1996: 193–4). At a new resort in north Sulawesi (Indonesia), though local people expected significant employment opportunities, few were employed, and then in menial positions where training was non-existent, and most employees – including 'traditional' dancers – came from other parts of north Sulawesi. Even some local workers who were employed quit because their wages were inadequate to meet needs (Simpson and Wall 1999). Sometimes, though, there may be little alternative to hiring distant workers, leading to employment hierarchies. In several parts of Malaysia local youths had migrated away before tourism began and now, without land in the tourist areas, they did not want to return merely to be employed as waiters or cooks; hence, imported workers were brought from Thailand and the Philippines (Hamzah 1997: 212). In the Cook Islands tourism workers have been brought from Fiji as Cook Islanders themselves have gone to New Zealand.

Gender

Partly because it is a labour-intensive industry, though more jobs are held by men, tourism provides formal employment opportunities for women where they are otherwise rare, which enables some 'empowerment and advancement' even where that employment is menial. Almost everywhere women are more likely to be involved in handicraft production, because it can be absorbed into domestic activities. In Bali informal employment was attractive since child care posed no problem (Cukier 1996), and informal selling was convenient for married women who could combine this with family and religious obligations (Cukier, Norris and Wall 1996). In such villages as Penglipuran men worked in agriculture, hence women learned the languages of overseas tourists, sold handicrafts and rented out rooms (Hitchcock 2004). The ability of women to gain employment and income from tourism is particularly important where women tend to be the victims of poverty.

Though tourism has sometimes brought substantial changes in gender relations, various incidences occur of women being reminded not to abandon their 'primary duties' as wives and mothers, or even as agricultural producers, as better-paid, more stable and sometimes less arduous jobs are taken by men (Kindon 2001). Even in the Asian informal sector men tend to hold better jobs. At Boracay (Philippines) women were the main vendors but there was a smaller market for the goods that they sold than those sold by men (Chant and McIlwaine 1995). Prostitution and gambling (in a formal casino sense) remain mainly urban phenomena (though drawing workers from rural areas) but in what are now small towns, such as Kuta and Koh Samui, massage parlours are common (Hall 1992; Garrick 2005; Green 2005). Pangandaran village (Java) had several enclaves of mainly female prostitutes: some were local women, though claimed to be migrants from elsewhere in Java. Cultural mores often exclude women from participation in the tourism industry. In Pangandaran women were prevented from becoming tourist guides, with such women being perceived as prostitutes because of their contact with foreigners, and also excluded from such 'modern' activities as driving. Several, though, had rental stalls and one had become the beach lifeguard (Wilkinson and Pratiwi 1995: 293–4; see also Fallon in this volume). Some social mobility was possible even in unpropitious circumstances. In the Pacific women are rather more likely to hold 'front-line' positions in hotels and elsewhere (Berno and Jones 2001). In the small tourist industry of Samoa 80 per cent of workers in the formal tourist industry are women, and the proportion of women in the informal sector is probably even higher. Taxi driving is the only component of the industry where men outnumber women (Fairbairn-Dunlop 1994). Transport is almost always a male arena.

On Beqa island (Fiji) the only available employment for women was in the single tourist resort, which gave many their first direct access to cash income and, since they sometimes worked late hours, increased personal mobility and social freedom (Burns 2003). Here, as elsewhere, tourism offers women new opportunities for social mobility, greater control over household incomes (because of their contribution to them) and, in some contexts, a break from patriarchal society. In both Nepal and Yunnan, the mere presence of educated Western women tourists was an inspiration to local women to fight for gender equality (Nyaupane *et al.* 2006). Women usually play an important role in cultural performances, though there is little evidence of the extent to which this may have generated new incomes and new social structures within communities, or even have contributed to the stereotyping of women's visual role. However even where, as in Namuamua village (Fiji), women played multiple roles in the tourist industry as guides, entertainers, food preparers and handicraft sellers, their incomes were less than half those of the men, who had different roles in the industry, notably in transport (Tokalau 2005). Here and elsewhere, women were more likely than men to use their incomes for community objectives rather than individual goals.

Informal sector employment

Much employment in the tourist industry is in the informal sector, with unregulated working conditions and wages and some ease of entry, especially where the tourist industry is relatively new. Controls over informal sector workers tend to increase over time as tourism becomes a major source of income. At Gili Trawangan (Lombok) local people were pioneers in meeting new demands for tourist facilities and 'they start home-stays, food stalls and transportation, or hire out snorkel gear, motorcycles or mountain bikes' (Kamsma and Bras 2000: 170). In Bali many women are employed in the small handicraft factories but more are informally engaged, as masseurs, hair braiders, drink sellers, traders and *losmen* (guesthouse) workers, whereas men are itinerant vendors and drivers. Partly because of competition, particular cultural groups may dominate parts of the informal sector. At Kuta migrants from Raas island almost completely dominate the trade in fake designer-label caps and watches and thus reduce local competition for a limited market (de Jonge 2000). Intense competition in the informal sector, sometimes between local and more distant entrepreneurs, characterises the informal sector.

Rigid regulation may exist, but informal sector workers often have the flexibility and initiative to avoid repression and regulation. Restrictions are often placed on informal sector workers, reducing their income-earning ability, as the informal sector is seen as the antithesis of up-market tourism or its workers considered to harass tourists. Rickshaws or their equivalents are excluded from parts of even small towns, on congestion grounds. Most informal sector workers are excluded from resorts, other than in carefully controlled areas where handicrafts may be made and demonstrated. Informal sector workers were banned from the planned resort area of Nusa Dua on Bali, though ubiquitous in the other main tourist centres there, though there too they experience severe government control (Bras and Dahles 1998; Dahles 2000). In Sapa (Vietnam), hilltribe vendors, once free to wander through the small town selling 'ethnic garments and trinkets', were forced into a featureless concrete market while paying a daily fee to the council (Michaud and Turner 2006: 801). In Port Moresby (Papua New Guinea) handicraft sellers, along with other street vendors, have been routinely attacked by police. In Koh Samui relatively poor villagers, who were never large landowners or engaged in commerce, participated at best in the tourist industry as wage labourers or informal producers and sellers of food sold from mobile carts, but their livelihood was constantly under threat from numerous foreign-owned restaurants, bungalow owners who kept them away from guests and village stores that offered similar goods (Williamson and Hirsch 1996: 198). The informal sector is constantly made unwelcome by the more powerful and opposed for what can be seen as the inappropriate imagery it offers and its 'threat' to formal sector economic activities.

Accommodation, local ownership and participation

The most rewarding economic activity for local people has usually been through the provision of accommodation, common in early phases of tourism, as 'new' tourists enjoy, or make do with, simple accommodation of local materials, in the search for cultural experiences. Uncertainty in marketing and low occupancy rates were of no particular significance where little capital was involved and tourist needs were few. However, to convert houses into accommodation, or build separate tourist accommodation, potential entrepreneurs usually required some capital, which was often hard to obtain other than through loans – sometimes difficult to acquire by those with small or non-existent landholdings. In Koh Samui (Thailand) 'bank loans are only given to those able to provide security; land is the most common security offered and this has enabled land-rich people to enter the tourist industry as entrepreneurs' (Williamson and Hirsch 1996: 194). Even developing basic tourist accommodation favoured those already relatively well off, with land in attractive areas. As tourism expanded, the need for even more capital made local access difficult.

Over time, much local-style accommodation has been replaced by 'modern' accommodation, and local owners displaced by distant owners. In the 1970s Batu Ferringhi beach (Penang island, Malaysia) was characterised by fishermen's cottages that were the main form of tourist accommodation; two decades later all had been replaced by hotels, and international tourists, mainly backpackers, replaced by domestic tourists. Many parts of Bali, from the 1970s in Kuta, were characterised by home-stays (*losmens*): family-owned houses with three or four rooms for tourists, usually owned by local Balinese, although by 1992 some 29 per cent were owned by other Indonesians, though none were foreign-owned or large enough to have foreign employees (Wall and Long 1996: 36). Many had little room to grow and in the larger centres were eventually bought out, as part of a familiar tourist cycle.

Characteristic of tourism as a labour-intensive industry is the manner in which, as places achieve success, people from elsewhere arrive as workers and become land, hotel and service owners. This is exemplified by the rapid arrival of hoteliers from Australia, Timor and Jakarta to the small town of Labuan Bajo in Flores (Indonesia) (Erb 2005: 166) and to such other Indonesian destinations as Kuta and Gili Trawangan (Lombok) (Cukier 1996), or the Thai islands of Koh Samui and Phuket (Cohen 1982; Williamson 1992; Parnwell 1993). Influxes of outsiders are normal and only exceptionally, as in Rapanui (see McCall in this volume), have local people been able to resist the incursions of others seeking to take advantage of valuable resources. Outsiders often displace local people, especially vendors, rather than work alongside them, a situation also true for accommodation and land ownership, producing resentment and tension. A characteristic of the tourist industry is the remarkable ease of entry, but also intense competition, especially for such informal sector activities as food vending, kiosks, guides, bicycle rental and prostitution.

Return migration of once local people may produce similar discontent. At Gili Trawangan questions about who were the 'local people' with rights to be beneficiaries of tourism development became critical. Relatively successful 'local' entrepreneurs tend to be marginal, liminal people, because of the necessity to have experience of a wider world. In Yunnan one of the principal guesthouse and tour operators was of Tibetan ancestry and had a Dutch wife; another was a local man with a university degree and fluent English. Each of them could purvey particular touristic versions of 'authenticity' (Ateljevic and Doorne 2005). Just as frequently middlemen, including entrepreneurs and guides, are from elsewhere (Toops 1992; Adams 1997; Michaud 1997) and are the main economic beneficiaries of tourism. Such migration poses problems. In Bali in the 1990s as many as 60,000–80,000 migrants a year were moving from other parts of the country and generating serious ethnic and religious tensions (Warren 1998: 238).

Characteristically tourism workers of perceived low status, such as prostitutes, are said to be from distant places, as in Bali where they are said to be Javanese, or in southern Thailand where they are 'from the north', Myanmar (Burma), or Yunnan where they were seen as different ethnic groups from 'poor, neighbouring villages' (Nyaupane *et al.* 2006: 1380). In various places where there has been migration tourists may perceive the migrants, whether hawkers, prostitutes or hotel workers, as 'locals' and 'hosts', but long-term residents see them as uninvited competitors, who may be subjected to violence and exploitation (McNaughton 2006). Many such migrant workers send or take remittances back to their home areas (Cukier 1996) and are also resented for this. Foreign entrepreneurs may develop small-scale tourism activities that are regarded as the province of local people. At Koh Samui expatriate Westerners (*farang*) were involved from the earliest days as bar and bungalow owners, diving instructors and more marginally as prostitutes and drug dealers. Though most sought to avoid direct competition with local workers, and brought innovations and new skills, their presence was resented and opposed, unless they aligned with local patrons (Williamson 1992). Less skilled workers came from as far away as Burma. In Bali too foreigners have been similarly involved in the ownership of the local tourist industry; almost all informal sector workers were from the neighbouring island of Java but Balinese retained many better and more formal positions (Cukier 1996).

As tourism grows local people are more likely to be displaced from their homes, from their land (required for hotel resorts, golf courses etc.) and even from jobs in the informal sector as tourism creates demands for higher, specialised standards of service and facilities. This occurred at Pangandaran (Java), but the more skilled local people retained jobs relative to the village poor who lost theirs (Wilkinson and Pratiwi 1995). Local people may even be displaced from cultural performances, as they have been in some hotels in New Caledonia, where indigenous Melanesians were replaced by Polynesian migrants from distant Tahiti (Gibson and Connell 2005: 146–7) if their repertoire was less exotic, vibrant and colourful than more distant groups.

In the early days of the tourism industry in upland Toraja, Sulawesi, local people were as marginalised and distanced from the industry as those of Komodo and Ban Suay, since the tourist economy was largely controlled by coastal people. However, unlike the examples above, tourism fostered local entrepreneurialism, with young Toraja taking up jobs in the business, initially as guides and hotel workers, and later as owners of travel agencies, at first in the Toraja area and later in Ujung Pandang, the provincial capital, thus reversing the more exploitative initial situation (Scarduelli 2005). Such reversals in favour of local people otherwise seem relatively rare, despite resorts and governments professing to be supportive.

Land

As tourism becomes successful land use and ownership may become contested, as distant interests seek to buy land and establish large hotels, which may have repercussions for marine tenure and for the exploitation and conservation of land and sea in adjacent areas. Like employment and accommodation, land has tended to leave local hands for more distant and non-local owners, or been expropriated by the state. Similarly local people have usually fought against land alienation, but have not always been averse to short-term profits from land sales.

In Samoa, local opposition to land alienation prevented forms of tourism that local people dislike (Fairbairn-Dunlop 1994), but such examples are rare. In several places, such as Koh Samui, though foreign ownership is illegal, land has still been bought (and also leased) under various schemes that use nationals as fronts (Williamson and Hirsch 1996: 196). Governments have frequently facilitated land acquisition by companies, not always legally, to enable large-scale tourism development. In Vanuatu, a high proportion of coastal land has been sold off to foreign investors, ensuring that the best land and the largest, most profitable tourist ventures are in foreign hands, while local small-scale ventures are marginalised (Slatter 2006).

Land losses may have other negative outcomes. In north Sulawesi a new resort took local land for what were seen as derisory compensation payments; although the new resort did purchase local foods in the nearby market, local villagers no longer had land to grow such products, which had to come from more distant villages (Simpson and Wall 1999: 247–9). At Ban Chaweng (Koh Samui) land was alienated to the extent that such 'natural features' as coconut plantations and pandanus trees became perceived as 'tourist attractions' rather than components of traditional life (Green 2005: 51–2). Land alienation was symptomatic of environmental change.

The increased value of land for tourism posed other problems. At Gili Trawangan tourist development was characterised by land disputes, as wealthy enterprises and entrepreneurs from other Indonesian provinces bought large parcels of land, and land disputes and lawsuits proliferated (Kamsma and Bras 2000). Similarly there have been disputes over who owns the land where tourist facilities have been situated, and thus over rights to compensation, employment and so on. Land ownership is rarely equitable, and community leaders have greater ownership and authority over land. In some contexts, as in attempts to develop ecotourism in Guadalcanal (Solomon Islands), local people have challenged project development because only leaders were consulted rather than the majority of the local people, and their priorities were different from those of the leaders, being concerned with local food production from the land and lagoon areas involved (Rudkin and Hall 1996).

Development of a tourist resort on the uninhabited island of Anuha (Solomon Island) in the 1980s was slowed by separate claims to both land and reef ownership by several groups from as far as 80 kilometres away and it eventually closed after local compensation claims escalated in response to the clearance of timber and other indigenous resources (Sofield 1996: 184). In Bintan island (Indonesia) villagers demonstrated against arbitrary and inadequate compensation for their resettlement and land alienation for tourism; they shut off power and blockaded a major resort in a context in which claims were complicated by speculators and new arrivals claiming to be inhabitants (Wong 2003). Somewhat similarly in western Fiji in the 1990s, Fijian landowners blockaded and occupied tourist facilities in protest at the late payment of land rent: 'the main problem with the luxury tourist industry, as experienced from the indigenous standpoint, is not that it commodifies Fijians and purveys an inauthentic culture in their name, but that it pays the former too little, too late, for the privilege' (Abramson 2004: 79). Similar tensions over land have threatened other tourism projects, especially where local people have limited direct participation.

That land is now much more valuable is only of short-term benefit to local people. Exceptionally, land of little local productive potential and value, such as low-lying sandy areas close to the coast, become of great significance for tourism. At Kuta, such land was leased early or sold off cheaply. In Koh Samui the boom in land prices that tourism brought prompted many to sell and generated 'instant millionaires', especially amongst traders who had already purchased cheap coastal land from indebted farmers and other creditors. For the near landless there was no benefit unless they wished to leave the island, hence higher land prices effectively encouraged out-migration of poorer islanders (Williamson and Hirsch 1996: 196). At Pangandaran, where those with land sold it and prospered, the main beneficiaries from elsewhere in Java bought the land and developed the tourist industry, while 'lower-class locals' became marginalised in their jobs, property and power (Wilkinson and Pratiwi 1995: 296). Similarly, at Mui Ne, where tourism boomed in the 1990s as Vietnam liberalised, land of little value suddenly became very valuable, so:

> The people who used to have fishing boats on the beach, they're rich now. Their speck of sand which used to be worth 10c a square metre is worth $380 a square metre and they can sell half, build a new house and get a new motorbike and still have half left.

In such boom circumstances tensions between the local people and newly arrived entrepreneurs were minimal (Prasso 2007: 24), though the massive profits being made by outsiders, and the displacement of local people, indicate emerging problems (Ledger 2007). In all three places, as migrant workers were moving in local people moved out. Tourism both highlights already existing land tenure inequalities, where those with greater areas of land and status may benefit to the exclusion or marginalisation of others, and introduces new inequalities between local people and migrants.

Incomes

Unsurprisingly, people welcome tourism primarily for the direct or indirect income-earning opportunities it provides, especially where there are few alternatives. This may also help retain culture and allow people to remain in their home areas without needing to migrate for work. The actual distribution of incomes in tourist contexts has rarely been examined, with most studies generalising benefits at a community level, but equitable income distribution is rare, and this may be a source of tension (and may also be veiled by those who have primary access to the income).

In south Pentecost (Vanuatu), where men perform a spectacular land dive, and most villagers are involved in related tourism construction or food production, virtually every person in the village receives some income, though money is distributed according to status and gender (men receiving more than double that of women), and for particularly valued contributions to the event, while some is also channelled into group items such as an outboard motor-boat (de Burlo 1996: 268). In Yunnan (China) villagers tended to contribute equally to village tourism projects and received equal shares of the profits (Nyaupane *et al.* 2006: 1381). In several places, including Tanna (see Robinson and Connell in this volume), tourist resources may be pooled for village projects such as water tanks or everyone's education fees, but practice may not always live up to theory.

In villages where the impact of tourism is slight and returns to tourism low, income might often be a significant proportion of all cash incomes in a remote place, as it was in south Pentecost, though few households made more than about US$10 from tourism in any year (de Burlo 1996). Similarly, on Beqa island, individuals might receive the equivalent of US$10 for four days' work (Burns 2003: 89) but where no other cash-earning opportunities existed on the island this was invaluable, and worth competing for. Where alternative employment exists such incomes are much less attractive; even in impoverished parts of southern China village cultural groups could earn only the equivalent of the average daily wage for a casual worker and even that was uncertain since there was considerable competition between groups (Luk 2005: 275).

Villagers establishing ecotourism projects in remote upland areas of Fiji were enthusiastic about the potential of shared, communal income to enable them to fulfil seemingly prosaic but important needs, such as visiting relatives, buying goods and gaining access to preventative medical care (Bricker 2001). In nearby Namuamua village, which tourists visited three days a week to see a cultural performance, tourist revenue constituted about 40 per cent of the incomes of those directly engaged in the tourism venture, the rest coming from market sales, and had enabled several of them to set up small village businesses such as stores and a boat operation, and to hire fellow villagers. However, not all villagers were involved in the venture. At the same time, the external company running the tour had invested in village development projects including concrete pathways, a diesel generator and a refrigerator for the health centre (Tokalau 2005). Other studies of ecotourism projects have shown that a significant proportion of the income generated remains in the local community, that it is often spent on basic needs, and that the use of local materials and expertise in ecotourism projects further concentrates income locally (Scheyvens and Purdie 1999).

In most contexts, however, incomes are predictably unequal, and particularly uneven as tourism develops and competition increases. Competition for tourist employment and income may be most intense where few alternatives exist. On Beqa, as more villagers sought to obtain contracts for performing fire-walking, they became more desperate to obtain contracts with hotels, and so undercut each other in the market and gained less income (Bigay *et al.* 1981: 149). Fiji Indians, with no tradition of fire-walking, have also sought to enter the market (Brown 1984). In one small Fijian village, debate over the distribution of income from a successful ecotourism project resulted in strong competition from another, opening up conflicts within the village (Scheyvens and Purdie 1999: 221). Elsewhere in Fiji there has been (sometimes violent) conflict between villagers over who has the right to ferry surfers, for significant incomes, to a popular surfing spot, a dispute centred on land and marine tenure (Harrison 2003: 14). Only skilful negotiation prevented Solomon Islands villagers blocking routes to distant villages where an ecotourism project operated, as they sought to gain some income from those who passed through (Russell and Stabile 2003: 54). Similar problems led to the demise of one of the most successful custom villages in Tanna (see Robinson and Connell in this volume). In the Solomon Islands, the Anuha resort was burned down by villagers irate over access to incomes from tourism on their land (Weaver 2002: 129). Frustrations are frequent.

Small-scale tourism, and thus ecotourism, are particularly prone to conflict because of the limited amount of income generated, the relatively few tourists, the marginality of many ecotourism destinations in terms of other sources of income and the often visible evidence of money changing hands. At Devokula village (Fiji) conflicts occurred over income distribution, with the tour organiser assumed to be holding most of the money rather than distributing it more evenly; villagers consequently refused to be involved in dances and argued over who should clear tourist access routes (Fisher 2003: 67). These disputes later led to the village tourist project closing. Similarly in Yap (Federated States of Micronesia), probably not atypical, one village chief simply kept all the entrance fees to the village for himself, with the predictable outcome that villagers believed that this was a specific case of 'money is making people stingy and therefore harming community spirit' and refused to participate further (Mansperger 1995: 90). As this suggests, it is often the relatively poor who tend to benefit least from tourism (see for example Zeng *et al.* forthcoming 2008), because their skills are less valuable, their land holdings are too small or they are simply ignored.

Inequality occurs where some local people are excluded. In upland Nepal traditionally wealthy families had the resources necessary to open lodges for tourists, whereas disadvantaged individuals from poor families and low-caste groups received few benefits from tourism; lower castes tended to have poorly paying jobs as porters or firewood collectors (Nyaupane *et al.* 2006: 1381). A project promoting small-scale home-stay tourism in Relau village, Kedah, originated from one affluent family's initiatives, and immediately excluded households with shabby houses and no money to make improvements. With the project under way, visitors were allocated to homes according to their 'comfort and quality level' and nepotism, further disadvantaging the relatively poor and resulting in considerable disparities in earnings (Liu 2006). Exactly the same was true for a home-stay tourism scheme in Paonangisu village, Efate (Vanuatu), where village households tended to be selected according to their wealth, and thus their 'appropriateness' for Western tourists, and their links to one of the chiefs, hence the income from home-stays accentuated existing inequalities. In Relau there were further complaints over income distribution as some 60 per cent of the income from home-stays was 'managed' by an urban coordinator for administration and marketing (Liu 2006). Similarly in the Chinese village of Hongcun the majority of income was retained by urban governments and companies, and little reached the local people, though in the nearby village of Xidi local control of the industry ensured that most income stayed in the village, with village incomes some three times as great as other villages in the region without tourist facilities (Ying and Zhou 2007). Unsurprisingly, as elsewhere, the level of local involvement in management and also the kind of tourists, whether package tourists or individual travellers, explains varying degrees of economic leakage, local control and inequality.

Where income gains are limited to only some people in a community, resentment may be considerable. In two small towns near the Komodo National Park, residents who personally benefited from tourism (through employment or sale of goods) were more positive about tourism than others, an unsurprising conclusion evident elsewhere. However, attitudes were most negative in the town that overall had derived the greater benefits, but where the income distribution was consequently the more skewed, and also amongst relatively poor local people, mainly farmers and fishers, who perceived benefits going to immigrants from other parts of Indonesia (Walpole and Goodwin 2000). The clear implication was that greater equity within a community was beneficial to sustainable tourism development. Similarly where income comes from joint activities, such as the performance of fire-walking in Fijian villages, or village stays in Flores (Indonesia), and is retained for village projects rather than a small sum being given to each household (Burns 2003: 89; Stymiest 1996: 11; Cole 1997: 225), the rewards are visible and long-term survival may be more likely. In Flores tourists sponsored a village water project rather than give individual payment to the villagers, who altruistically recognised that the whole community would benefit, including those who had played no part in the tourism but had 'still had their lives interrupted and were generally inconvenienced by the presence of guests' (Cole 1997: 225). Especially in the Pacific region, islanders may forgo what might elsewhere be seen as capitalist rationality in favour of ways of life and development strategies that maintain harmony and equity rather than maximise profit for some. Such altruism and broad community spirit is rare; the Yap situation may be rather more familiar.

More rarely, the local benefits from tourism have been largely positive and relatively equitable. Tap Mun, an island off the northeast coast of Hong Kong, saw its population decline rapidly from the 1960s, from around 5000 to 100, with the decline of agriculture and fishing, and emigration or migration to the city. Recent tourism has brought large numbers of weekend tourists, who spend money on food (in restaurants and grocery stores), souvenirs of dried fish and seaweed, and water taxis. Although the principal beneficiaries are the package tourism operators who bring the tourists, the local population has gained enough income to ensure an island livelihood, retain a 'cultural connection with the sea', maintain 'their traditional lifestyles for most of the week without interference or interruption by tourists' and have an 'acceptable balance between meeting modest economic needs and optimal lifestyle opportunities' (McKercher and Fu 2006: 521). As at Mui Ne in Vietnam, this kind of explosive growth tends to be beneficial in terms of incomes, but short-term income may not equate with long-term livelihood.

Benefits from tourism are usually uneven, as certain local people and groups with better connections and education, more land or entrepreneurial skills have advantages. Tourism becomes one more means of local socio-economic differentiation. It may also contribute to marginalisation and loss of local autonomy as distant outsiders take over the critical components of the industry, such as hotels, restaurants and car hire. Local people never gain more than a proportion of tourist expenditure, and intermediaries, or overseas companies, may often be the key beneficiaries.

Environmental change

Tourism has diverse environmental impacts, usually perceived as negative. Change is inevitable where facilities must be constructed, but, given tourist interests, is less visually intrusive and damaging to the environment than other forms of development, such as industry, mining or logging. In some cases it may even result in a subsequently improved environment, where governments and hotels undertake various forms of conservation and land and marine management. In Fiji, for example, several hotels are in advance of the government in stimulating environmental conservation. In Thailand, coral reefs were managed when it became evident that their viability was crucial to the tourist industry (Wong 2001). Too often these are rare and somewhat self-interested examples. Otherwise scarce resources may be diverted 'to fuel tourist development' (Richter 1993: 110). As Pholpoke argued for northern Thailand, even tourism projects that are labelled 'ecotourism' often 'reproduce the same contradictions as other forms of capital-intensive development . . . exacerbating economic and social disparities, diverting resources and alienating the majority of people from their resource base' (1998: 262). Whether this is a 'majority' or a minority, and the extent of alienation, vary.

The gradual recognition of the value of resources, tourists' demands for pristine sites and their willingness to pay have resulted in better management practices in many destinations, and the declaration of parks and wildlife preserves. In some cases, such as Moalboal (Philippines), it was a 'revelation' to local people that tourists would pay for access to reefs; the subsequent collection of entrance fees enabled the reefs to be more effectively managed and a relatively poor village to enhance its income (White and Rosales 2003). However, where incomes from tourism are low and alternative income sources few, local people are reluctant to engage in conservation with no economic value to them. The conservation activities needed to maintain parks, reefs and similar facilities may disrupt local people, if traditional uses of such areas are banned, and park conservation may be disrupted by villagers illegally taking animals or wood from the parks, planting crops or grazing cattle there (see for example Cochrane 2006a; Rugendyke and Nguyen Thi Son 2005). In Bhutan, farmers near the Jigme Singye Wangchuck National Park were allowed to remain in the area but restrictions were placed on shifting cultivation, hunting and the collection of timber products, and wild animals, protected in the park, damaged their crops and attacked their domesticated animals, hence they resented the park and wildlife conservation, and implicitly state intervention (Wang *et al.* 2006; see also Wong in this volume). Integrating conservation and community development in and around parks remains a source of contention and frustration (Novelli and Scarth 2007). Tourists themselves may disrupt conservation management practices, animal behaviour and the long-term future of some parks. Overall, using tourism to support conservation has failed to fulfil expectations, in large part because of cultural differences that are suspicious of wilderness rather than supportive of conservation and management (Cochrane 2006a). Tourists may be more supportive of environmental stability and management than some local people (see Rajaratnam *et al.*, and Rugendyke and Nguyen Thi Son, both in this volume), though their expectations and patterns of visiting vary enormously (see Backhaus in this volume).

Environmental changes may have complex outcomes. In southern Thailand, restrictions were imposed on a considerable part of the marine habitat, to conserve a relatively pristine environment for the benefit of tourists. Conservation meant that the local Moken people were unable to fish there; since fishing was their main income source they were forced to turn to tourism as a last resort (Cohen 1996: 247). In the waters off the resort hotel on the tiny island of Ulithi (Federated States of Micronesia) tourist divers were confronted and threatened by local fishermen, concerned at their impact on marine resources (Rubinstein 2003), averting the kind of situation that faced the Moken. Elsewhere, as at Yanuca island (Sigatoka, Fiji), resorts have taken care not to alienate local people or their land and marine resources, for fear of conflict and disruption.

The correlation between rapid tourism growth and environmental degradation is usually close. By the end of the 1980s the rapid growth of tourism at Kuta had outpaced the development of tourist infrastructure so that drainage, sanitation, traffic congestion and air, water and noise pollution were all problems (Wall and Long 1996: 43), alongside the visual pollution of poles, posters, neon lights and garbage. Several similar coastal areas, such as Koh Samui (Thailand), have experienced degradation, where inadequate environmental planning and management, alongside sand mining, land clearance and other deleterious activities, including golf courses, have resulted in the loss of coastal mangroves and fisheries habitats, coral reef damage, coastal erosion and pollution from solid and liquid waste. Tourist pressure on water resources may literally drain water from irrigation systems. In Koh Samui the greatest constraint to further tourism development was access to fresh water, even intermittently brought from the mainland, but it also brought unexpected income to local people with access to productive aquifers (Williamson and Hirsch 1996; see also Cushman *et al.* 2004). Golf courses, with their demands on land and water and fertiliser run-off, have created localised problems in Bali and elsewhere in Asia (Warren 1998; Pleumarom 2002). At places such as Denarau (Fiji) and Panagsama (Philippines) accelerated coastal erosion has actually followed attempts to stabilise beaches using dikes, seawalls and even coastal cottages (White and Rosales 2003). World heritage sites such as Borobudur (Indonesia) and Angkor Wat (Cambodia) have simply become 'loved to death'. Environmental stresses are often greater where the tourism industry is not locally owned, as in parts of the Philippines, Thailand and Fiji, though this is also a function of size. However, in quite remote places, like villages associated with trekking in Nepal and upland Thailand, bamboo groves used for raft making are depleted and non-biodegradable garbage is dumped indiscriminately, often by local people and guides rather than tourists. The task of removing it or burying it is considerable (Dearden 1996).

The environmental degradation (and social transformation) of Pattaya (Thailand) as it grew from a small village into an expanding city has contributed to the quintessential tourist-directed dystopia: 'the most extreme example of the touristic transition of a seaside resort' (Cohen 2001: 159). Vung Tau (Vietnam), with its bars, karaoke parlours and casinos aimed at Chinese tourists, and Olongapo (Philippines) have acquired similar reputations. At Koh Samui tourist 'strips' dominate beaches and local people no longer have easy access for their own recreation or fishing: 'locals are typically barred access to hotels that front the beach, and are even discouraged from being on the beach in front of hotels' (Green 2005: 52). Such environmental costs of tourism merge with social costs. Although in only a handful of places have tourists been accused of being 'unpleasant guests' who are 'loud, lecherous, drunken and rude' (Boissevain 1996: 5) and/or contribute to overcrowding and environmental stress, a common complaint concerns the dress sense, or lack of it, of tourists and their intrusion into private areas (Mansperger 1995; Cole 1997; Hamzah 1997; Liu 2006). Fears that these provide demonstration effects for local youth increase these concerns (Macnaught 1982). Other concerns have been price rises following tourism (for example, Walpole and Goodwin 2001), the breach of cultural taboos on recreational activities on Sundays (Ringer 2004) and crime, often linked to outsiders.

Much of the region has remained distant and distinct from the hedonistic mass tourism of parts of the Mediterranean and Caribbean. Backpacker trails have however grown in some parts of Asia, hence a few places such as Khao San Road in Bangkok and Kuta have become centres of a particular kind of low-key, capitalist hedonism. The islands of Koh Phangan and Koh Samui (Thailand) are rare examples of the excesses of party tourism. One journalist has described this well:

What is Koh Phangan if it isn't the embodiment of the globalisation we all profess to despise? Twenty years ago before the full moon parties, Hat Rin Nok was a tiny fishing village unchanged in millennia. A generation later – our generation – and the streets are paved with Internet cafes and the fishing boats conduct all you can smoke ganja cruises . . . Equitable distribution of wealth? Not in this place. Many of the bars are owned or leased by farangs, foreigners who came for a holiday and never left. The only locals who can get jobs are those who can speak English.

(Smith 2002)

Such seemingly limited economic gains have been at some social and environmental cost.

Linkages

Tourism stimulates development in other economic sectors, notably transport, agriculture, fisheries and handicrafts. Tourist income multiplier effects are least in small island states where consumer goods, and even service industries such as hire cars, are more likely to be imported, and where tourism exists in enclaves. They are greatest where tourism is on a large scale (Brohman 1996). Leakages are also substantial from small islands in larger states, such as Malaysia (for example, Hamzah 1997). Where tourism markets are small, consistent linkages with agriculture and especially fisheries have proved difficult to develop, and neither has grown in response to tourism. In the island state of Fiji, fruit and vegetable producers were apparently unresponsive to the higher prices but uncertain demands of the tourism industry (Varley 1978) and preferred guaranteed export crop production. Today, foreign-owned hotels usually have significant global chains of food and drink supply independent of local producers. Nonetheless resorts have usually sought to stimulate local production, to encourage good relations with local people and gain a convenient, regular supply of fresh food. In Lombok the Sheraton Senggigi Beach Resort initiated programmes with local fishermen and a farmer, but the former principally benefited one local middleman and the latter failed because occupancy rates varied and the hotel did not require the volume produced by the farmer, though it continued to purchase food from local markets (Telfer and Wall 1996). Elsewhere hotels have had a commitment to both regularity and high quality that local suppliers have not always been able to meet. In contrast local guesthouses import little and depend on local production, though they may also prepare few meals.

Tourism may compete with agriculture for labour, with tourism tending to be favoured, and resorts may occupy agricultural land or disrupt coastal fisheries, hence tourist development may result in reduced food production where it would be most useful. In Bali some tourist workers have become divorced from village life, even paying fines for the non-performance of expected village activities, or selling off their village land (Cukier 1998), and no longer contributing to local production. In such places as Koh Samui, agriculture was declining prior to significant tourist development, because of world prices, increased competition and emigration; hence, tourism was 'a timely boost to a troubled island economy' (Williamson and Hirsch 1996: 188), which hastened the near demise of the agricultural economy. In some hilltribe villages the distractions of tourism employment reduced time spent on agriculture so that food had to be bought and cash crops went unharvested though the income generated by tourism was inadequate to replace these losses (Toyota 1996; Michaud 1997). Although the implication of such conclusions is that villagers should revert to their more sustainable economic system there is no evidence that this has actually occurred.

Handicrafts

Typically tourism stimulates handicraft industries, which may involve many people in local communities. Few areas where tourism is of even slight importance have failed to develop handicrafts, some, such as tee-shirt production or silverwork, initially demanding imports, some based on traditional products and some innovative creations and manifestations of 'airport art' (Parnwell 1993), fridge magnets and so on that are virtually mass-produced for tourists (evident in the many small-scale highly specialised handicraft industries of the villages between Ubud and Denpasar in Bali), or artifacts, such as didjeridus in Bali, quite unconnected to local culture. Wooden swords, hustled by itinerant carvers, are emblematic of tourism in Fiji yet never existed in the past, and the hard sell of the vendors has even been claimed to have deterred tourism in Suva, the capital city (Miyazaki 2005).

Declining trade production has sometimes been revitalised by tourism, as in the case of brassware production in the Thai village of Pa Ao (Parnwell 1993: 2521–2). In several hilltribe villages of Thailand, Vietnam, Laos and China, handicrafts, especially sewn items, have become the main sources of village income, ahead of agriculture. In Chuuk (Federated States of Micronesia) even limited tourism has helped promote handicraft production, reinvigorated artistic skill and raised awareness of the artistic merit of traditional artifacts (Nason 1984), a situation not unusual elsewhere. Thus the 'art form of tivaevae [quilts, designed and made by the women of the Cook Islands] was in danger of being lost to the world, but it has flourished since it was initially discovered by visitors to the islands and is now sold on the Internet' (Ayres 2002: 153). Ironically, though, whereas tourism has revitalised the art form, tivaevae are today rarely found for sale to tourists in the Islands – their cost, based on labour-intensive production, means they are beyond the budget of most tourists, who are more likely to buy cheaper, mass-produced artifacts, often brought to the Islands from elsewhere. Thus, even regions distant from the tourist industry may benefit from handicraft production. Many carvings from West Papua are sold elsewhere in Indonesia, especially in Bali, though intermediaries are the principal beneficiaries, rather than West Papuan villagers.

Collectively, linkages into production, accommodation and handicrafts, and both the formal and informal sectors, boost household incomes and add diversity and flexibility to household survival strategies. As with employment, the small-scale nature of some of the linkages may change local power structures and, sometimes, increase the economic role of women. Given the particular bias towards women in travel brochures and travelogues, and in the Pacific even the feminisation of destinations, there is a certain irony in this gradual change.

The complexities of culture

Throughout these processes of broadly economic change, society and culture also evolved, but in an even more complex manner. Culture has never been static, but evolved long before the advent of tourism, yet under the scrutiny of the tourist gaze visual culture at least has often changed more rapidly. Tourism is, however, just one influence on such changes, alongside trade, migration, missionisation, education, television and so on. Social change can therefore be exceptionally complicated, and neither 'tradition' nor 'change' can easily be defined or recognised.

Visual culture (art, dance and music) may be transformed in quite different ways: it may be reinforced and strengthened, embellished and changed in form (by shortening or adapting more lively and exotic components), abandoned or even invented (for example, Harnish 2005), with simultaneous conservation and dissolution. Various forms of performance have been altered in China, Bali, Fiji and elsewhere, according to their links to other non-visual components of culture and to the tourist industry. The production of material goods has likewise changed. Traditional techniques may be degraded in mass production, but artistic skills may also be invigorated and salvaged from oblivion (Crick 1989). The village of Nakabuta (Fiji), near Sigatoka on the Coral Coast, has both constructed houses within the village as tourist accommodation (a rare phenomenon in Fiji) and become the only village in the country to have retained a pottery industry, though its products have little similarity to the utilitarian goods of pre-tourist times (see also Silverman in this volume). Similarly in the northern Thai village of Ban Thawai, known as a 'carving village', three-quarters of village households produce handicrafts, but with Buddhist, Christian and Disney motifs (Nimmonratana 2000). In Suva, swords, alongside 'cannibal forks', are simply crudely invented traditions. In each of these contexts, tourism has enabled handicraft production to continue but introduced wholly new forms, and in an age of global markets socially significant meanings, and even multiple meanings of 'authenticity' may be retained by village handicraft producers (Wherry 2006).

What was once in the course of being abandoned may be revitalised for tourist consumption. Being engaged in what outsiders, specifically tourists, perceive as somewhat exotic practices, or simply being colourfully dressed (or undressed), can confer advantages within the tourism scene and sometimes enable those who may be seen locally as somehow 'backward' to gain greater benefits from tourism. Thus the Sa people of Pentecost island (Vanuatu), regarded as more 'traditional' than others on the island, are the main recipients of tourist income on the island because of their retention of spectacular customs that involve land-diving (de Burlo 1996; see also Robinson and Connell in this volume). Amongst the Toraja, migration and remittances enabled the flourishing of impressive funeral ceremonies, ironically staged by those who had converted to Christianity and embraced 'modernity', the exoticism of which brought larger numbers of tourists. Balinese cremations similarly became marketed to tourists and drew in thousands of visitors. Such efflorescences of culture, however transformed, were functions of more effective participation in a modern cash economy and resulted in the somewhat incidental ability to stimulate tourism.

In a rather similar manner, but at a vastly different scale, physical heritage, from great sites like Angkor Wat and Borobudur, or recent colonial endeavours such as the French concession in Shanghai, to more prosaic temples and shrines, is preserved at least in some part for the tourist gaze. In some contexts, such as the great walled city of Lelu on Kosrae (Federated States of Micronesia), only tourism has enabled the site to be conserved, as younger generations have ignored it (Ringer 2004) and local people have even pillaged the walls for building material, as they have done at Angkor and the Great Wall of China.

Just as with material goods, invariably there is a local perception that tourism has helped keep culture and heritage alive, though this equally invariably refers to culture's visible expressions, despite occasional concerns that kinship relations have been disrupted by commodification (for example, Nyaupane *et al.* 2006) and emerging individualism. In Makira (Solomon Islands), in a very early and short-lived phase of ecotourism, local pride and local performances of culture were both enhanced; moreover local youth, hitherto often disaffected with village life, gained roles through being tour guides, pan pipe band members or carvers (Russell and Stabile 2003: 48–9). Yet such new roles were themselves rapid responses to the arrival of the tourism.

Even in very early stages of tourism local cultures, such as that of the Moken ('sea gypsies') of southern Thailand, are adapted, embellished and staged at particular times for the tourist gaze; tourist guides on the island of Alor (Indonesia) actively promoted it as an island of black magic, and Sepik villagers were caught up in Cannibal Tours (Cohen 1996: 246; Adams 2004). Conversely local people also engage with outside imagery; in Alor it was not only tourists who created images of the 'other' but 'the Alorese themselves (drawing on anthropological writings and encounters with travellers) who were actively sculpting self-images in the hopes of luring tourists' (Adams 2004: 129). Such local agency is rarely absent (see Silverman in this volume). In almost every context local people 'traditionalise' some aspects of their lives and environments in order to enhance tourism and reduce the disappointment of those who have come invariably looking for something 'different' (see for example Hoskins 2002; Volkman 1990). At a national level, tourism agencies also shape images that almost always imply cultural (and scenic) distinctiveness, such as Papua New Guinea's slogan 'Every Place You've Never Been', hinting at the cultural capital that might be acquired from observing difference.

Difference may be enhanced. In Bali the frog, *barong* and *kecak* dances have been invented and much modified for the tourist gaze (Picard 1996; Dunbar-Hall 2001, 2003); the Short-skirt Miao of southern China transformed their dress style to be the 'sexy exotic Short-skirt Miao' and integrated up-tempo music and even breakdancing into traditional dances (Luk 2005), and fire-dancing, characteristic of Samoan performances for tourists, was actually invented by Samoans working in Hollywood in the 1960s. In Yunnan (China) Tibetan architecture, commodities and food now embellish tourist precincts (Ateljevic and Doorne 2005). Fijian fire-walking, initially confined to part of one small island in Fiji, has now spread to most significant tourist areas and is rarely if ever performed outside a tourist context (Stymiest 1996). Hilltribe girls who make a living from being photographed with tourists on the Thai–Burmese border 'willingly don a contrived tribal costume' invented from elements of different tribal attire, but marketed as 'traditional' (Cohen 2001: 164).

Akha hilltribe villages in northern Thailand perceived that offering opium to tourists was 'easy money', and 'jungle guides', who took tourists to the villages, stressed that this was traditional Akha behaviour and they should try an 'authentic' experience of Akha culture by smoking opium. The Akha themselves largely disdained the practice but, ironically, by demonstrating the practice some became addicted to it, while Akha people were looked on unfavourably by other nearby social groups (Toyota 1996). Similarly the guides described Akha women as 'traditionally obedient to men' and adept at massage, rather than explain that massages were given for necessary income, and thus deflate tourists' expectations of the 'noble savage' (Toyota 1996: 233). In multiple ways therefore local people and guides have colluded in offering 'authentic' experiences.

Collusion may become external pressure. Even in upland Makira 'women are under some pressure to be (authentically) bare-breasted when performing cultural shows, which seems to make all but the very oldest of them uncomfortable' (Russell and Stabile 2003: 52–3). Where tourism intermediaries are not villagers or local people, and are detached from village life, as in Sarawak, they pose threats to particular tourist destinations, unless certain conditions are met, which demand alterations to Iban longhouses that make them 'more authentic and traditional' and similar changes to people's dress, appearance and community roles that make them more 'exotic, primitive, untouched by the modern world' (Yea 2002: 175). Construction of 'human zoos', contemporary versions of early twentieth-century colonial exhibitions, have been notorious in the case of Padaung women in Burma (Myanmar), illegally lured or forced across the border into Thailand, to be exhibited in a 'model village' (Cohen 2001: 163–5; Parnwell 2001: 242–4). Universally, local people wish to focus on only some elements of the past, and are selective about the 'authentic', the 'traditional' and the 'primitive' in their own or others' perceptions.

Where visitors have become frequent, numerous social and cultural activities, staged for the tourist gaze, merely amount to 'pseudo-events' (MacCannell 1976) drawn away from their identification with either locality or culture. Rituals undergo reinvention and commodification, and are disentangled from cultural requirements and divorced from both time and space, to fit in with tourist cycles and space requirements. In Toraja funeral ceremonies were truncated to meet the needs of tourists, which resulted in community resentment, to the extent that several Toraja communities temporarily refused to accept tourists, but had to relent in order to sell the souvenirs and the accommodation on which they had become dependent (Adams 1990). Elsewhere, but in a different context, local people have managed to differentiate performances for tourists from those for local consumption from which tourists are excluded (Dunbar-Hall 2001). The Naxi people of Yunnan (China) took advantage of state-initiated tourism to revitalise Dongba culture, suppressed during the Cultural Revolution years, but under the tourist and government gaze it was transformed and purified of so-called 'superstitious and unscientific' elements of local religion, so alienating the tourist version from what remained everyday Dongba practice (Chao 1996). Marginality and difference are therefore marketable and thus transformed as relationships between tourists and local people become more formal, impersonal and commercialised.

Authenticity is valued by tourists and locally promoted even if it cannot be identified. Tourists perceive Toraja funerals as 'authentic', and not staged for tourist consumption, hence they are both more 'authentic' and 'real' than shows performed in Tahiti or Bali (Scarduelli 2005: 397). Local people themselves recognise the utility of 'authenticity' and its potency in drawing tourists, as in the Sepik (Papua New Guinea), where Chambri villagers stress that the presence of tourists in their villages is a testimony to the persistence and strength of tradition, and even Padaung women are proud of their cultural distinctiveness (Errington and Gewertz 1989; Parnwell 2001; Allerton 2003). Yet, like others, 'the Chambri were of value to tourists only because they were different and unequal. They would remain of interest only as long as they remained primitive, only as long as they remained a vanishing curiosity in the modern world' (Gewertz and Errington 1991: 56). In Sarawak (Malaysia) tourists have stopped visiting certain areas because they are not 'traditional' enough (King 1993: 114) though this may not herald the end of all forms of tourism in such areas. Moreover, though tourists may recognise and reject obvious travesties of local culture, they may have little sense of what 'traditional' culture might once have been, in some imagined historical past time, and, as post-tourists, be content to enjoy the 'show' for what it is (rather that what it might be supposed to be). Gewertz and Errington thus point to the wider paradox whereby 'tourists were drawn to Chambri to see those less developed whereas the Chambri sought to attract tourists so that they could be more developed [and so] tourism, the source of their future strength and autonomy in the world, was premised on continuing inequality' (1991: 28), but their participation in the tourist industry aimed at reducing that inequality.

In almost every context 'tradition' is mediated by various intermediaries, by the villagers themselves, who are selective about what variants of 'tradition' are appropriate, and by tourist reception of previous performances. Tourists sometimes had narrow, if predictable, visions of what tradition entailed, as in Papua New Guinea, where villagers were expected to dress up and act as 'savages', staging tribal fights and so on (Kulick and Willson 1992), in their quest for coherent and unchanged cultures and environments from 'the world we have lost'. There is considerable selectivity over what tourists experience and see, structured by the local people, by the promoters and cultural gatekeepers/entrepreneurs, and ultimately by the 'tyranny' of the guidebook (see McGregor 2000). The past 'is another country', which may perhaps become a 'living museum', valued by tourists and local people in quite different ways.

Perceptions therefore vary enormously; in the highlands of Papua New Guinea, Huli men, elaborately dressed for tourist dance performances, provided 'the authentic culture of a timeless present', through appropriate signs and spectacles, while themselves acquiring a 'sense of superiority and empowerment in the face of foreign strength and wealth satisfying their own desire and agency in the modern world system' (Timmer 2000: 121). Tourists and villagers are rarely anthropologists or historians, and self-deception may satisfy everyone. As Xie has put it, 'tourists are not cultural anthropologists seeking authentic experiences. Rather they are consumers looking for purchasable versions of culture' (2006: 133). As tourists shift towards 'post-tourism' – the recognition and acceptance that all is somehow fake but pleasurable – imagination triumphs and the show goes on.

Local people have often effectively become their own cultural entrepreneurs, mediating contemporary village life and 'traditional' performances and carefully balancing front and back stages, to satisfy both tourists and the villagers themselves, as a means of securing continued tourism and thus income. Village life goes on in a 'back zone', away from the 'stage', and takes very different forms from that portrayed to tourists. Even so the maintenance of tradition, even in artificial form, poses problems for the sustainability of tourism when residents seek to use their tourist income to purchase elements of modernity. One Sarawak longhouse, newly constructed of concrete and bricks, rather than with traditional materials on wooden stilts, was excluded from the tourist itinerary. Both here and in Toraja, villagers have consequently made considerable adjustments to their homes, even replacing iron with thatched roofs, to ensure the continuity of tourism, but were frustrated by their inability to either modernise or influence the itineraries of tourists (Adams 1984; Yea 2002). In Naxi, local people have 'traditionalised' guesthouses to create 'customised' authenticity (Wang 2007). In Flores, by contrast, villagers resented the government attaching heritage status to megaliths in the village, generating tourism but consigning them to being a 'primitive' and unchanged society (Cole 2003). In such contexts especially, the simple dichotomy of hosts and guests, even where it seems most evident, is simply absent: middlemen, brokers and the state play crucial roles.

In some places tourists may even be 'quarantined' to minimise their direct impact or maintain illusions. Near the southern Vanuatu island of Anatom (Aneityum) huge cruise ships periodically disgorge literally thousands of tourists on the small unpopulated offshore island of Inyeug (known to the tourists only as Mystery Island). Tourists purchase goods in an artificial market specifically set up for them, and there is restricted contact with the islanders, who were initially concerned about possible disease transmission. Islanders cross from Anatom solely to set up the market and have been able to make significant economic gains from handicrafts and food sales to pay school fees that would otherwise have been difficult to raise (Miles 1988: 175; Slatter 2006). This is effectively a cordon sanitaire to avoid the physical disruption of village life that the presence of so many tourists would entail. At Bunlap in Pentecost (Vanuatu) tourists are not allowed to see the preparations involved in staging the ritual land-diving, nor are they allowed to participate, which would render the event profane. Indeed, there as elsewhere, tourists are held in low regard by local people because of their inability to understand the meanings of local ritual and their fixation on the most colourful visual elements, despite their willingness to travel to see it (de Burlo 1996). More generally tourists are encouraged to participate in some elements of local culture, usually dancing, and even to be ridiculed, rather than be merely passive observers; local bystanders may even outnumber tourists. Typically local people enjoy such performances through their own direct participation, or simply by watching the antics and performances of other villagers or the tourists (Fisher 2003; Russell and Stabile 2003) as momentarily 'hosts' and 'guests' reverse roles.

Over time, as in Bali, local people may even shift towards a 'touristic culture' in which, through a 'circle of representation', their culture has increasingly taken on the characteristics expected of them rather than those of their own past. In Chambri local people may well be 'engaging in a form of indigenous ethno-Orientalising – portraying themselves in response or, perhaps, in resistance to images they think we have of them' (Gewertz and Errington 1993: 652). More prosaically, in Bali and elsewhere dances commissioned and developed for tourists have been absorbed back into rituals (Picard 1993; Bruner 2005; McKean 1976). In a sense they had deliberately moved towards the 'exotic other' of the tourist brochures. As in Toraja 'the tourist gaze compels the Toraja to look at their own traditions through alien eyes and enhances their consciousness of their own cultural heritage . . . which leads to the development of a new ethnic identity' (Scarduelli 2005: 394). In a world of syncretism and hybridity there is no reason why this should matter, or that culture loses meaning because it is partly commodified. After all, Balinese ritual performances have three separate audiences – the divine, the local and the touristic – and the last function does not negate the first two (McKean 1976).

Devastating critiques have been made of the impact of tourism on local cultures, not least by Picard, who has written of 'cultural tragedy' (1993: 71; see also Wood 1998). However, such critiques have tended to be the superficial legacy of intellectual hostility to rising mass tourism rather than recognition that social change was often highly acceptable. Social change has brought positive and negative consequences. Income from tourism that enables access to education, water supplies and so on is welcomed. Yet, in Mustang, within the first few months of tourism, some children had already begun begging for money and pens and had learned sexual swear words from visitors (Shackley 1994; see also Burns 2003). In large parts of the Pacific tourism was seen by many governments and villagers alike as a 'last resort' because of the threat it was thought to pose to local cultures (Macnaught 1982). In Ban Suay many Hmong actively resisted tourism since it challenged concepts of Hmong identity that centred on productive agriculture and associated community rituals (Michaud 1997). Generally local people have been much less cautious about tourism, often perceiving agriculture to be dirty, dull and difficult and merely producing 'slow money' compared with the easy and 'fast money' of tourism. More negative consequences were rarely immediate. Critiques of the social impact of tourism were often a response to the 'loss of visual culture' rather than a recognition that the most cherished values, centred on kinship relations, were usually resilient to outside influence and that tourism enabled new and positive means of self-representation. Moreover social change has usually been at least as much a consequence of education, and the myriad facets of globalisation, while, over time, national governments, such as that of China, have revalued distinctive ethnicity (albeit often as a tourist resource) and many local communities have gained pride in cultural distinctiveness, in whatever form that now takes. Change and resilience have been contemporaneous as tourism has brought a series of interconnected and sometimes paradoxical outcomes, as culture and tourism become increasingly intertwined.

New directions

Tourism usually excites local interest. It smacks of relatively easy incomes compared with agriculture, as richer outsiders pass briefly through to depart with handicrafts, photographs and memories. In many parts of the region villagers have constructed guesthouses, often way off any beaten track, in anticipation and vain hope that tourists will somehow materialise, only usually to be disappointed. Distance denies participation on any terms.

Where local people have participated in tourism it has never been easy or equitable. Capitalism and competition characterise the industry. Even at the local level tourism engenders competition as much as cooperation in the quest for success; altruism is rare and conflict not unusual. Control by local capital may not necessarily be superior to development by distant capital. Moreover not all local communities have been willing participants in the industry: they may be tourees rather than hosts. Participation in the tourist industry is also complex where local people both stage exotic cultural events that hark back to a distant past and provide services in nearby modern hotels. There is an inherent ambivalence about cultural marginality and economic incorporation, and local cultural entrepreneurs are poised in between. Yet across the Asia-Pacific region the variety of experiences and outcomes is enormous.

Communities (and districts and regions) were never homogeneous and the uneven development that has sometimes followed tourism has tended to build on existing inequalities where these relate to power, land tenure, access to resources and so on. The old social order, with its inequalities, can also be replaced by new resentments and obligations. In a sense most local communities actually fit quite well with Anderson's (1991) notion of 'imagined community', formulated in a quite different context, in glossing over the fragmentation, fissures and tensions that exist within them. In almost every place, clearly evident in Kuta and Koh Samui, there are both local winners and losers, challenging any simple notions of change.

Villages like Korotogo, where all seem to have benefited and none have been significantly disadvantaged or excluded, and which can revert to a more 'traditional' economy should tourism collapse, are rare. Indeed within the Asia-Pacific region (for example, Tokalau 2005), and more generally (see Stronza 2001; Azarya 2004), new forms of social stratification, conflict and inequality at the local level have been a more familiar outcome of tourist development. Villages like Kampung Komodo, in an impoverished part of the country, where few have benefited, are not unusual. Moreover, at Kampung Komodo a government training programme for the local cooperative, devised to train villagers in woodcarving techniques, was given no support by local tourism operators, thus undermining its marketing and distribution strategy (Walpole and Goodwin 2000). In the nominally socialist state of Vietnam at Sapa, local cultural minorities, despite representing about 85 per cent of the local population, 'are basically left to watch and hope for beneficial effects to trickle down, deprived as they are from access to economic success and political power in the state apparatus due to their cultural distinctiveness, their lack of formal education, and their limited economic capital' (Michaud and Turner 2006: 803). Even at the most local level particular interests thus shape inequitable structures. Although such conclusions have been reported in other world contexts (see Juarez 2002) there is no necessary reason why inequality, the loss of cultural autonomy and the subordination of local people in global cultures and economies are the inevitable outcome of tourist development.

However, where tourist numbers have been considerable, few have been able to retain autonomy and any real degree of control over an industry that originates and is stimulated so far from home. Indeed in some contexts governments have discouraged small-scale tourism in favour of resort tourism, effectively preferring national revenues to more equitable local development, and putting local participation under pressure, as in Lombok, where local enterprises were literally destroyed (Kamsma and Bras 2000). In Flores the local tourist board was extremely anxious to create a tourism industry that resembled that of the elite tourism of Bali, centred on extravagant facilities, even discouraging local small-scale entrepreneurs, despite tourists to Flores mainly being those who were seeking to avoid extravagance (Erb 2005). Governments can stifle local initiative as much as assist it.

Indeed it is no accident that pro-poor tourism has become of some philosophical and practical importance at the time that tourism within the rural areas of the Asia-Pacific region has become of much greater significance, and its ability to reduce inequality has been questioned. Nonetheless, it is readily evident that many areas have benefited dramatically from the rise of tourism. Indeed, Hong Kong islands and the once remote Yasawa island group in Fiji have survived only because of it; where emigration was depleting island populations, tourism has brought new development and enabled villagers to remain on their home islands. Other than in such fortunate places, where tourism has transformed island life in positive ways and negative consequences are few, without some degree of social and environmental regulation the achievement of equitable and sustainable development through tourism is rare.

Although there have been social costs from tourism its cultural impacts are almost impossible to assess by local people or outsiders. The potential economic rewards from tourism have generated intense local interest in tourism. Out of some degree of necessity, and interest in modernity whatever that might entail, many people have been willing to change their livelihoods and their cultural practices for the tourist gaze. But then, the income that comes from tourism is a major enticement. Changes are always risky. At Sapa in highlands Vietnam, where barely a decade of tourism has brought significant changes:

> Sa Pa town is perceived by the majority of backpackers as noisy, unsightly and ultimately an infringement on nature. Modernity shocks them, urban sprawl drives them away, karaoke excesses and rampant prostitution are judged sickening; they came here precisely to get away from it all . . . the popularity of Sa Pa for backpackers is likely to decline. There is already a tendency in their discourse to . . . label it as increasingly 'worn out', a 'spoilt' destination where the damages of 'bad tourism' have made interactions too 'commercial' and rendered a visit less appealing.
>
> (Michaud and Turner 2006: 799)

Tourism is always poised for a potential fall. Yet, what may here be unappealing to backpackers is admirable for the growing number of urban Vietnamese tourists; a new tourist cycle takes over as Sapa goes from small town to mass tourism destination. Ultimately it is valuable to constantly question whether tourism is sustainable, given its obvious environmental impacts, but also in social, cultural and economic terms, and in whether it is equitable. The following chapters elaborate on these themes through detailed case studies from diverse contexts, where local society and environment and the nature and impact of tourism have taken quite different forms.

References

Abramson, A. (2004) 'A Small Matter of Some Rent to be Paid: Towards an Analysis of Neo-Traditional Direct Action in Contemporary Fiji', in T. van Meijl and J. Miedema (eds), *Shifting Images of Identity in the Pacific*, Leiden: KITLV Press.

Adams, K. (1984) ' "Come to Tana Toraja, Land of the Heavenly Kings": Travel Agents as Brokers of Ethnicity', *Annals of Tourism Research*, 11: 469–485.

—— (1990) 'Cultural Commoditization in Tana Toraja, Indonesia', *Cultural Survival Quarterly*, 14: 31–34.

—— (1997) 'Touting Touristic "Primadonnas": Tourism, Ethnicity and National Integration in Sulawesi, Indonesia', in M. Picard and R. Wood (eds), *Tourism, Ethnicity and the State in Asian and Pacific Societies*, Honolulu, HI: University of Hawai'i Press.

—— (2004) 'The Genesis of Touristic Imagery: Politics and Poetics in the Creation of a Remote Indonesian Island Destination', *Tourist Studies*, 4: 115–135.

Allerton, C. (2003) 'Authentic Housing, Authentic Culture? Transforming a Village into a "Tourist Site" in Manggarai, Eastern Indonesia', *Indonesia and the Malay World*, 31: 119–128.

Anderson, B. (1991) *Imagined Communities*, New York: Verso.

Ateljevic, I. and Doorne, S. (2005) 'Dialectics of Authentication: Performing "Exotic

Otherness" in a Backpacker Enclave of Dali, China', *Journal of Tourism and Cultural Change*, 3: 1–17.

Ayres, R. (2002) 'Cultural Tourism in Small-Island States: Contradictions and Ambiguities', in Y. Apostolopoulos and D. Gayle (eds), *Island Tourism and Sustainable Development*, Westport, CT: Praeger.

Azarya, V. (2004) 'Globalization and International Tourism in Developing Countries: Marginality as a Commercial Commodity', *Current Sociology*, 52: 949–967.

Baker, K. and Coulter, A. (2007) 'Terrorism and Tourism: The Vulnerability of Beach Vendors' Livelihoods in Bali', *Journal of Sustainable Tourism*, 15: 249–266.

Berno, T. (1996) 'Cross Cultural Research Methods: Content or Context? A Cook Islands Example', in M. Hitchcock, V. King and M. Parnwell (eds), *Tourism in South-East Asia*, London: Routledge.

Berno, T. and Jones, T. (2001) 'Power, Women and Tourism Development in the South Pacific', in Y. Apostolopoulos, S. Sonmez and D. Timothy (eds), *Women as Producers and Consumers of Tourism in Developing Regions*, Westport, CT: Praeger.

Bigay, J., Green, M., Rajotte, F., Ravuvu, A., Tubanavau, M. and Vitusagavulu, J. (1981) *Beqa: Island of Firewalkers*, Suva: Institute of Pacific Studies.

Boissevain, J. (1996) 'Introduction', in J. Boissevain (ed.), *Coping with Tourists*, Oxford: Berghahn.

Bras, K. and Dahles, H. (1998) 'Women Entrepreneurs and Beach Tourism in Sanur, Bali: Gender, Employment Opportunities and Government Policy', *Pacific Tourism Review,* 1: 243–256.

Bricker, K. (2001) 'Ecotourism Development in the Rural Highlands of Fiji', in D. Harrison (ed.), *Tourism and the Less Developed World*, Wallingford: CABI Publishing.

Briedenhann, J. and Wickens, E. (2004) 'Tourism Routes as a Tool for the Economic Development of Rural Areas – Vibrant Hope or Impossible Dream', *Tourism Management*, 25: 71–79.

Brohman, J. (1996) 'New Directions in Tourism for Third World Development', *Annals of Tourism Research*, 23: 48–70.

Brown, C. (1984) 'Tourism and Ethnic Competition in a Ritual Form: The Firewalkers of Fiji', *Oceania*, 54: 223–244.

Bruner, E. (2005) *Culture on Tour: Ethnographies of Travel*, Chicago: University of Chicago Press.

de Burlo, C. (1996) 'Cultural Resistance and Ethnic Tourism on South Pentecost, Vanuatu', in M. Hitchcock, V. King and M. Parnwell (eds), *Tourism in South-East Asia*, London: Routledge.

Burns, G. (2003) 'Indigenous Responses to Tourism in Fiji', in D. Harrison (ed.), *Pacific Island Tourism*, New York: Cognizant Communication Corporation.

Butler, R. (1980) 'The Concept of a Tourist Area Cycle of Evolution: Implications for Management of Resources', *Canadian Geographer*, 24: 5–12.

—— (1991) 'Tourism, Environment and Sustainable Development', *Environmental Conservation*, 18: 201–209.

Chant, S. and McIlwaine, C. (1995) *Women of a Lesser Cost: Female Labour, Foreign Exchange and Philippine Development*, London: Pluto.

Chao, E. (1996) 'Hegemony, Agency and Re-presenting the Past: The Invention of Dongba Culture among the Naxi of South West China', in M. Brown (ed.), *Negotiating Ethnicities in China and Taiwan*, Berkeley, CA: University of California, Institute of Asian Studies.

Cochrane, J. (2006a) 'Indonesian National Parks: Understanding Leisure Users', *Annals of Tourism Research*, 33: 979–997.

—— (2006b) 'The Sustainability of Ecotourism in Indonesia: Fact and Fiction', in M. Parnwell and L. Bryant (eds), *Environmental Change in South-East Asia: People, Politics and Sustainable Development*, London: Routledge.

Cohen, E. (1982) 'Marginal Paradise: Bungalow Tourism on the Islands of Southern Thailand', *Annals of Tourism Research*, 9: 189–228.

—— (1996) 'Hunter-Gatherer Tourism in Thailand', in M. Hitchcock, V. King and M. Parnwell (eds), *Tourism in South-East Asia*, London: Routledge.

—— (2001) 'Thailand in "Touristic Transition"', in P. Teo, T. Chang and K. Ho (eds), *Interconnected Worlds: Tourism in Southeast Asia*, Oxford: Elsevier.

Cole, S. (1997) 'Anthropologists, Local Communities and Sustained Tourism Development', in M. Stabler (ed.), *Tourism and Sustainability: Principles to Practice*, Wallingford: CAB International.

Cole, S. (2003) 'Appropriated Meanings: Megaliths and Tourism in Eastern Indonesia', *Indonesia and the Malay World*, 31: 140–150.

Connell, J. (1993) 'Bali Revisited: Death, Rejuvenation and the Tourist Cycle', *Environment and Planning D*, 11: 641–661.

Crick, M. (1989) 'Representations of International Tourism in the Social Sciences: Sun, Sex, Sights, Savings and Servility', *Annual Review of Anthropology*, 18: 307–344.

Cukier, J. (1996) 'Tourism Employment in Bali: Trends and Implications', in R. Butler and T. Hinch (eds), *Tourism and Indigenous Peoples*, London: International Thomson Business Press.

—— (1998) 'Tourism Employment and Shifts in the Determination of Social Status in Bali: The Case of the "Guide"', in G. Ringer (ed.), *Destinations: Cultural Landscapes of Tourism*, London: Routledge.

Cukier, J., Norris, J. and Wall, G. (1996) 'The Involvement of Women in the Tourism Industry of Bali, Indonesia', *Journal of Development Studies*, 33: 248–270.

Cushman, C., Field, B., Lass, D. and Stevens, T. (2004) 'External Costs from Increased Island Visitation: Results from the Southern Thai Islands', *Tourism Economics*, 10: 207–219.

Dahles, H. (2000) 'Tourism, Small Enterprises and Community Development', in D. Hill and G. Richards (eds), *Tourism and Sustainable Community Development*, London: Routledge.

Darma Putra, I. and Hitchcock, M. (2006) 'The Bali Bombs and the Tourism Development Cycle', *Progress in Development Studies*, 6: 157–166.

Dearden, P. (1996) 'Trekking in Northern Thailand: Impact Distribution and Evolution over Time', in M. Parnwell (ed.), *Uneven Development in Thailand*, Aldershot: Avebury.

Dunbar-Hall, P. (2001) 'Culture, Tourism and Cultural Tourism: Boundaries and Frontiers in Performances of Balinese Music and Dance', *Journal of Intercultural Studies*, 22: 173–187.

—— (2003) '*Tradisi* and *Turisme*: Music, Dance and Cultural Transformation at the Ubud Palace, Bali, Indonesia', *Australian Geographical Studies*, 41: 3–16.

Erb, M. (2003) '"Uniting the Bodies and Cleansing the Village": Conflicts over Local Heritage in a Globalizing World', *Indonesia and the Malay World*, 31: 129–139.

—— (2005) 'Limiting Tourism and the Limits of Tourism: The Production and Consumption of Tourist Attractions in Western Flores', in C. Ryan and M. Aicken (eds), *Indigenous Tourism: The Commodification and Management of Culture*, Amsterdam: Elsevier.

Errington, F. and Gewertz, D. (1989) 'Tourism and Anthropology in a Post-Modern World', *Oceania*, 60: 37–54.

Fairbairn-Dunlop, P. (1994) 'Gender, Culture and Tourism Development in Western Samoa', in V. Kinnaird and D. Hall (eds), *Tourism: A Gender Analysis*, Chichester: John Wiley.

Fisher, D. (2003) 'Tourism and Change in Local Economic Behavior', in D. Harrison (ed.), *Pacific Island Tourism*, New York: Cognizant Communication Corporation.

Garrick, D. (2005) 'Excuses, Excuses: Rationalizations of Western Sex Tourists in Thailand', *Current Issues in Tourism*, 8: 497–509.

Gewertz, D. and Errington, F. (1991) *Twisted Histories, Altered Contexts. Representing the Chambri in a World System*, Cambridge: Cambridge University Press.

Gewertz, D. and Errington, F. (1993) 'We Think, Therefore They Are?', in A. Kaplan and D. Pease (eds), *Cultures of United States Imperialism*, Durham, NC: Duke University Press.

Gibson, C. and Connell, J. (2005) *Music and Tourism: On The Road Again*, Clevedon: Channel View.

Green, R. (2005) 'Community Perceptions of Environmental and Social Change and Tourism Development on the Island of Koh Samui, Thailand', *Journal of Environmental Psychology*, 25: 37–56.

Hall, C. (1992) 'Sex Tourism in South-East Asia', in D. Harrison (ed.), *Tourism and the Less Developed Countries*, London: Belhaven.

Halvaksz, J. (2006) 'Becoming "Local Tourists": Travel, Landscapes and Identity in Papua New Guinea', *Tourist Studies*, 6: 99–117.

Hamzah, A. (1997) 'The Evolution of Small-Scale Tourism in Malaysia: Problems, Opportunities and Implications for Sustainability', in M. Stabler (ed.), *Tourism and Sustainability: Principles to Practice*, Wallingford: CAB International.

Harnish, D. (2005) 'Teletubbies in Paradise: Tourism, Indonesianisation and Modernisation in Balinese Music', *Yearbook for Traditional Music*, 37: 103–123.

Harrison, D. (2003) 'Themes in Pacific Island Tourism', in D. Harrison (ed.), *Pacific Island Tourism,* New York: Cognizant Communication Corporation.

Hitchcock, M. (1993) 'Dragon Tourism in Komodo, Eastern Indonesia', in M. Hitchcock, V. King and M. Parnwell (eds), *Tourism in South-East Asia*, London: Routledge.

—— (2004) 'Margaret Mead and Tourism: Anthropological Heritage in the Wake of the Bali Bombings', *Anthropology Today*, 20: 9–14.

Hoskins, J. (2002) 'Predatory Voyeurs: Tourists and "Tribal Violence" in Remote Indonesia', *American Ethnologist*, 29: 797–808.

Howe, L. (2005) *The Changing World of Bali*, London: Routledge.

de Jonge, H. (2000) 'Trade and Ethnicity: Street and Beach Sellers from Raas on Bali', *Pacific Tourism Review*, 4: 75–86.

Juarez, A. (2002) 'Ecological Degradation, Global Tourism and Inequality: Maya Interpretations of the Changing Environment in Quintana Roo, Mexico', *Human Organization*, 61: 113–124.

Kamsma, T. and Bras, K. (2000) 'Gili Trawangan – from Desert Island to "Marginal" Paradise: Local Participation, Small-Scale Entrepreneurs and Outside Investors in an Indonesian Tourist Destination', in D. Hill and G. Richards (eds), *Tourism and Sustainable Community Development*, London: Routledge.

Kindon, S. (2001) 'Destabilizing "Maturity": Women as Producers of Tourism in Southeast Asia', in Y. Apostolopoulos, S. Sonmez and D. Timothy (eds), *Women as Producers and Consumers of Tourism in Developing Regions*, Westport, CT: Praeger.

King, V. (1993) 'Tourism and Culture in Malaysia', in M. Hitchcock, V. King and M. Parnwell (eds), *Tourism in South-East Asia*, London: Routledge.

Kontogeorgopoulos, N. (2004) 'Ecotourism and Mass Tourism in Southern Thailand: Spatial Interdependence, Structural Connections and Staged Authenticity', *GeoJournal*, 61: 1–11.

Kulick, D. and Willson, M. (1992) 'Echoing Images: The Construction of Savagery in Papua New Guinean Villages', *Visual Anthropology*, 5: 143–152.

Ledger, H. (2007) 'Pressure Put on Paradise', *Sunday Telegraph Escape*, 2 December, 6–7.

Liu, A. (2006) 'Tourism in Rural Areas: Kedah, Malaysia', *Tourism Management*, 27: 878–889.

Luk, T. (2005) 'The Poverty of Tourism under Mobilizational Developmentalism in China', *Visual Anthropology*, 18: 257–289.

MacCannell, D. (1976) *The Tourist*, New York: Schocken.

McGregor, A. (2000) 'Dynamic Texts and Tourist Gaze: Death, Bones and Buffalo', *Annals of Tourism Research*, 27: 27–50.

McKean, P. (1976) 'Tourism, Culture Change and Culture Conservation in Bali', in D. Banks (ed.), *Changing Identities in Modern Southeast Asia*, The Hague: Mouton.

McKercher, B. and Fu, C. (2006) 'Living on the Edge', *Annals of Tourism Research*, 33: 508–534.

Macnaught, T. (1982) 'Mass Tourism and the Dilemmas of Modernization in Pacific Island Communities', *Annals of Tourism Research*, 9: 359–381.

McNaughton, D. (2006) 'The "Host" as Uninvited "Guest": Hospitality, Violence and Tourism', *Annals of Tourism Research*, 33: 645–665.

Mansperger, M. (1995) 'Tourism and Cultural Change in Small-Scale Societies', *Human Organization*, 54: 87–94.

Mauer, J.-L. and Zeigler, A. (1988) 'Tourism and Indonesian Cultural Minorities', in P. Rossel (ed.), *Tourism: Manufacturing the Exotic*, Copenhagen: IWGIA.

Michaud, J. (1997) 'A Portrait of Cultural Resistance: The Confinements of Tourism in a Hmong Village in Thailand', in M. Picard and R. Wood (eds), *Tourism, Ethnicity and the State in Asian and Pacific Societies*, Honolulu, HI: University of Hawai'i Press.

Michaud, J. and Turner, S. (2006) 'Contending Visions of a Hill-Station in Vietnam', *Annals of Tourism Research*, 33: 785–808.

Miles, W. (1988) *Bridging Mental Boundaries in a Postcolonial Microcosm: Identity and Development in Vanuatu*, Honolulu, HI: University of Hawai'i Press.

Miyazaki, H. (2005) 'From Sugar Cane to "Swords": Hope and the Extensibility of the Gift in Fiji', *Journal of the Royal Anthropological Institute*, 11: 277–295.

Nason, J. (1984) 'Tourism, Handicrafts and Ethnic Identity in Micronesia', *Annals of Tourism Research*, 11: 421–449.

Nimmonratana, T. (2000) 'Impacts of Tourism on a Local Community: A Case Study of Chiang Mai', in K. Chon (ed.), *Tourism in Southeast Asia*, New York: Haworth.

Novelli, M. and Scarth, A. (2007) 'Tourism in Protected Areas: Integrating Conservation and Community Development in Liwonde National Park (Malawi)', *Tourism and Hospitality Planning and Development*, 4: 47–73.

Nyaupane, G., Morais, D. and Dowler, L. (2006) 'The Role of Community Involvement and Number/Type of Visitors on Tourism Impacts: A Controlled Comparison of Annapurna, Nepal and Northwest Yunnan, China', *Tourism Management*, 27: 1373–1385.

Parnwell, M. (1993) 'Tourism and Rural Handicrafts in Thailand', in M. Hitchcock, V. King and M. Parnwell (eds), *Tourism in South-East Asia*, London: Routledge.

—— (2001) 'Sinews of Interconnectivity: Tourism and Environment in the Greater Mekong Subregion', in P. Teo, T. Chang and K. Ho (eds), *Interconnected Worlds. Tourism in Southeast Asia*, Oxford: Elsevier.

Pholpoke, C. (1998) 'The Chiang Mai Cable-Car Project: Local Controversy over Cultural and Eco-tourism', in P. Hirsch and C. Warren (eds), *The Politics of the Environment in Southeast Asia*, London: Routledge.

Picard, M. (1993) '"Cultural Tourism" in Bali: National Integration and Regional Differentiation', in M. Hitchcock, V. King and M. Parnwell (eds), *Tourism in South-East Asia*, London: Routledge.

—— (1996) *Bali: Cultural Tourism and Touristic Culture*, Singapore: Archipelago Press.

Pleumarom, A. (2002) 'How Sustainable is Mekong Tourism?', in R. Harris, T. Griffin and P. Williams (eds), *Sustainable Tourism: A Global Perspective*, Oxford: Butterworth Heinemann.

Prasso, S. (2007) 'The Vietnamese Way', *Australian Financial Review/Life and Leisure*, 2 March: 20–25.

Ranck, S. (1987) 'An Attempt at Autonomous Development: The Case of the Tufi Guest Houses, Papua New Guinea', in S. Britton and W. Clarke (eds), *Ambiguous Alternative: Tourism in Small Developing Countries*, Suva: University of the South Pacific.

Richter, L. (1993) 'Tourism Policy-Making in South-East Asia', in M. Hitchcock, V. King and M. Parnwell (eds), *Tourism in South-East Asia*, London: Routledge.

Ringer, G. (2004) 'Geographies of Tourism and Place in Micronesia: The "Sleeping Lady" Awakes', *Journal of Pacific Studies*, 26: 131–150.

Rubinstein, D. (2003) 'A Tale of Two Islands', in T. Aoyama (ed.), *Social Homeostasis of Small Islands in an Island Zone*, Occasional Paper No. 39, Kagoshima: Kagoshima University Research Center for the Pacific Islands.

Rudkin, B. and Hall, C. (1996) 'Unable to See the Forest for the Trees: Ecotourism Development in Solomon Islands', in M. Hitchcock, V. King and M. Parnwell (eds), *Tourism in South-East Asia*, London: Routledge.

Rugendyke, B. and Nguyen Thi Son (2005) 'Conservation Costs: Nature-Based Tourism as Development in Cuc Phuong National Park, Vietnam', *Asia Pacific Viewpoint*, 46: 185–200.

Russell, D. and Stabile, J. (2003) 'Ecotourism in Practice: Trekking the Highlands of Makira Island, Solomon Islands', in D. Harrison (ed.), *Pacific Island Tourism*, New ork: Cognizant Communication Corporation.

Scarduelli, P. (2005) 'Dynamics of Cultural Change among the Toraja of Sulawesi', *Anthropos*, 100: 389–400.

Scheyvens, R. and Purdie, N. (1999) 'Ecotourism', in J. Overton and R. Scheyvens (eds), *Strategies for Sustainable Development: Experiences from the Pacific*, London: Zed Books.

Shackley, M. (1994) 'The Land of Lo, Nepal/Tibet', *Tourism Management*, 15: 17–26.

Simpson, P. and Wall, G. (1999) 'Environmental Impact Assessment for Tourism: A Discussion and an Indonesian Example', in D. Pearce and R. Butler (eds), *Contemporary Issues in Tourism Development*, London: Routledge.

Sinha, C. and Bushell, R. (2002) 'Understanding the Linkage between Biodiversity and Tourism: A Case Study of Ecotourism in a Coastal Village in Fiji', *Pacific Tourism Review*, 6: 35–50.

Slatter, C. (2006) *The Con/Dominion of Vanuatu? Paying the Price of Investment and Land Liberalisation – A Case Study of Vanuatu's Tourism industry*, Auckland: Oxfam New Zealand.

Smith, C. (2002) 'Backpackers Inc', *Sydney Morning Herald*, 7 July: 1, 6.

Sofer, M. (1990) 'The Impact of Tourism on a Village Community: A Case Study of Votua Village, Nadroga/Navosa, Fiji', *Journal of Pacific Studies*, 15: 107–130.

Sofield, T. (1996) 'Anuha Island Resort: A Case Study of Failure', in R. Butler and T. Hinch (eds), *Tourism and Indigenous Peoples*, London: International Tourism Business.

Stronza, A. (2001) 'Anthropology of Tourism: Forging New Ground for Ecotourism and Other Alternatives', *Annual Review of Anthropology*, 30: 261–283.

Stymiest, D. (1996) 'Transformation of Vilavilairevo in Tourism', *Annals of Tourism Research*, 23: 1–18.

Tapari, B. (1988) *Problems of Rural Development in the Western Papuan Fringe*, Occasional Paper No. 9, Port Moresby: University of Papua New Guinea Department of Geography.

Telfer, D. and Wall, G. (1996) 'Linkages Between Tourism and Food Production', *Annals of Tourism Research,* 23: 635–653.

Timmer, J. (2000) 'Huli Wigmen Engage Tourists: Self-Adornment and Ethnicity in the Papua New Guinea Highlands', *Pacific Tourism Review*, 4: 121–135.

Tokalau, F. (2005) 'The Economic Benefits of an Ecotourism Project in a Regional Economy: A Case Study of Namuamua Inland Tour, Namosi, Fiji Islands', in C. Hall and S. Boyd (eds), *Nature-Based Tourism in Peripheral Areas. Development or Disaster?*, Clevedon: Channel View.

Toops, S. (1992) 'Tourism in Xinjiang, China', *Journal of Cultural Geography*, 12: 19–34.

Toyota, M. (1996) 'The Effects of Tourism on an Akha Community: A Chiang Rai Village Case Study', in M. Parnwell (ed.), *Uneven Development in Thailand*, Aldershot: Avebury.

Tuinabua, L. (2000) 'Tourism and Culture: A Sustainable Partnership', in A. Hooper (ed.), *Culture and Sustainable Development in the Pacific*, Canberra: Asia Pacific Press.

Ulack, R. (1993) 'The Impact of Tourism in Fiji: A Comparison of Two Villages', *Focus*, 43, Summer: 1–7.

Varley, R. (1978) *Tourism in Fiji: Some Economic and Social Problems*, Bangor Occasional Papers in Economics No. 12, Bangor: University of Wales Press.

Volkman, T. (1990) 'Visions and Revisions: Toraja Culture and the Tourist Gaze', *American Ethnologist*, 17: 91–110.

Wall, G. and Long, V. (1996) 'Balinese Homestays: An Indigenous Response to Tourism Opportunities', in R. Butler and T. Hinch (eds), *Tourism and Indigenous Peoples*, London: International Thomson Business Press.

Walpole, M. and Goodwin, H. (2000) 'Local Economic Impacts of Dragon Tourism in Indonesia', *Annals of Tourism Research*, 27: 559–576.

Walpole, M. and Goodwin, H. (2001) 'Local Attitudes towards Conservation and Tourism around Komodo National Park, Indonesia', *Environmental Conservation*, 28: 160–166.

Wang, S., Lassoie, J. and Curtis, P. (2006) 'Farmer Attitudes towards Conservation in Jigme Singye Wangchuck National Park, Bhutan', *Environmental Conservation*, 33: 148–156.

Wang, Y. (2007) 'Customized Authenticity Begins at Home', *Annals of Tourism Research*, 34: 789–804.

Warren, C. (1998) 'Tanah Lot: The Cultural and Environmental Politics of Resort Development in Bali', in P. Hirsch and C. Warren (eds), *The Politics of the Environment in Southeast Asia*, London: Routledge.

Weaver, D. (2002) 'Perspectives on Sustainable Tourism in the South Pacific', in R. Harris, T. Griffin and P. Williams (eds), *Sustainable Tourism: A Global Perspective*, Oxford: Butterworth Heinemann.

Wherry, F. (2006) 'The Social Sources of Authenticity in Global Handicraft Markets: Evidence from Northern Thailand', *Journal of Consumer Culture*, 6: 5–32.

White, A. and Rosales, R. (2003) 'Community-Oriented Marine Tourism in the Philippines', in S. Gossling (ed.), *Tourism and Development in Tropical Islands*, Cheltenham: Edward Elgar.

Wilkinson, P. and Pratiwi, W. (1995) 'Gender and Tourism in an Indonesian Village', *Annals of Tourism Research*, 22: 283–299.

Williamson, P. (1992) 'Tourist Developers on Koh Samui, Thailand', *Journal of Cultural Geography*, 12: 53–64.

Williamson, P. and Hirsch, P. (1996) 'Tourism development and social differentiation in Koh Samui', in M. Parnwell (ed.), *Uneven Development in Thailand*, Aldershot: Avebury.

Wilson, D. (1993) 'Time and Tides in the Anthropology of Tourism', in M. Hitchcock, V. King and M. Parnwell (eds), *Tourism in South-East Asia*, London: Routledge.

Winter, A. (2007) 'Rethinking Tourism in Asia', *Annals of Tourism Research*, 34: 27–44.

Wong, P. (2001) 'Southeast Asian Tourism: Traditional and New Perspectives on the Natural Environment', in P. Teo, T. Chang and K. Ho (eds), *Interconnected Worlds: Tourism in Southeast Asia*, Amsterdam: Pergamon.

Wong, P. (2003) 'Tourism Development and the Coastal Environment on Bintan Island', in S. Gossling (ed.), *Tourism and Development in Tropical Islands*, Cheltenham: Edward Elgar.

Wood, R. (1998) 'Touristic Ethnicity: A Brief Itinerary', *Ethnic and Racial Studies*, 21: 218–241.

Xie, P. (2006) 'The Development of Cultural Iconography in Festival Tourism', in P. Burns and M. Novelli (eds), *Tourism and Social Identities: Global Frameworks and Local Realities*, Oxford: Elsevier.

Yea, S. (2002) 'On and Off the Ethnic Tourism Map in Southeast Asia: The Case of Iban Longhouse Tourism, Sarawak, Malaysia', *Tourism Geographies*, 4: 173–194.

Ying, T. and Zhou, Y. (2007) 'Community, Governments and External Capitals in China's Rural Cultural Tourism: A Comparative Study of Two Adjacent Villages', *Tourism Management*, 28: 96–107.

Zeng, B., Carter, R., De Lacy, T. and Bauer, J. (forthcoming 2008) 'Effects of Tourism Development on the Local Poor People – Case Study of Taibai Region, China', in V. Jauhari (ed.), *Global Cases on Hospitality Industry*, New York: Haworth Press (in press)

2 Another (unintended) legacy of Captain Cook?

The evolution of Rapanui (Easter Island) tourism

Grant McCall

Rapanui Islanders take a certain bitter delight in how the name for their home can be translated as either 'the navel of the world' or 'the place at the end of the world'. Tepito o te Henua is a name that figures in old songs, chants and later documents produced by Rapanui and others: the world's most remote inhabited place (Cristino and Izaurieta 2006). Small size, small population and remoteness have shaped Rapanui's history for the millennium and a half that the place has had people on it. The nearest occupied island is Pitcairn, some 1900 kilometres away. That Polynesians managed to find and settle Rapanui is a tribute to their navigational prowess. That almost completely locally owned tourism is now the most important component of the Rapanui economy, and that large numbers of tourists visit Rapanui, may be almost as remarkable.

Once on the island, the settlers who became Rapanui did their best to adapt to its sub-tropical climate and variable rainfall, coupled with high winds and rocky soil, developing an agricultural system around lithic mulching, which, if not on the modern visitor's itinerary, permitted Rapanui to develop a considerable surplus to finance the construction of a remarkable megalithic building complex, unique in the world, but still the subject of research and debate (Hunt 2006). Rapanui is at once recognisable as a Pacific island, but also central to theories about lost continents, space alien visitors and, of late, environmental morality tales (Bahn and Flenley 1992; Flenley and Bahn 2002; Rainbird 2002; Diamond 2005). 'Mysterious' Easter Island, a resonant phrase in world languages, has given way to its being a lesson for the planet.

Unlike many other accounts of the island's dramatic history (Ebensten 2001; Fischer 2001; Streining 2001; Sierra 2002; Peteuil 2004; Fischer 2005), a theme of Rapanui agency runs throughout this chapter. Rapanui suffered a number of depredations at the hands of outsiders, but they also manipulated (when they could) these outsiders and their desires. That is how they survived and eventually prospered.

Contact and colonial history have merged into a tourism narrative. There is sufficient evidence that Rapanui were aware of their island's attractions for outsiders and began an intermittent tourist trade from Captain James Cook's 1774 visit, which included the production of portable artifacts to supplement their meagre trade in foodstuffs. This controlled Rapanui reception of visitors was subsequently curbed and then curtailed almost entirely from its incorporation into the Chilean state in 1888, which began a long period of restrictions on the access of outsiders, until the birth of contemporary tourism from the 1950s.

Isolation and a constrictive political scene eventually favoured Islanders' control of their own tourism enterprise in the last four decades of rapid development. Unlike other parts of the world, the Rapanui control almost all their tourism infrastructure, from accommodation and guiding to transport and support for those activities. This unusual dominance is both an unintended consequence of their remoteness and isolation, and a tribute to the Islanders' persistence and inventiveness in the face of considerable odds, both environmental and historical.

Contact and colonial history as tourism narrative: visitors and souvenirs

Two accounts record the first visiting expeditions, of the Dutch who arrived on Easter Sunday 1722 and named it Easter Island and the Spanish who were the first to stake a formal claim. In accounts of both the Dutch and Spanish visits, a Rapanui came out to one of the ships and seemed to understand its size and importance. Typically, the outsiders presented trinkets, even food and drink, but no return objects are recorded. The only portable artifacts observed by the Dutch and Spanish were items of clothing, weapons and tools. If the visitors received anything at all in exchange for the goods they gave, it was island food and little of that (McCall 1990)

Whereas the Dutch visit lasted just five days and was terminated by an inadvertent burst of violence – resulting in the death of a dozen Rapanui, including the first man to make contact with the outsiders – the Spanish had much more agreeable relations. Over the seven-day visit, a pair of officers travelled around the entire inhabited island, staying with and near Rapanui households. In their account, they lamented the lack of material culture, apart from survival basics such as food and tools. There was very little that the Rapanui had to exchange.

Yet, just four years after the Spanish visit, Captain James Cook in 1774 and his crew found that the Rapanui had a considerable quantity of inventive and attractive carvings, which they pressed on their visitors, like any artisan at a tourist site today. Indeed, Cook was so angered at being pestered to trade, that he returned disgusted to his ship shortly after landing, and the crew was later cheated by being sold baskets of sweet potatoes that were mainly stones. The first Rapanui to swim out to the Cook expedition even carried carvings tied on a rope belt around his waist, as did other subsequent and hopeful traders. The Rapanui probably used some of the tools (tomahawks and knives are mentioned) from the Spanish to produce the artifacts urged on the Cook expedition.

The souvenir trade and tourism – the receiving of outsiders according to a plan for profit – was effectively born with the 1774 Cook expedition. Visitors liked mementos and local people profited by making them. That some were very finely made does not detract from their status as souvenirs manufactured for trade, part of a nascent commercial drive: an impulse to produce tradeable materials to obtain the valuables that outsiders seemed willing to supply. After all, on their sparse island, on the brink of a new ecological disaster with resources vastly depleted (Fischer 2005: 45–6), they had little else to entice visitors to part with attractive and useful goods.

Following Cook, all visitors reported the artifact trade, often producing drawings of Rapanui engrossed in that work, such as the famous illustration from the La Pérouse expedition of 1786 (Figure 2.1). Here, the visitors are avidly studying the monuments whilst their hats are being stolen, or members of the crew are enticed by women whilst their pockets are picked. Even the illustrator figures in the tableau: guiding trade (to get to the monument), while sexual trade and a mirror are shared by the couple to the left. Given that a dozen years had passed between Cook and La Pérouse, no wonder that they were keen to exploit as many aspects of trade with new outsiders as they could.

The format established during the Cook visit was repeated dozens of times with ship visits from 1774 until the twentieth century, when air travel supplanted ocean voyages as the main source for visitors. A single scout, or group of scouts, visited the ship, bearing items for trade. On returning to shore, more Rapanui appeared, similarly supplied. Until the early twentieth century, though ship visits were still few and unpredictable, Rapanui had no idea when the next group of visitors/tourists would appear, so they set to carving when they could, to have supplies always ready, produced for their exchange rather than use value (McCall 1994a). Yet then, as now, not all Rapanui found entertaining visitors and making artifacts to their taste. Some remained aloof and distant, watched from a distance, but had no apparent desire to engage in trade (Loti 1998: 30).

INSULAIRES ET MONUMENS DE L'ÎLE DE PÂQUE.

Figure 2.1 Lithograph of the La Pérouse visit of 1786.

Similarly commencing with Cook's visit was the discovery that female sexual services were a trade item and many visitors commented on this in subsequent years. Cook, who was very concerned about the transmission of disease by his crew, tried to prevent such liaisons, but they still took place. The development of a sex trade for visitors is hardly a surprise; lonely men were engaged on long voyages. Even though most visitors recorded the sexual trade, none failed to comment on how few women seemed to be in evidence. Again, probably, this was a specialisation practised by just some on Rapanui, but evidence of the early development of this abiding element of the tourism business.

Rapanui curbed and curtailed: the island of innocent inmates

After 140 years of sporadic contact, 1862 heralded a major change in the Rapanui way of transacting with outsiders, as slave raids impressed on the Islanders the vulnerability of their homeland to attack (McCall 1976a). Although this labour trade was of short duration, it was all-encompassing, with disease being a considerable feature for the first time. Aside from population decline, through death and emigration, which had taken the indigenous population to fewer than 200, the island society acquired for the first time a few resident outsiders whose role in mediation with later outsiders was to be considerable for more than a century.

The Rapanui continued to adapt in terms of their internal social order, making five attempts between 1868 and 1886 to have France declare a protectorate over their island; a succession of French governments denied all these petitions. Confidence gradually returned and people returned to tourism and souvenir production. Apart from tourism, a French settler on Rapanui, Jean-Baptiste Dutrou-Bornier, hit on the idea of transforming the island into a boutique farm with European products for what he saw as the growing influence of his country in tropical Oceania. His trading ships supplied mutton and wool, along with grapes, apples and peaches, to the hungry European settlers of Papeete, but alongside assorted artifacts produced by the islanders.

Museums from Britain, Chile, France, Russia, the USA and Germany sent ships to Rapanui to collect artifacts. Collectors came with ideas of what they wanted, sometimes bringing illustrations and showing them to carvers (McCall 1994a). A member of one 1883 German expedition wrote with some disappointment that artifacts were displayed on organised shelves with prices in various European currencies. He thought he was at 'the end of the world' but was actually visiting what was by then a population with considerable experience of receiving visitors and catering for their needs (Ayres and Ayres 1995). Some carvings from this era indicate lesser skill than exists on the island today, unsurprising since in 1877 the population had declined from about 3500 to only 110 people: there simply were not enough people to run the sheep ranch, produce souvenirs and take visitors about when they arrived.

Beginning in 1868, with the visit of a British warship, collectors wanted something larger than the figurines so popular in a previous age. Father Roussel, the self-appointed head of the Catholic mission on Rapanui from 1866 to 1870, remarked dryly that the captain of the ship liked 'souvenirs' so, apart from figurines, Rapanui assisted the crew to remove two full-sized *moai* stone figures, one of considerable iconic value today, on display in the British Museum for decades, and regarded as one of that institution's most treasured objects. In 1870, a French ship wanted a *moai*, but removed a head instead, owing to lack of cargo space. Later Chileans, Belgians, Americans and (again) the French removed the larger souvenirs that the Rapanui literally had lying around!

Chile acquired Rapanui in 1888. The first Chilean administrator failed to impose Chilean rule, though Chile did succeed in ending ties between Rapanui and the rest of Oceania, except as mediated through Chile. Chileans, developers and the Chilean state were poorly resourced, so there was little contact during the 1890s. The occasional non-Chilean ship managed to call, but the souvenir trade dwindled to a Chilean-controlled dribble.

With the tightening grip of the sheep ranch and Chilean control, the Rapanui were enclosed in their single village and prevented from visiting any other part of their island. Contacts with ships became more restricted by naval authorities and commercial interests, though expeditions of various sizes, some lasting over a year and others a few months, provided opportunities for souvenir sales, tour guiding and even other services such as the occasional accommodation request. Chileans who arrived on company or naval ships reported how people offered them meals, accommodation, guiding and transport during the week or so they stayed. Perhaps such visits could be cast as an early form of adventure tourism, since no special structures were built to receive the tourists: they were accommodated in private Rapanui homes.

Payment for these services was rarely in cash: there was no commerce on Rapanui, so goods were the most highly valued form of payment, though Rapanui sometimes used money to purchase things from passing ships. Late nineteenth- and early twentieth-century photographs show how widespread European clothes were among the population, and Islanders even ordered such things as tea sets from passing ships. Attending to the occasional visitor was the only business open to Rapanui, and the only means of acquiring goods; the little commercial life was dominated by the ranch on what was now a 'company island' (Porteous 1981). Resident officials and their families would be taken on tours, or offered fishing and other exploratory expeditions, if they were not interested in the island's dramatic past. In return, Rapanui hoped for special access to the visiting ships, trade items, even money that could be used to obtain goods. Ship officials always received favoured treatment, in the hope of profitable exchanges.

The first long-term visitors who lived in Rapanui accommodation and who were guided about the island were archaeologists in the 1950s and 1960s, before civil government brought freedom. Heyerdahl's famous account of his own expedition to Easter Island in 1955 recorded his first meeting with islanders who had swum out to his ship:

> One by one the ragged fellows jumped down on the deck, shook hands with everyone they could get near and produced bags and sacks full of curious

things. The most bizarre wooden carvings began to circulate from hand to hand and soon attracted more attention than their owners . . . All the carvings were masterly, faultlessly executed and so highly polished they were like porcelain to the touch . . . I had never seen such productivity on any of the other Polynesian islands whose inhabitants prefer to take life very calmly . . . The wood-carvers pointed with apologetic smiles to their ragged trousers and bare legs and wanted to barter their goods for clothes and shoes. In a few seconds business was in full swing all over the deck.

(Heyerdahl 1958: 39–40)

Rapanui who had worked for the company as shepherds knew the island well. Some had made it their business to collect stories from old people and retail these to interested visitors: a specialist kind of cultural tourism. The sites visited by Cook and others a century or so before had not changed and Rapanui were eager and mainly knowledgeable guides. Through newspaper reports and personal recommendation, some Islanders, such as Juan Tepano, a guide for the archaeologist Katherine Routledge in 1914–16 and the Franco-Belgian expedition of 1934, became known for their expertise.

Visitors were still few, the Chilean regime was brutal and dogmatic, there was considerable hardship until the 1960s, and many Rapanui sought out-migration. Yet hardship and isolation prepared the Islanders well for the next stage of freedom and development. Since it was so cut off from the rest of the world and had only a small non-Rapanui resident population, language and folklore still flourished and were practised in performance for locals, visitors and resident non-Rapanui. Any visitors, in whatever capacity, could be confident of being given an expert tour of the major sites, depending upon the amount of time available. Tours, now by four-wheel drives, followed the paths taken by Captain Cook's crew more than two centuries earlier.

Turning disadvantage to advantage

After 1953, when the foreign-owned ranch closed down, the entire operation of Rapanui became military. Governance and commerce fell under the rule of the Chilean Navy. Administration became even more authoritarian until it eventually produced revolt in 1964. The revolution was not so much anti-Chilean as anti-Navy and brought many changes in Rapanui society and culture, especially in terms of visitors, their numbers and needs. The so-called 'Easter Island Law' of June 1966, intended for 15 years but effectively the governance statute of Rapanui in 2006, had several important provisions, including exemption of all activities on the island from Chilean tax of any kind, a stipulation that contributed to the development of tourism on the island, allowing local capital accumulation. Relative isolation and the ignorance of Chilean authorities of what the island was actually like meant that Islanders were largely free to develop in their own way as numerous distant and irrelevant plans had minimal impact.

Resident non-Rapanui numbers were boosted considerably by the establishment of a joint Chilean and United States Air Force 'secret' base in 1966, probably a listening post for imagined and real Soviet naval movements in the Pacific. From the Islander point of view, it was a boon to the local economy. Military development was extravagant and bountiful (Porteous 1981: 192–3). Young men stationed on Rapanui were looking for entertainment, with considerable material and monetary resources; governments shipped great quantities of expensive, specialist equipment, not all of which they used immediately and much of which was discarded. The first airstrip in 1966 was made from high-quality steel plates, which eventually served entrepreneurial Rapanui as excellent fencing and building materials. Military supply markets, called 'Post Exchanges', provisioned not only the base personnel but all their Rapanui friends with a variety of goods, including the beginnings of consumer commodities such as gramophones, domestic appliances and even motor vehicles.

The Americans took the lead in all practical terms at the 'joint' base; they supplied the facilities and material and maintained the link with the outside world. The base hosted a regular 'Open House', to which resident Rapanui and Chileans were invited, where the beginning of training in catering was introduced to Islanders who worked in those facilities. Some refer still to the country and western recordings of the time as 'Open House' since that was the music normally played there.

Rapanui, in return, developed 'bars' and 'discotheques' in their lounge rooms or in special adjacent buildings built, again, from base cast-off materials, using recordings on players traded for various services. One early discotheque was made entirely of flattened 44-gallon drums, used to bring fuel to the island. As the military staff were male and young – including the Chileans there and in the navy – these Rapanui facilities became very popular, and an entertainment industry came into being, complete with Tahiti-inspired shows providing additional employment in the growing industry.

The first organised tourism also came about because of American dominance. The luxury adventure tourism operator Lindblad (influenced by the Heyerdahl expedition of 1955–6) initiated their inaugural tour in 1967, with the first civilian but not yet scheduled flight. Clients travelled by a LAN-Chile chartered DC-6 propeller-driven aircraft that had Pullman train seats, but only 40 persons, and usually arrived monthly. Lindblad imported Dodge pick-up trucks from the USA, fitting them with timber benches and tarpaulin protective tops. The hotel was simply a cluster of tents around a central communal area housing lectures, meetings and dining. Rapanui were always the drivers, the guides being mostly outsiders hired by Lindblad, including archaeologists who did the work in exchange for support for their research. The manager was an American. Beyond employment the impact was limited since all the food had to be flown in from the mainland (Fischer 2005: 221).

Lindblad clients typically arrived with a copy of Heyerdahl's (1958) popular *Aku-Aku*, where the places on the tour were described and explained. The Heyerdahl volume served as Rapanui's first tour guide and visitors would ask Rapanui to take them to places mentioned there, such as 'secret' – actually fabricated – caves.

The itinerary always involved a Sunday church service (mainly in Tahitian) and usually a barbecue lunch at Anakena, the one white sand beach on the island, where coconut palms had been brought and planted by the Chilean Navy in the 1950s. In 1967 there were 444 tourists but this number increased when flights became twice weekly in 1971. Easter Island's first summer festival, later called Tapati, was held in 1969, eventually becoming the island's main social attraction. Most tourists were from North America, with smaller numbers from Asia, Europe and Australasia.

The three years that Lindblad operated there resulted in Rapanui gaining skills in modern tourism organisation on a continuing basis. It was one thing to receive the occasional visitors but quite another to actually develop a programme of return visits and a sense of planning for visitor satisfaction. The Lindblad period served as on-the-job training for Rapanui and those who Lindblad employed became the first tourism and accommodation operators.

The resident military conducted short- and long-term liaisons with Rapanui women, some resulting in offspring, and some in marriage and long-term residence in the USA. Some tourists, despite the short visits, became acquainted with Rapanui men and women, and ultimately took them to their homes in Europe and America. With the education of Rapanui in Chile, there were a number of marriages between Rapanui and Chileans. Rather dramatically, the island was joining the rest of the globe. Whereas some outsiders had known of and visited Rapanui since 1722, only after the 1960s were Rapanui able to experience the world with the same freedom.

What had been a small number of resident non-Rapanui before 1966 ballooned into several hundred, then a thousand, as government services proliferated. The socialist government of Salvador Allende (1970–3) appointed three young educated Rapanui to head public services, but most were Chileans from the mainland and they also brought their families, which produced a considerable expatriate population to service and supply, resulting in the growth of a local entrepreneurial Rapanui population. The 'Easter Island Law' established that only Rapanui could own land on the island, except that used for government and military purposes. Sometimes Rapanui, from their privileged landholding position, would employ Chileans as managers or other staff. Occasionally, a partnership might be formed.

With the election of the socialist Allende government in 1970, American investment in Chile fled, including that of Lindblad on Rapanui. Those who had worked for Lindblad found themselves with a tourism plan and vehicles, even some accommodation in parts of their homes, and they were now in charge. The Allende regime favoured communal development of Rapanui, which never caught on. The Lindblad tent hotel became part of a chain of government-organised accommodation in remote Chilean places, and imported a 120-bed prefabricated hotel from Florida which replaced the tents.

In late 1971 LAN-Chile leased a Boeing 707 transcontinental jet from Lufthansa, along with a servicing agreement, and started their Frankfurt to Papeete itinerary. Once a week, the plane arrived from Germany, via Santiago, bringing new tourists to Rapanui, where they stopped briefly. After a refuel, which allowed passengers to take a rush tour of some of the archaeological sites close to the airport, the flight continued with the bulk of those on board going to Papeete. There new passengers were collected, and the return flight, fuel stop and short tour repeated. Each leg between Santiago, Rapanui and Papeete took about five hours. Despite such limited airline capacity, tourism slowly developed, with more duty-free vehicles being imported through Chile to carry passengers to tourist sites. Rapanui entrepreneurs organised their kin resources into providing labour and materials, including food, and the front rooms of Rapanui homes became guesthouses, then small hotels, as properties were enlarged and more services added. Indeed in the 1970s and 1980s the tourist rooms were sometimes the first rooms to be built in new houses to enable an early start to income generation.

The Chilean military coup on 11 September 1973 shocked not only the world, but also the Rapanui, who soon adapted to the situation and profited from it. The new military dictator, Augusto Pinochet, became the first President of Chile to visit the island, a visit that he and his top officials repeated several times, showering the island with more facilities including improved roads and other infrastructure that contributed to better visitor facilities. Ironically it was one of the few parts of Chile where Pinochet and his conspirators could feel safe, as Islanders treated these distant guests like any others.

Rapanui maintained their dominance of commerce, including the tourism industry. People learned English and French, some extending to other languages. Rapanui married visiting outsiders who possessed language skills and they joined the family business: looking after tourists. By the mid-1970s annual tourist numbers had reached around 2000, mainly from the USA, France (including French Polynesia) and Germany; it was too costly for Chileans (Porteous 1980). The island population was itself still below 2000, almost half of whom had come from the Chilean mainland.

Towards the modern era

Tourism began to grow steadily in the 1980s, once the impact of the coup had faded, and by 1994 tourist numbers had reached 8000. Tourists increasingly outnumbered the local population, despite its having grown following substantial migration from Chile. Though LAN-Chile now scheduled four flights a week, most tourists were cruise ship passengers ashore for less than a day. Despite their limited spending on the island, tourism (alongside growing government employment) had resulted in as many as 1200 cars on the island, let alone motorcycles, jamming the few streets of Hanga Roa. About ten small hotels and 28 guesthouses were now evident – alongside two supermarkets and new stores that accepted bank cards, and restaurants. The Tapati Festival had become a week-long event, and planting of flowers, shrubs and imported tropical trees had transformed a rather barren landscape on Hanga Roa into a more evidently Polynesian township (Fischer 2005: 241–3).

During the 1990s tourist numbers continued to increase, so that by 1999 there were no fewer than 21,434 tourists who stayed overnight on the island – by then about seven times the local population. Some 8500 came from western Europe,

7350 from South America (almost all from Chile – as it had become an economic possibility), 2460 from North America, 1600 from Oceania (mainly Australia) and 1500 from Asia (almost entirely Japan). The number of small hotels and guesthouses had increased to 46 (Fischer 2005: 250). Yet Rapanui experienced even faster growth in the twenty-first century, indeed one of the most spectacular increases in the world. Tourist numbers grew to 45,000 in 2005 and a reported 48,000 in 2006 – much the highest number of tourists per capita anywhere in the Pacific (Gonschor 2007: 244) and comparable only with a few exceptional Caribbean islands.

Tourism is Rapanui's only industry of any consequence. Some 91 per cent of the economy is now dependent on tourism (Gonschor 2007) and it employs more people than the once dominant public service. Early on, income was small as few visitors stayed long, if at all, and in the 1970s the lone state hotel received perhaps two-thirds of the island's food and lodging income. Moreover, the benefits were even more limited because agriculture and fishing had both declined already and residents, as well as tourists, had to import most of their food (Porteous 1980: 137). Through family links, tourism touches everyone associated with the island, but people do not share equally in this wealth. It was not until the larger numbers of tourists of the 1990s and the rapid increase in numbers of guesthouses and transport businesses that Islander incomes significantly grew. A nascent middle class was present at the end of the 1970s (Porteous 1980: 138) and by 1993 most of the relatively few wealthy local families were the proprietors of small hotels. For many families, however, the cost of maintaining, servicing and renovating facilities was barely covered by tourist income (Stanton 2003: 116), but there were few other alternatives. By then tourism was

> clearly the economic mainstay of many of the island's residents – especially the Rapa Nui. Individuals rent rooms, vehicles, horses, and even snorkeling equipment to tourists. Paid entertainers in the hotels, discotheques and res-taurants are normally locally born, and most crafts and souvenirs are made by Rapa Nui artisans. Nevertheless most Rapa Nui in tourism work at the lower level (e.g. in room service, food preparation, grounds maintenance or desk clerks) while better jobs with a more dependable salary (accountants, secretaries, interpreters and tour guides) are predominately [*sic*] taken by non-islanders.
>
> (Stanton 2003: 117)

Traces of the barter economy that Cook would have been familiar with remained into the 1990s but by the present century the money economy dominated all trans-actions. As that occurred the more traditional extended Polynesian family took on a more nuclear form (Fischer 2005: 252). However, certainly until the 1990s, the small size of the island and the extensive kin relationships there 'encouraged more egalitarianism and sharing than elsewhere would have been the case' (Porteous 1981: 226). While incomes might still be limited and skewed, tourism dominated the island economy.

During the 1990s, growing numbers of tourists actually stayed on the island. Of the just over 30,000 visitors in 2000 some 17,791 (58 per cent) used island accommodation for at least one night, and that number was three times what it had been at the start of the decade. On average, tourists from Chile stayed for six nights and those from other destinations three or four nights (Escuela de Turismo 2002), hence the tourist impact on the island had become considerable. About half of these tourists came by sea and half by air.

Accommodation, food and travel costs (for tours and taxis) dominate tourist expenditure, with the main tour companies on Rapanui all being run by local people. Handicrafts remain important, whether as scrupulously accurate copies of older works or tawdry 'airport art'. Woodcarvings generate significant incomes and shell necklaces, entirely the preserve of groups of women, were introduced in the 1970s. At least into the 1980s carvings were exchanged and bartered for such goods as jeans, radios, sunglasses and watches that were hard to obtain on the island, while before the arrival of a passenger ship 'people plan[ned] their carving programme for several weeks, often working long hours to build up a large stock of figures to sell or barter' (McCall 1994b: 131–2). By 2002 some 20 shops specifically sold tourist artifacts of various kinds.

The actual size of this economy is difficult to measure. As no taxes are collected, there are no official figures. One way to judge this might be to assume that every visitor spends an average of US$50 per day (the currency in which much tourism business is conducted) for not less than a three-day visit. This estimate would yield US$7.2 million based on 2006 figures. This is not an unreasonable assumption, since many tourists spend much more than this and stay longer. Despite a major loss of income to Rapanui (because of the extent of imports, especially food and fuel, from the mainland) it constitutes a massive income for a small island.

New incomes brought new family structures and a rise in wealth. Most Rapanui gradually came to target a more Chilean middle-class ideal of regular employment, higher education for their children, at least one car, regular overseas travel and such goods as colour televisions, computers, microwaves and telephones. The local standard of living had dramatically risen, alongside what has been described as 'a run-away materialism' where 'individual avarice and virulent consumerism are undermining social cohesion' (Fischer 2005: 257, 260).

Just as the publication of *Aku-Aku* in 1955 brought the first hesitant and elite tourism, and the opening of the airport in 1967 stimulated a second wave of tourism, so the filming of Kevin Costner's US$20 million epic *Rapa Nui* in 1993 brought a new phase of tourism (despite dismal response to the film). For many Europeans much of that was to see the spectacular and mysterious human landscape. Pico Iyer's explanation for visiting is not so unusual:

> We had come here to step out of a century we knew too well and to ground ourselves in what belonged to the collective past, as it seemed: something totemic, ancestral, all that we associated with Easter Island's stone figures, or moai, which move across the island after dark, the locals believe.
>
> (2006: 8)

For growing numbers of South Americans it was to see the island touted as a Spanish-speaking Tahiti or Hawai'i (Stanton 2003: 118). The title of one article in Qantas' in-flight magazine suggests what might be valued: 'Lost World: Mysterious Stone Monoliths, Ancient Hieroglyphics and Extinct Birdmen – Experience the Timeless Culture of Remote Easter Island' (Banks 2003). Mystery, intrigue and isolation are all selling points.

Rapanui's most obvious attraction is cultural tourism: its *moai* of universal renown. It became a UNESCO World Heritage site in 1995. People come to the island in increasing numbers by air and sea. Easter Island, in spite of not having a suitable harbour, is on the cruise routes for a number of lines, each visit landing several hundred passengers for a day's stay. Flights are still the monopoly of the Chilean national airline, LAN, but are more frequent than in the past, allowing visitors to choose either a long or a short stay. The idea that Rapanui's Mataveri airport would become sort of a South Pacific hub was never realised. The Government of Chile prefers visitors to arrive through Santiago as part of a tour of South American cultural sites such as Machu Picchu and the Galapagos. Most LAN flights to Rapanui come from Santiago, not Papeete.

In the midst of grassroots control by Rapanui of the tourism industry is one outstanding exception – the Hotel Hanga Roa, which had begun as the Lindblad tent accommodation. When the Pinochet regime came to power in 1973, neo-liberal economic advisors set about divesting the Chilean state of its assets, including the national hotel chain. Along with other properties, the Rapanui hotel was sold, despite this being illegal under the 'Easter Island Law', but because of this is now the only freehold land on the island. The property has been bought and sold several times since the 1970s and was acquired in 2005 by the German Chilean Schiess family, which had plans for its considerable expansion. The Rapanui family who originally allowed the hotel construction thought that it was the government who was going to run it and, when the transaction took place, a complete Chilean manufactured kit house seemed good compensation: it was the only private one on the island. Today, the 'payment' – what people then thought was compensation for temporary use – is run down and shabby, whilst the hotel continues to grow and prosper next door. The Hotel Hanga Roa employs Rapanui in menial capacities, but the managerial staff are outsiders. In that respect, the Hotel Hanga Roa is little different from any other implanted tourism development in the Pacific and elsewhere. But in Rapanui it is an exception that proves the rule. Even so, some wealthy and influential Rapanui plan to rectify this anomaly in an otherwise Islander-controlled local economy.

Autonomous futures in modernity

Rapanui has moved towards autonomy since the 1950s and Chilean government plans are to grant the island more freedom in the future. Partly this derives from Chilean consciousness of the island, but at a popular level rather than in government policy. As many Rapanui comment, neglect always was the order of the day as far as official government action was concerned. The moves towards internal autonomy derive from the 1993 'Indigenous Law' and are part of long-term national policy rather than being distinctive to Rapanui. Rapanui are the smallest of Chile's ethnic minorities, and vocal Mapuche groups in the south and others have made a greater impact through protest and politics. A succession of Chilean governments never knew what to do with their Pacific island possession, apart from its use for a brief period as a prison for political prisoners in the 1920s and 1930s. Moreover a small, poor country must administer a 6435-kilometre coastline, with Rapanui constituting a 'far west' some 3600 kilometres away.

Autonomy can only benefit the islanders and their tourism development. When people discussed autonomy in 2001–2, one of the great attractions was that planning could be done locally: local services could be used and local suppliers brought in to projects. Instead of more people being employed from Chile, Rapanui residents could be engaged. Sometimes taken as anti-Chilean, it was a distinct preference for local involvement and talent, whatever the ethnicity. Tourism planners from Chile often come with incomplete local knowledge, although they have the power to access resources and direct central government funding. One enticing element of autonomy is that it would permit Rapanui to regulate their own air space. Rapanui for over three decades have decried the monopoly of LAN, the Chilean national carrier, to carry passengers and freight. In 2002, many eagerly embraced a doomed plan for a specialist charter service.

Coupled with issues of autonomy, the viable carrying capacity of the island for residents as well as visitors has emerged in government reports since the 1971 *Plan regulador*, alongside concern about environmental damage from rapidly expanding volumes of garbage and raw sewage (Campbell 2006). Newspaper reports in early 2006 documented the lack of water, space, food, even electricity, for the growing number of tourists (Molina and Saona 2006). A recurring theme of visiting journalists (along with those promoting tourism) is the environmental and social impacts of residents and tourists. The municipal tip is over one of the island's main aquifers, rubbish collection is a problem, and heavily visited sites are deteriorating. Electricity blackouts occur and there are concerns over future water supply (Gonschor 2007). Graffiti were present on *moai* by the mid-1970s and ancient rock carvings have been scraped with stones to make them easier to photograph, cruise ship visitors have stood in such numbers on old stone houses that some have collapsed and stone walls have had to be erected around sites to keep out rental vehicles. Shorelines have been washed away as sand has been removed for building. Bulldozing land has removed irreplaceable archaeological sites: 'the island's economic lifeblood' (Fischer 2005: 252). No coherent management plan exists for the island.

Other kinds of pressures remain. The Sociedad Agricola y Servicios (SASIPA), which provides water and electricity to the island, and operates a cattle ranch, has sought to exploit their valuable land by building a golf course, a luxury resort and a botanical garden (Stanley 2002: 302–3; Fischer 2005: 261). An especially intense controversy over tourism erupted in 2005 when a Chilean company announced plans to build a casino on the island, in a joint venture with a Rapanui entrepreneur. It was argued that this would generate 150 jobs and 10 per cent of the income would go to the island, but many Rapanui signed a protest against the project on the grounds that it would 'increase the problems accompanying mass tourism and would contribute to the erosion of Rapanui culture' (Gonschor 2007: 245). Nonetheless these projects may never materialise; many Rapanui 'have seen mass tourism . . . on the continent or while visiting friends and relatives in Tahiti and do not like what they see' (Stanton 2003: 124) though dislike may not prevent changes that increase the extent of overseas ownership. Even in 2006, many recent Chilean immigrants were working as taxi drivers and selling cheap imitation woodcarvings, undermining the Rapanui handicrafts trade (Gonschor 2007: 245).

Some Rapanui urge a reduction in tourist numbers, but rarely for long before their relatives silence them with scorn, pointing out how widespread the benefits are: school fees paid, children sent for education elsewhere, enabling a standard of living higher than in most of Chile and, probably, much of the island Pacific. Whether selling a room, an artifact or a tour, all Rapanui get their cut of the tourism pie, through family or personal connections. Local fishermen and agri-culturalists sell their produce to people who accommodate tourists and there is always casual work available for young people who wish to work in the Rapanui hotels, guesthouses and restaurants. Thus 'young Rapanui are among the most fortunate in the Pacific islands . . . because they need not leave their home island to prosper' (Fischer 2005: 228).

Rapanui has developed a local industry in visitor reception since 1774 – even then tourism perhaps – supplying tours and local crafts of increasing diversity and sophistication, from small crude carvings to high-quality replicas of museum pieces and dramatic contemporary artworks. As the 'Easter Island Law' protects Islander land, few outsiders have managed to survive in the island's commercial environment. Just one hotel symbolises a tiny alien presence. Outsiders who have succeeded have done so invariably by joining Rapanui families as spouses and in-laws. All this has kept commerce localised (despite food and other imports) to a degree difficult to find in the rest of the Pacific, and perhaps elsewhere, with so many tourists and so few local people. Rapanui visitor numbers have become substantial and will probably continue to grow as transport complexities ease, despite occasional misgivings about tourist carrying capacity.

Necessarily therefore there is some concern about a possible future without tourism, should that falter, or should LAN-Chile – still the lone airline – face tough times. A few, thoughtful Islanders wonder what might be the next stage for Rapanui. Is there anything beyond tourism? Most Rapanui emphasise the educa-tion of their children. Families have regularly made great sacrifices to support the schooling of their offspring hoping that *something* must result. In the first genera-tion of education, Rapanui left the island in 1956, the sons and daughters of people with little or no formal schooling, owing to the island's remoteness and Chilean neglect. Parental expectation is for ever more sophisticated levels of education and training, moving from trades to the professions. Indeed tourism offered and ena-bled social mobility: 'driver to vehicle owner to office manager to tour operator to hotelier and entrepreneur – many Rapanui of the "junta generation" experienced just such a career' (Fischer 2005: 228), though this played its part in eroding the egalitarianism that had hitherto characterised Rapanui. Tourism provides a place for such people to work, if autonomy enables controls on the employment of non-Rapanui. Some second- and third-generation educated Rapanui are networking as professionals with an eye to higher levels of achievement. As this pool of educated talent becomes a major force in community leadership, there will be more demand and more opportunities for locals to fill island jobs.

More visitors, more education and greater sophistication in the world will bring about the paradox that, just as people are easily able to visit the place, the people who live there will have become increasingly cosmopolitan. Cultural identity has already changed substantially as Rapanui retain minority status on their own small island and few, if any, have purely Polynesian ancestry. With a census population of just 3837 in 2002, and tourist numbers around 12 times that, external influence is enormous. With autonomy and increased tourism, Rapanui will continue to marry non-Rapanui outsiders, which will lead to the loss of indigenous cultural patterns, for example in exchange, ritual and beliefs. Most particularly, these advances into modernity contribute to a decline in that most precious cultural commodity, language (Santa Coloma 2006).

Cultural change on Rapanui will not prevent the island remaining a remote place seething with outrageous mystery, and a lure for new generations of tourists. Chileans will continue to see the place as their Hawai'i (or Tahiti) and the rest of the world will see it as a site of scientific or esoteric knowledge. For many visitors, it is the island and not the islanders that they come to visit: it is the island and not the islander that is the teacher. Many Rapanui are happy with this impenetrability. Indeed, it was a characteristic that most had perfected some time ago owing to their extensive experience in receiving visitors. The last two sentences in my doctoral thesis remain true some three decades later:

> These modes of conduct are evidence rather of the sophisticated methods of the Rapanui of capturing for their vulnerable little land resources in goods and people which it otherwise lacks. Within the family, the fiction of sharing cloaks the reality of give and take and in the wider world the mask of *chilenidad* covers the unperturbed Rapanui soul.
>
> (McCall 1976b: 286)

James Cook might thus still recognise island ways. Tourism to the 'dream island' at 'the end of the world' will probably continue to increase and, with autonomy and continued Chilean national government backing, it will remain a unique case of a global industry remaining firmly in local, grassroots hands. Fame has created island tourism; isolation and neglect have thus far enabled Rapanui to remain in control.

Note

Fieldwork on Rapanui was carried out in 1972–4, 1985–6 and 2001–2. I am grateful to the Australian National University, the Australian Research Council and the University of New South Wales at various times for supporting the fieldwork and research.

References

Ayres, W. and Ayres, G. (1995) *Geiseler's Easter Island Report: An 1880s Anthropological Account*, Asian and Pacific Archaeology Series No. 12, Honolulu: Social Science Research Institute, University of Hawai'i.

Bahn, P. and Flenley, J. (1992) *Easter Island: Earth Island: A Message from our Past for the Future of our Planet*, London: Thames and Hudson.

Banks, J. (2003) 'Lost World: Mysterious Stone Monoliths, Ancient Hieroglyphics and Extinct Birdmen – Experience the Timeless Culture of Remote Easter Island', *Qantas Magazine*, January, 26–31.

Campbell, P. (2006) 'Easter Island: On the Verge of a Second Environmental Catastrophe', *Rapa Nui Journal*, 20: 67–70.

Cristino, C. and Izaurieta, R. (2006) 'Easter Island: Total Land Area of Te Pito o Te Henua', *Rapa Nui Journal*, 20: 81.

Diamond, J. (2005) *Collapse*, London: Penguin/Allen Lane.

Ebensten, H. (2001) *Trespassers on Easter Island*, Key West: Ketch and Yawl Press.

Escuela de Turismo (2002) *Informe Final Estudio de Demanda por Evaluación y Certificación en Subsectores del Area Turismo*, Santiago: Escuela de Turismo, DuocUC.

Fischer, H. (2001) *Sombras sobre Rapa Nui: Alegato por un pueblo olvidado*, trans. Luisa Ludwig, Santiago: Editorial LOM.

Fischer, S. (2005) *Island at the End of the World: The Turbulent History of Easter Island*, London: Reaktion Books.

Flenley, J. and Bahn, P. (2002) *The Enigmas of Easter Island: Island on the Edge*, Oxford: Oxford University Press.

Gonschor, L. (2007) 'Rapa Nui', *The Contemporary Pacific*, 19: 240–7.

Heyerdahl, T. (1958) *Aku-Aku: The Secret of Easter Island*, London: Allen and Unwin.

Hunt, T. (2006) 'Rethinking the Fall of Easter Island', *American Scientist*, 94: 412–19.

Iyer, P. (2006) 'Like Nowhere Else', *The Australian*, 21 October: 7–8.

Loti, P. (1998) *Isla de Pascua*, Santiago: Libros del Ciudadano, LOM Ediciones.

McCall, G. (1976a) 'European Impact on Easter Island. Response, Recruitment and the Polynesian Experience in Peru', *Journal of Pacific History*, 11: 90–105.

—— (1976b) *Reaction to Disaster: Continuity and Change in Rapanui Social Organisation*, unpublished doctoral thesis, Australian National University.

—— (1990) 'Rapanui and Outsiders: The Early Days', in B. Illius and M. Laubscher (eds), *Circumpacifica: Festschrift für Thomas S. Barthel*, vol. 2, Frankfurt am Main: Peter Lang.

—— (1994a) 'Rapanui Images', *Pacific Studies*, 17: 85–102.

—— (1994b) *Rapanui: Tradition and Survival on Easter Island*, 2nd edition, Sydney: Allen and Unwin.

Molina, J. and Saona, T. (2006) 'Colapso energético por alta demanda turística', *El Mercurio* (Santiago), 18 February.

Peteuil, M.-F. (2004) *Les Évadés de l'île de Pâques*, Paris: L'Harmattan.

Porteous, J.D. (1980) 'The Development of Tourism on Easter Island', *Geography*, 65: 137–8.

—— (1981) *The Modernization of Easter Island*, Western Geographical Series No. 19, Victoria, BC: University of Victoria.

Rainbird, P. (2002) 'A Message for our Future? The Rapa Nui (Easter Island) Ecodisaster and Pacific Island Environments', *World Archaeology*, 33: 436–51.

Santa Coloma, M. (2006) *Guardianes de la tradición: Mestizaje y conflicto en la sociedad rapanui*, Isla de Pascua: Rapanui Press.

Sierra, M. (2002) *Rapanui: Náufragos del planeta*, Santiago: Editorial Persona.

Stanley, D. (2002) *South Pacific*, 8th edition, Emeryville, CA: Avalon Travel.

Stanton, M. (2003) 'Economics and Tourism Development on Easter Island', in D. Harrison (ed.), *Pacific Island Tourism*, New York: Cognizant Communication.

Streining, H.-D. (2001) *Das Osterinsel Syndrom*, Düsseldorf: Metropolitan.

3 Moderate expectations and benign exploitation

Tourism on the Sepik River, Papua New Guinea

Eric Kline Silverman

MTS Discoverer. PNG's Ultimate Travel Experience . . . designed and built specially to suit the unique conditions along the Sepik River and north coast of PNG . . . can cruise for 30 days without the need to take on fuel and water. It is fast, comfortable and is equipped with five tenders and a helicopter (subject to conditions). The vessel is entirely air conditioned and its facilities include a library, in-house television system, cocktail bar, luxurious lounge, covered deck space, a dive shop, laundry, gift shop and restaurant. Satellite telephones are located in each cabin and all twenty-two cabins have private bathroom and PA system.

(www.mtsdiscoverer.com/)

The SEPIK SPIRIT accommodates 18 passengers in 9 deluxe and spacious twin bedrooms each with private bathroom. Facilities aboard also include a dining room, lounge, bar, video and covered upper observatory deck. The entire vessel is airconditioned. Because the SEPIK SPIRIT is the ideal size for exploration, it provides intimate exposure to the area without the mass tourism often experienced elsewhere.

(www.pngtours.com/lodge4.html)

Introduction: the obvious

The luxurious *Melanesian Discoverer* cruise ship regularly sets sail from the coastal town of Madang in Papua New Guinea (PNG) to meander along the Sepik River, one of the premier and exclusive tourist destinations in Melanesia. The popular allure of the Sepik is at once mysterious, artistic, scenic, and tribal. Like the Amazon and Congo, the Sepik has long served as a metaphor for Western desires and conquests. The river offers a legendary glimpse of raw, pristine nature and, especially, simple villagers unsullied by the conventions and conveniences of the modern world. Not incidentally, the Sepik is also valued by the West for the many opportunities it provides for purchase of souvenirs along this journey into the mythic infancy of humanity.

Typical American tourists who visit the Sepik fly from New York to Sydney, then on to Port Moresby, the capital city of Papua New Guinea, and finally to the coastal town of Madang. There they board the *Melanesian Discoverer* ship for a characteristic five-night cruise, then climb into a small charter airplane on a grass airstrip beside the river at Timbunke village and visit Mount Hagen or Goroka, the two main towns in the PNG Highlands. After a few more days sightseeing, they return to New York, perhaps stopping at Cairns or Sydney. The fare for the five-day cruise is US$1925. (The total cost of the entire journey is approximately US$6000.) Maximum capacity on the *Melanesian Discoverer* is 42 passengers. A full voyage thus represents a minimal collective expenditure of about $80,000. Although this figure pertains solely to the river portion of the entire trip, and so excludes airfare and any other sightseeing, it nonetheless probably exceeds the typical annual income of any village along the Sepik.

The American Museum of Natural History in New York City offers an extensive programme of upscale adventure travel throughout the world. The 18-day 'Faces of Melanesia' voyage in October–November 2006 is scheduled to charter the *Clipper Odyssey* expedition vessel to visit Papua New Guinea, the Solomon Islands, Vanuatu and New Caledonia. The fare for the trip is US$8780 per person for the least expensive berth. Those preferring more sumptuous accommodation can spend as much as $15,880. The pre-trip expedition to the Sepik and Highlands costs an extra $2780 per person (with a surcharge of $480 for a single supplement).[1] The 'Faces of Melanesia' brochure estimates roundtrip airfare from the United States at US$2685. Flights for the Sepik and Highlands excursion will add an additional $850. Minimally, then, a US tourist interested in the entire package can expect to spend about $15,000. A full ship of American passengers would represent a total amount of disposable income in excess of $1,770,140. Thus framed, Sepik River tourism unquestionably expresses and sustains the marginal position of local Melanesian communities in a grossly unequal global economy.

By almost any standard, Papua New Guinea plays a peripheral role in the world system. The World Bank, for example, ranks the United States as fourth of 171 countries in terms of gross national product (GNP). The rank of Papua New Guinea is 133. In 2004, the GNP per capita for PNG was US$550; the US exceeded this figure by a factor of 75, or $41,400. Here, again, the presence of Western tourists in the Sepik would seem best analysed in terms of vast inequalities in wealth and power.[2]

Typically, social scientists and theorists construe tourism in peripheral communities as essentially a form of post- or neo-colonialism that forces local folks into self-humiliation and cultural prostitution for the lurid gaze of wealthy Westerners. What is so attractive about marginal communities is precisely their marginalisation (Azarya 2004). Local people are said to transform themselves into little more than commodities. Staged performances and tawdry trinkets pass for cultural authenticity. To add insult to injury, many tourists feel good about themselves for having so magnanimously passed a few dollars on to the poor primitives for their dances and souvenirs.

In an earlier era, dark-skinned savages were displayed for the almost pornographic pleasure of a gawking Western public. Visitors at the 1904 St Louis World's Fair stared at dog-eating, head-hunting Igorots. The Monkey House at the Bronx Zoo encaged a young man, Ota Benga, from the Belgian Congo. Sarah

Baartman, otherwise known as 'The Hottentot Venus', was paraded across Europe, even in death. Contemporary tourism is simply a more palatable form of the same brutality. Tourists often justify this dehumanising exploitation as a simple-minded curiosity about the world. But there is no dignity here; it is tourism, to draw on Ness (2005), as terror.

The celebrated 1987 film *Cannibal Tours*, directed by Dennis O'Rourke, offers the most vivid example of touristic gall and immorality for Melanesia (Silverman 2004). Most tourists who visit the Sepik, as *Cannibal Tours* so powerfully portrays, yearn to connect with the naked essences of humanity that the West long ago swapped for civilisation. They wish, in other words, to rejoice in how far we have come since our meaner days, or to celebrate those who still live in simple harmony with nature. Tourists also seek brief refuge from the angst of modernity – a key draw of ecotourism (West and Carrier 2004). MacCannell (1992) interprets the motivation behind contemporary tourism as the popular search for an antidote to the artificiality and superficiality of Western culture. Tourism offers tourists something real – a reality, of course, entirely staged. But despite the image of escape, tourism is thoroughly implicated in the ongoing colonial project by virtue of its relentless drive to inspect and scrutinise the natives. Tourists, argue MacCannell (1992) and Root (1996), and not the proverbial natives, are the true cannibals who consume the Other.[3]

This chapter does not seek to diminish the moral intent of those who view tourism through the lenses of colonialism, economic exploitation and racism – lenses ground to focus on the cruelties of the contemporary world system. Instead, it aims to complicate this view, not to apologise for tourism, but to illuminate what is far less obvious: the ways that local people harness tourism for their own subjective purposes. An analytic framework that focuses *only* on the exploitative aspects of tourism, however ethically virtuous, further contributes to the erasure of indigenous agency by preempting the possibility of identifying authentic creativity and genuine meaning in a world that is ineluctably hybrid – a hybridity partly formed and often exemplified by tourism.

The ethnographic focus is the Eastern Iatmul village of Tambunum, a fishing and horticultural community of about a thousand people, most of whom are now engaged in part-time entrepreneurship or petty capitalism. Tambunum is the largest Iatmul-speaking village in the middle Sepik. It is also one of the most successful tourist destinations in the region, in large measure thanks to its bustling size, spectacular architecture and extensive repertoire of art. And it is in the area of tourist art that we can identify what is perhaps the most authentic aspect of contemporary culture in the Sepik.

Sepik tourism

Tourism is no new phenomenon in Melanesia. As far back as the 1880s, Burns Philp was advertising full-scale 'excursion trips' to New Guinea (Douglas 1996). In the same era, Sepik villagers were producing wooden shields for external consumption (D. Newton, personal communication 1987). Tambunum villagers concur, and claim that their great-grandparents carved all sorts of objects specifically to exchange for steel tools with the earliest German colonists and traders.

Although tourism in the Sepik did not begin in earnest until the 1970s, Western interest in local art remained as strong as it was nearly a century earlier. One of the initial purveyors of tourist houseboats along the river, Wayne Heathcote, is today a key participant in the upscale tribal art market in Europe, the United States, and Australia – albeit someone recently implicated in the illicit trade in 'national cultural property' of Papua New Guinea (Dalton 2006). The sale of tourist art and artifacts remains, as it was in the 1970s (May 1977) and perhaps earlier, a central source and promise of income along the Sepik. At the very least, art continues to mediate many, if not most, interactions between Sepik villagers and Western travellers.

In gross economic terms, the tourism industry in PNG remains underdeveloped, but whether to applaud or deplore this situation pivots on perspectives. Industry supporters such as Levantis (1998) bemoan that PNG is 'by far the worst performing nation in the region in terms of tourist growth'. The Papua New Guinea Tourism Promotion Authority (www.pngtourism.org.pg) reports only 69,250 visitors in 2005, most of whom arrived primarily for business. The holiday or leisure-oriented component amounted to a paltry 18,115. Not surprisingly, most of these visitors hailed from Australia (approximately 7100). Japan ranked second, with about 4300 tourists, followed by the US (approximately 2800). Despite the allure of the Sepik, the East Sepik Province itself is not one of the premier destinations in the country. Nevertheless, tourism is a vital and often vibrant aspect of contemporary life along the river, though one which hinges on a world system over which local people can exercise little control.

Today, three general categories of tourists visit the Sepik. The first and most affluent group enjoys the comforts of two lavish vessels that dominate the touristic landscape of the river. Melanesian Tourist Services operates the *Melanesian Discoverer*, a plush and state-of-the-art catamaran cruiser (complete with helipad) that in 1988 replaced the *Melanesian Explorer* steamship featured in the film *Cannibal Tours*. The ship often visits Tambunum and is berthed at the Madang Resort Hotel. A rival tourist craft, the *Sepik Spirit*, arrived on the river in 1990 and is operated by Trans Niugini Tours. The two competitors generally ply different regions of the Sepik. They also maintain their own chains of resorts, hotels and lodges throughout the country.[4]

The second and third groups of tourists, or 'travellers' as they prefer to be called (Errington and Gewertz 1989), seek adventure rather than comfort, or simply have smaller budgets. They tend to arrange tours with small operators, such as Sepik Adventure Tours, or travel solo with no fixed itinerary. They, too, gain access to the river through small, regularly scheduled airplanes that land on grass airstrips. More commonly, 'travellers' climb aboard the ubiquitous PMV (public motor vehicle) at the provincial capital town of Wewak and suffer (or enjoy) a long, uncomfortable, sometimes dangerous ride on a potholed dirt road to the river. These travellers sleep in village homes, small tourist lodges and their own tents.

Most tourism companies in PNG are relatively small and nationally owned or co-owned. The major global hotel chains and resort companies have not invested significantly in touristic ventures in the country. Unjustly or not, Papua New Guinea has a pervasive reputation for lawlessness. The single greatest hindrance to the growth of tourism in the country is a 'colossal crime problem' that, using data from the United Nations, Levantis (1998: 100) deems perhaps the worst in the world. This, coupled with the remoteness of most places in the country from the accoutrements of modern life (which represents both a draw and a hindrance to many potential visitors), severely curtails the tourism industry.

Indeed, the US and Australian governments issue frightening warnings on their foreign travel websites about violent crime in PNG. Visitors are hardly encouraged by these reports of carjackings, armed robberies, pickpockets, bag-snatchers, robbery and gang rape, and admonitions against using public transportation, hiking in remote areas, travelling alone and visiting isolated beaches. So severe is the perceived (or real) law-and-order situation that the US State Department offers a 'Primer on Personal Security in Papua New Guinea' that mentions a weak police force, urban gangs, tribal fighting and rape (http://travel.state.gov/travel/ cis_pa_tw/cis/cis_1757.html). One can read these cautions as a sober warning about travel to a so-called 'failed state' or an exaggerated rendition of colonial anxiety over 'native rule'.

Periodically, news items about tourism and crime appear in the Papua New Guinea dailies. In a well-publicised incident, a group of Australian tourists were mugged while hiking the Kokoda Trail in late December 2000 (see also Foster 2002). Numerous op-ed pieces, policy statements and letters to the editor appeared over the next few months in *The Post-Courier* to berate the criminals and bewail the adverse effects of crime on the fledgling PNG tourism industry. The then-Minister of Culture and Tourism, Andrew Bain, pointed to a broader need for the entire nation to engage in moral self-reflection, and introduced 'a force of tourist police' in 2005. But some writers accused the Australian media of exaggeration and pandering to stereotypes about PNG lawlessness. Throughout this incident, tourism emerged as a window for assessing the overall civic and economic status of the country.

A different sort of moral evaluation

Upon greeting tourists, who typically arrive on the *Melanesian Discoverer*, the Eastern Iatmul people make no effort to conceal the paraphernalia of modernity, despite their recognition that tourists travel to the Sepik precisely to view a different culture and custom. Nor do villagers simulate activities. What does happen is that many people cease their everyday activities to paint tourists' faces, to prepare for a brief dance performance and, most significantly, to fetch art and artifacts. Additionally, the few elder women who sometimes forgo shirts will quickly don a blouse.

Neither audiences nor performers in Tambunum seem especially enthralled by the dances. They appreciate them for what they are, namely, staged performances imbued with an ethos of vague obligation rather than genuine or sincere authenticity. They are something that must be done and seen in order for everybody to adhere to a broader, unstated and vague script of tourism or expectation. Still, as Otto and Verloop (1996) discuss for the famous Asaro Mudmen of the PNG Highlands, even the most obvious of touristic performances in Melanesia often entail complex histories and notions of local ownership that are invisible to the audience. Tourist performances, too, often symbolise local and national identity precisely because they are enacted and highlighted during the touristic encounter. Additionally, it is important to realise that the tourist operators along the Sepik exercise little formal or informal authority that overrides local initiatives and agency. Quite the opposite. Tourism is one of the few arenas of cross-cultural interaction in the Sepik where local people do not feel silenced and pacified by Western power.

Of far greater importance to tourism along the river than the dance performances, and to the thesis of this chapter, is the huge, impromptu bazaar of art and material culture that is quickly assembled on either side of the main village pathway. This somewhat spontaneous marketplace offers the greatest possibilities during the touristic setting for Westerners and local people – men and women, youth and elders – to interact freely. Some Eastern Iatmul are passive, refusing to speak or gesture until first approached by a tourist; others, especially younger men, assume playfully aggressive and animated postures. Conversation, questions, comments, banter, laughter and a great deal of miscommunication are common. The encounter is largely enjoyed by all present.

The economic inequalities of this market, as in Melanesian tourism more generally, require little underscoring. But an economic perspective alone fails to offer a full understanding of tourism in the middle Sepik River. An alternative or complementary framework begins with Pratt's (1992: 7) notion of a 'contact zone,' defined as 'the spatial and temporal copresence of subjects previously separated by geographic and historical disjunctures, and whose trajectories now intersect'. The idea of the 'contact zone' stresses not simply hegemonic forces of 'conquest and domination,' or Western power, but also the 'interactive, improvisational dimensions' of the encounter, that is, local creativity. It is this idea that is so useful for analysing Sepik tourism.

Dialogues with the wider world

Among the Chambri, another Sepik society, men modify ceremonial spears for sale to tourists. Chambri peddle these souvenirs inside the village as well as outside a hotel in Wewak. Some modifications to the spears are practical: smaller items are more appealing to tourists since they can stuff them in their suitcases. But, as Gewertz and Errington (1991) discuss, the transformation from ritual object into souvenir involves far more than physical appearance. For example, whereas men imbue ceremonial spears with a totemic spirit, tourist spears are mystically inert.

A similar process occurs in Tambunum. Eastern Iatmul, like Chambri, attribute both personhood and mystical agency to ritual objects. This magical animation partly arises from the social organisation of artistic creation. A clan's sacred objects gain spiritual efficacy when carved by hereditary ritual partners or sisters' children in exchange for food. Then, too, the owners must sponsor a totemic chant that intones the name and spirit into the ceremonial object. Neither practice applies to tourist art. In this sense, ceremonial art is group-oriented, or collective, and named. By contrast, tourist art is largely individualistic, or carved by one man, and nameless. Men often sign their tourist objects; but the objects fail to possess, in a traditional sense, any significant identity.

Gewertz and Errington (1991) discuss another difference between ritual and touristic objects among Chambri. Ceremonial objects entail social obligations within a wide-ranging moral system of exchange. Tourist spears adhere to the logic of capitalism, not reciprocity. They elicit no long-term relationships, much to the dismay of the Chambri, who see the tourist trade as the only viable means of creating enduring bonds with Westerners. Additionally, ceremonial spears circulate among men and groups who see themselves as more equal than not. Local political and ritual hierarchies are 'commensurate', that is, unstable, negotiable, and potentially reversible. They are also defined by knowledge rather than material wealth. By contrast, differences between tourists and villagers are 'incommensurate'. For all practical purposes, these differences are insurmountable, permanent and marked by a dramatic inequality in access to wealth and commodities.

Gewertz and Errington report that Chambri modify male initiation ceremonies in order to accommodate a paying audience of tourists. From one angle, this commodification dilutes the authentic meaning of the rite and further evidences how local Melanesian societies must submit to a more powerful world system. From another angle, however, this transformation maintains the relevance of the ceremony. To the extent that masculine competence among the Chambri today entails successful interaction with the agents and forces of modernity, a traditional initiation rite would fail to transform boys into capable men. The contemporary authenticity of the ceremony thus arises partly from the inauthentic presence of tourists. Through tourism, in other words, Chambri ritual 'dialogue[s] with the wider world', to borrow a phrase from Adams' (1993: 70) study of Tana Toraja tourism in Indonesia, and helps forge a sense of local identity that neither wholly ignores nor wholly embraces the dominant global system.

The economics of tourist art

Tambunum villagers – men, women, and children – probably create a greater quantity of tourist art, in a wider range of styles, than any other community along the Sepik. In the late 1980s, upwards of 5000 objects were sometimes displayed for art dealers and tourists. Additionally, several men in the village regularly send small consignments of carvings to dealers and shops in Port Moresby, Australia, America and Europe – often through local middlemen who migrated to town. These commercial affiliations wax and wane with considerable frequency, and it would be false to portray the village as some sort of 'factory' that exports large quantities of art in return for immense sums of cash. But other Sepik communities see Tambunum as an affluent village arising from a high traffic in tourists and dealers who purchase large amounts of art. Eastern Iatmul generally agree with this assessment, while nonetheless aspiring to a richer level of material wealth and 'true' development that tourism is unable to provide or sustain. In the heyday of tourism in the late 1980s, art sales grossed the village approximately 30,000 kina (about US$25,000) per year. It is reasonable to assume that this sum was unmatched elsewhere along the river. But this status offered little consolation to Eastern Iatmul who yearned for greater material comforts.

The relationship between tourist art and local economic aspirations is decidedly ambivalent. The creation of art is fuelled by the desire – really, the need – for money to pay for, among other things, outboard motor fuel, clothing, health supplies, radios, school fees, batteries, food and public transportation. At this point in history, it would be nothing short of absurd to view money as enabling Sepik villagers simply and solely to purchase prestige items. The local economy is now thoroughly monetary. Eastern Iatmul would greet the question 'Why do you need money?' – a question asked by the filmmaker in *Cannibal Tours* – with the same bafflement as would most Americans or Australians. Eastern Iatmul have no interest in regressing to premodern subsistence horticulture.

That the sale of tourist art is unable to satisfy the full spectrum of modern desires was made clear by one man who grumbled, while finishing a crocodile-shaped coffee table, 'We carve because we do not have real development here.' His anger also belied resignation. No amount of artistry, he continued, would bring to the village electricity, modern housing, and the other material and symbolic privileges associated with the rising middle class in the towns of PNG. This urban elite, as Gewertz and Errington (1999) show, often defines itself against less sophisticated village kin – the very folks tourists wish to see.

Most Eastern Iatmul, it is worth noting, attribute their monetary plight to bureaucratic incompetence and petty corruption rather than global history and the rise of the modern world system. Village men regularly exclaim that Papua New Guineans, but not tourists, languish under an inept government. Eastern Iatmul, to draw on van Beek (2003) in relation to Africa, view the arrival of tourists (like that of anthropologists) through two lenses. For some, the presence of external visitors shows evidence of local cultural vitality and significance. To others, however, tourism betokens a profound marginality.

In the late 1980s, many people in Tambunum pinned their hopes for the elusive goal of development on the construction of a tourist guesthouse (Silverman 2000). A decade later, this enthusiasm waned for two reasons. First, tourism itself declined with the rising perception of PNG as a dangerous country with severe law-and-order problems, a view partly fuelled by the crisis on Bougainville. Second, most people in Tambunum and throughout the region had tied their aspirations for economic success on the 'vanilla boom' that swept the Sepik. In local perceptions, cash-cropping replaced tourism as the economic strategy most likely to succeed; the village guesthouse contained drying vanilla beans rather than tourists.

Tourism only partly satisfies modern desires; likewise tradition (Silverman 2005). Elder men complain that most youth are far more interested in money than ritual skills and esoteric knowledge. But it is precisely local perceptions of the village as economically viable that prevent the degree of out-migration that has depleted many other villages in the Sepik. Tourism, in this sense, enables tradition as much as it erodes it. Tourism, too, contributes to a wider sense of uncertain identity in Tambunum. Villagers know well that they are, as one man said in the film *Cannibal Tours*, 'living between two worlds'. They are not fully Eastern Iatmul in the mould of their grandparents, but neither are they fully modern in the cast of the tourists. Eastern Iatmul identity today is one in transition.

The rise of the individual

The popular understanding of tourist art as in some sense meaningless or value-less is illustrated by a recent search of eBay, a venue for the sale of Sepik objects, that yielded 72 items (30 March 2006). Most of the labels and descriptions were inaccurate; a few seemed downright deceptive. For example, an 'antique' wooden lime container, sculpted to resemble a bird, was labelled Iatmul. This attribution is correct, but its antiquity was dubious, and hardly warranted the shocking 'Buy It Now' price of US$2300. A more accurate description of the object would place its date of manufacture within the past 15 years or so and perhaps indicate an original price of about $10. These objects, moreover, are now made exclusively for sale to tourists and dealers; villagers prefer to store their lime powder for chewing with betel nut in discarded plastic rice bags, tobacco tins or photographic film canis-ters. But none of the Sepik objects on eBay were described with any hint about their touristic nature.

The label of another object, 'Powerful Iatmul Mwai Mask 1970s; Oceanic Tribal', presented an honest date of manufacture. But, lest this attribution deter potential buyers, the item was further described as 'powerful . . . a great mask with real presence'. But why not a great *tourist* mask? The answer is obvious: tourism to most dealers, buyers, collectors and museums nullifies aesthetic authenticity. Tourist art, in other words, is not art but something less significant, such as mere craft, an exhibition of skill but not genuine aesthetic creativity.[5]

This view is naïve, ethnocentric and erroneous. Several genres of tourist art in Tambunum express the complexities of contemporary identities in the Sepik and Papua New Guinea. Many touristic works diverge considerably from traditional styles, motifs and forms. Tourism promotes the emergence of a hybrid aesthetic or what Causey (1999a) dubs, writing about Toba Batak wood-carvers in North Sumatra, 'conflation'. Aesthetic conflation in Tambunum reflects various aspects of modernity, especially a heightened sense of individualism.

In the Sepik, as Gewertz and Errington (1991) show, young men and women who embrace the modern ideals of romantic love and self-destiny often flee to towns such as Wewak to escape traditional marriage expectations and the authority of elders. Similarly, capitalism fosters a more heightened sense of the individual than normally permitted by traditional village morality. An effective capitalist must shun reciprocity and kinship obligations and embrace marketplace values of autonomy and self-reliance. The mundane act of shopping, Foster (2002) shows, stresses free choice. So, too, does the Christian notion of the afterlife, whereby each person is ultimately responsible for the fate of their soul. In multiple ways, then, modernity in Melanesia is inextricably linked to individualism.

Tourist art and identity

Tourist art in Tambunum aesthetically expresses the profound shifts in personhood and identity that now shape Sepik lives. Most obviously, tourist art evidences what might be called aesthetic individualism (Causey 1999b). Eastern Iatmul carvers clearly state that they aspire to create unique objects that will attract the attention of tourists and buyers. Consequently, individual styles have emerged over the past decade or two within the village, and there is considerable competition to create distinctive objects.

Traditional art also evidenced some degree of individual creativity and innovation. But the intent behind traditional artistry is to reproduce faithfully a spirit or motif that represents a collective descent group. Carvers thus act on behalf of their clans or lineages rather than self-interest. Tourist art dramatically differs. The goal is individual innovation, not fidelity to tradition and the representation of collective identity. Otherwise, men say, nobody's works would stand apart. In tourism, too, men carve for themselves or for a small number of kin, usually wives and children. They also carve for money and commodities, not totemic power and ritual prestige. All told, tourist art represents several key institutions of modernity, including the individual, capitalism and the nuclear family.[6]

A typical tourist artwork exhibits motifs and forms that were once consigned to distinct objects and displayed, moreover, in unrelated settings. This stylistic blurring befits the contemporary context in which traditional features of social life are often blurred with modern aspirations. Some art dealers counsel village men to carve only within the traditional repertoire. But Eastern Iatmul resist this request. They also carve styles that fail to sell with any great regularity. The local appeal of these objects cannot therefore be explained with reference to market forces. Rather, they resonate with local sensibilities, specifically the expression of modern individualism.

Two types of aesthetic authenticity are evident through tourism. On the one hand, unique works that diverge from tradition seem genuine precisely because they are plainly novel and therefore do *not* copy tradition. Yet, as Steiner (1999) argues for Africa, stylistic redundancy based on traditional forms and styles communicates an opposite but equally valid sense of authenticity. These works are genuine precisely because they *do* copy tradition. If some tourist works evidence the authenticity of modern individualism, others evidence the authenticity of an enduring heritage.

For all of its conformity with modern individualism, tourist art in Tambunum also sustains traditional modes of identity. Men generally say that they refrain from carving the totemic creatures of other lineages and clans without receiving permission lest they capitalise on, and consequently transgress, another group's esoteric patrimony. Most men also claim to disperse any income from the sale of tourist art within the kin group. Such comments reflect moral principles rather than actual practice. Nonetheless, the evidence in Tambunum shows that tourist art, however much it recreates identity, fails to erase entirely the force of traditional sentiment.

Sepik tourist art also reflects the prismatic, transformative and emergent aspects of contemporary identity. Many masks confound the distinctions between inside and outside, figure and ground. Three-dimensional animals recede into two-dimensional faces. Works exhibit no clear visual focus. The viewer is unable fully to see the object and its motifs from any single angle or perspective. Animals curve beneath masks and crawl through the surface. Mouths ambiguously devour and expectorate – give birth or even excrete – fish, frogs and snakes. This genre of tourist art defies any stable imagery and so reflects the instability of contemporary social life in Melanesia.

The most common creature to crawl through a maw or mask is the crocodile. Traditionally, crocodiles represented senior, clan-specific spirits who governed, among other processes, human fecundity. These spirits were, and remain, impersonated by men during ritual through the musical guise of bamboo flute melodies. But in the context of tourism, crocodile motifs and figures represent something entirely different. They signify the relatively new form of regional identity – 'Sepiks' as opposed to 'Islanders', say, or 'Highlanders'. This sense of regionalism was non-existent prior to the colonial era and the development of long-distance travel between disconnected places.

Literacy communicates the same message about regional identity, and also expresses individualism (Silverman 2004). Eastern Iatmul men commonly sign their tourist carvings with their Christian and sometimes totemic names. They may incise or paint the acronym ESP or its referent, East Sepik Province. Women similarly weave 'PS' into baskets, an alphabetic slogan for *Pikinini Sepik* (Sepik Child). Although these slogans by themselves might appear somewhat insignificant, they constitute yet further evidence for the overall thesis of this chapter that tourist art visualises wide-ranging processes in Melanesia that are profoundly reshaping what it means to be a person.

For many persons and communities in PNG, the idea and ideal of the nation-state remains tenuous and inchoate (Gewertz and Errington 1991; Foster 2002). The concept of anonymous citizenship, too, as a foundational legal principle that must replace kinship in order for the state to endure, is no less challenging. Eastern Iatmul carve myriad variations of PNG's national emblem, often accompanied by Christian phrases such as 'God Bless This House' or 'Mama Mary'. These objects represent efforts by individual local people to 'think through' the elusive notions of nationhood, citizenship and Christianity. The latter slogans also communicate the message that success in capitalism and modernity is inextricably linked to Christian morality (Smith 1994).

Woodcarving and artistry more generally is a traditional Sepik practice. For the tourist trade, these skills are frequently applied to modern materials such as shoe-polish and writing. Similar cross-cultural heterogeneity is in evidence throughout the village. Every now and then, a battery-operated clock ticks away the modern hours next to the wooden figure of Tuatmeli, the originary ancestor of the cosmos, inside the men's house. A garish print of the Holy Family was tacked atop, but unable to conceal, the painting of a spirit on a house-post. These visual cues point to the inter-cultural hybridity of contemporary personhood and culture in the Sepik.

Tourist art in Tambunum often exhibits a proliferation of multiple, sometimes hidden, faces unknown in premodern art. Several Sepik societies traditionally emphasised the self as a multiple construction. In some contexts, the self was masked by a spirit or magical preparation to avoid direct responsibility for certain kinds of social action, especially now-extinct homicide vendettas (Harrison 1993). Even today, Iatmul ritual partners (*tshambela*) may act on a man's behalf in situations likely to result in a quarrel – as when acquiring a second wife – or when someone demonstrates certain forms of clumsiness (Silverman 2001a). From one angle, the multiplicity of faces on tourist art expresses the traditional 'partibility' of the self. From another angle, this recent style portrays a dialogue between traditional and modern identity that is unlikely to achieve resolution. Indeed, we can identify a type of debate over personhood within the genre of tourist art itself. Whereas some works, as suggested above, represent the modern individual, works exhibiting multiple faces cast this contemporary ideal into doubt.

Anxiety and cognition

The 'visual redundancy of tourist artworks', suggests Steiner insightfully for Africa, communicates an intent to rise above the 'noise' of cross-cultural exchange and interaction (Steiner 1999: 101). From this angle, when a man carves and displays ten copies of the same object in Tambunum, he seeks to create a sense of aesthetic equilibrium amid the usual cacophony of cultural clash. Artistic reduplication thus offers visual stability and security to Westerners, who are more likely than local people to experience some unease in the unstructured, sometimes chaotic and noisy touristic encounter. Stylistic redundancy also allows carvers to create an illusion of permanence within, and control over, an era of extraordinary historical change and cultural dissonance. From this angle, tourist art soothes anxiety.

From another angle, tourist art induces anxiety. Often, contemporary artworks in Tambunum evidence a type of *horror vacui*, the avoidance of empty spaces. Most touristic works are filled with ornate patterns and motifs; there is little in the way of what might be called touristic minimalism. The lack of unadorned spaces on tourist art is often noted (for example, Abramson 1976). In Tambunum, though, this ornamental exuberance tends to remain ordered and restrained rather than chaotic and hectic. Several scholars have attempted to identify an unconscious relationship between the organisation of artistic designs, patterns of cognition and the underlying structures of social life (for example, Abramson 1990). Fischer (1961) saw isolated motifs as representing individual persons, and thus artistic style correlated with political organisation. Adams (1973) tied the growing intricacy of Attic vase painting – specifically the complexity of styles, the reduction in empty spaces and the use of enclosed figures – to the evolution of the Athenian city-state. From this angle, tourist art in the Sepik would seem to reflect the increasing complexity of social life and the basic features of the local worldview. It might also represent an effort by local people to challenge the Western assumption that Sepik societies are simple rather than complex. Tourist art in Tambunum conveys a sense of dynamism rather than sociocultural stasis. In other words, some genres of tourist art in the village refuse to accede to the touristic expectation of a simple primitivism.

From another angle, however, one that builds on Bateson's (1973) brilliant analysis of a Balinese painting, some touristic styles in the Sepik cognitively 'correct' the chaotic pattern of contemporary PNG culture. Male discourses about crime, politics, youth, the male cult, women, the state and post-contact history tend to emphasise cultural disorder and the erosion of authority. Events no longer abide solely by a localised, admittedly masculine, logic. Consequently, the restrained quality of the touristic *horror vacui* can be seen as an aesthetic effort to impose or re-impose local, male control over the upheavals and conflicts of everyday life. Stylistic complexity induces, but then contains, social anxiety.

Touristic art objects tend, much more than traditional works, to display prominent symmetries. This style may correlate with traditional cognitive patterns: the binary organisation of social structure, a layered view of reality and knowledge, and the use of pairings to organise phenomena (Silverman 1999, 2001b). Alternatively, and in line with prior interpretation of *horror vacui*, the proliferation of symmetries in tourist art may express the duality of contemporary life in the Sepik as a clash, synthesis and disjunction between the traditional and the modern.

A traditional hybridity

It would be mistaken to ascribe cultural hybridity or conflation in the Sepik, whether in art or any other cultural realm, exclusively to recent decades. Long ago, Margaret Mead (1938) flagged the tendency of Sepik societies to continuously traffic in objects, magic, ritual and other forms of culture. Iatmul villages in particular, she continued, were so saturated with external ideas and objects that many alien cultural forms never achieved full integration into the local system. Tourism builds on the scope and scale of these premodern processes. Indeed, the cultural fluidity noticed by Mead probably predisposed Iatmul to their innovative success with touristic art.

Today, Tambunum villagers procure various items such as shell necklaces and masks from nearby villages and town markets. After slight modification, they peddle these wares to tourists for a slight profit (Silverman 1999). Tambunum therefore sometimes serves as a small-scale, spontaneous entrepôt for villagers from remote communities in the area who wish to ply the tourist trade, but are too far away, despite their yearnings to attract tourists (Smith 2002). The touristic marketplace is not limited, as in more traditional commerce, just to men. Women offer their own touristic items, namely, baskets, netbags and rattan animals.

Often, as Mead might have predicted, Eastern Iatmul model their tourist objects after artwork associated with other regions of the country. Yet Tambunum villagers always seem to vary slightly these exogenous forms, styles and patterns, thus effectively making them their own. Many dealers and some tourists view these objects scornfully as forgeries, and in some sense they are correct. But, as Mead observed, the authenticity of Sepik art and objects often arises from precisely this process of importation. It thus seems far more accurate to view tourism as reproducing rather than destroying the process of traditional artistic creation.

Rediscovering and reinventing culture

The tourism trade in Tambunum, especially the drive to create distinctive objects, encouraged one local man to draw on marriage ties with a non-Iatmul group to obtain 'copyright' permission to reproduce their art for sale to tourists. In this instance, traditional morality joined rather than clashed with tourism. Other men reproduce objects they view in art catalogues, buying guides and books that circulate in the village. That is, they rely on Western institutions such as museums to provide examples of what tourists expect from traditional Iatmul art (see also Causey 2000; Chibnik 2003).

In the 1970s, the village of Kambot, located on the Keram River, a tributary of the Sepik, devised one of the most distinctive styles of Sepik tourist art, the storyboard, and this style is now said to be traditional (Dougood 2005: 260). Storyboards are even used to teach myth to youth. In Kambot, the boundary between the touristic and the indigenous, between the authentic and the fake, blurs. A similar process occurs in Tambunum.

In the late 1980s, a middle-aged man named Gamboromiawan was viewing photographs in Tambunum of Sepik art collected long ago during the era of German colonisation. These objects are now stored and displayed at the Museum für Völkerkunde in Basel. So captivated was Gamboromiawan by one forgotten mask from the village that he personally set about reintroducing the style into the repertoire of village art, only now for sale to tourists rather than ritual display. Nevertheless, the discovery of this object filled Gamboromiawan with no uncertain pride in the artistic achievements of his ancestors and allowed him to enact a small instance of cultural revival. In the absence of tourism, the photo of the mask would have passed with little notice or relevance.

Local resistance

Sepik men concoct imaginative tales about the antiquity and ritual significance of objects hewn recently and expressly for external sale. On the surface, these men are simply telling lies to earn the equivalent of a few dollars, thus illustrating the immorality of capitalism and tourism. But anthropologists have argued for years that Melanesian epistemologies differ significantly from Western ideas about truth, falsehood and the fabrication of knowledge. The veracity of an assertion in Melanesia often pertains more to the social implications of the claim than to some abstract notion of ultimate right and wrong (Weiner 1995). Consequently, a false belief may be true when it serves, at least for some people, a positive and often practical social function. Inventive creativity is often valued over convention. Fictions of touristic authenticity are thus culturally truthful.

Touristic performances in the Trobriand Islands often mock tourists through verbal barbs and threats, but these challenges are veiled since tourists obviously lack competence in the vernacular (Senft 1999). Tambunum villagers too mock tourists, while also joking with them in the vernacular as well as in a fast-paced rendition of pidgin (*tokpisin*). The joking seeks both to challenge asymmetries of power manifest in tourism and to create a sense of shared understanding and experience.

In Tambunum, too, men refer to their touristic carvings as *tumbuna*, the *tokpisin* word for ancestor. This linguistic convention accedes to Western expectations since tourists and art dealers usually look to purchase objects that adhere to stereotypically generic notions of primitive religion. But if tourist art is contextualised in the pervasiveness of the totemic system, wherein the sensible world is the materialisation of an invisible reality constituted by ancestral spirits, then the classification of tourist art as *tumbuna* rings true. This assertion, too, like the joking, can be seen subtly to challenge Western hegemony by endorsing a local reality even as it seems to adhere to touristic expectations.

During one touristic encounter in Tambunum, a young man, perhaps in his twenties, displayed to female tourists a small wooden replica of a vulva (Silverman 2000). With unabashed cheekiness, he called the object a '*tumbuna* cunt'. Later, several women from the group of tourists were visibly unnerved by the young man's brashness and aggressively sexual gesture. The decorative carvings on the tourist guesthouse that sits on a spit of land across the river from the main village similarly display exaggerated genitals and a serpentine act of intercourse (ibid.). Many touristic figures likewise display large, erect penises, images that refer to moments of mythic sexuality, although these referents are unknown by nearly all Westerners. Additionally, artistic sexuality represents how local men understand one side of modernity – namely, the overt eroticism of Western advertisements. (The other side, of course, consists of Christian notions of renunciation.) More importantly, though, these sexual images offer muted challenges to Western economic and cultural dominance through a traditional framework that might be called the erotics of rivalry (ibid.). Touristic renditions of sexuality might also exemplify the internalisation of Western fantasies about savage eroticism. Local people, in this view, adhere to a Western script about what 'natives' are supposed to do. But it is far more likely, at least in the Sepik, that these images intentionally seek to unnerve tourists and thus to subvert an encounter that, in many respects, favours and sustains Western power.

Conclusion

At DePauw University in the United States, it is common to show the film *The Mission*, starring Robert De Niro and Jeremy Irons, to students who elect to participate in three-week-long 'service' trips to the developing world. The main historical narrative of the film focuses on a tragic clash in South America after the

Treaty of Madrid in 1750, when Spain agreed to cede part of its Latin American claims to Portugal. In the film, Spanish Jesuits endeavour to protect innocent Christianised Indians against a brutal alliance between the Catholic Church and Portuguese slave traders after the papacy orders the closure of the Jesuit missions. But the Jesuits are no match for the Portuguese military, and the missionaries and Indians are slaughtered. Ultimately, the film is a tale about the wanton destruction of indigenous peoples as a result of Europeans' pursuit of god, gold and glory. 'Though I knew that everywhere in Europe the States were tearing at the authority of the Church', writes the Cardinal in the film to the Pope, 'and though I knew well that to preserve itself there, the Church must show its authority over the Jesuits here, yet I still couldn't help wondering whether these Indians would not have preferred that the sea and wind had not brought any of us to them.' Most viewers agree.

Likewise, most anthropologists and scholars tacitly endorse the same view of tourism. It would be better if the natives were simply left alone. There is much truth in this view. But moral condemnation, however vital and ethical, does not enable a full understanding of the local experience of tourism in the Sepik or beyond, because moral disgust often and ironically denies the very type of 'voice' to local people that critics so often attribute to tourists. This is not to justify or defend tourism. But simplistic claims about economic benefits and moral apprehension are often naïve and, precisely because they are simple, devoid of nuance and sensitivity. As this chapter showed, there are numerous moments of genuine creativity, aesthetic expression and meaning in the touristic encounter. Moreover, as Adams (1998) argues for tourism among the Toraja of Indonesia, several aspects of tourist art in the Sepik serve as a mild form of resistance to Western hegemony. Tourist art does, to say the obvious, cater to tourists and represent the gross inequalities of the modern world system. But tourist art also evidences the vitality and validity of local cultures and the strivings of Sepik people to make sense of their lives by transforming global processes into local experiences. Through tourist art, Eastern Iatmul villagers literally carve meaning out of plight imposed from afar. Tourism may be unjust. But it is no less meaningful.

Acknowledgements

I would like to thank, naturally, the kind people of Tambunum village who tolerated my presence in 1988–90 and 1994. Additionally, I am grateful to for assistance from the Fulbright Program, the Wenner-Gren Foundation for Anthropological Research, the Institute for Intercultural Studies and DePauw University.

Notes

1 See www.discoverytours.org/tours/search.php?sear=103 and www.discoverytours.org/tours/uploads/103custom4melanesiafinal.pdf.
2 Tourism receipts in Papua New Guinea hover at around 2 per cent of GNP or 6 per cent of GDP, although the World Travel and Tourism Council forecasts that tourism in 2006 will account for 9 per cent of GDP (www.wttc.org/2006TSA/2006individual%20pages/papuanewguinea.htm).

3 For a criticism of the common metaphor of cannibalism in cultural critique see King (2000).
4 In late 2006, Melanesian Tourist Services announce the sale of the *Melanesian Discoverer*, partly because, according to the company website, of 'increases in the cost of fuel, the reduction of tourist arrivals to PNG due to lack of competitive airfares', and competition from foreign-owned vessels that occasionally visit Papua New Guinea.
5 For further elaboration of the problematic High Art/Tourist Art dichotomy, focusing on the New Guinea Sculpture Garden at Stanford University, see Silverman (2003).
6 This is not to deny the presence of individualism in traditional personhood (see Silverman 2001b).

References

Abramson, J.A. (1976) 'Style Change in an Upper Sepik Contact Situation', in N.H.H. Graburn (ed.), *Ethnic and Tourist Arts: Cultural Expressions from the Fourth World*, Berkeley, CA: University of California Press.
—— (1990) 'Structural Aspects of Visual Art Design and their relation to Broader Sociocultural Contexts', *Empirical Studies of the Arts*, 8: 149–91.
Adams, K.M. (1993) 'Club Dead, Not Club Med: Staging Death in Contemporary Tana Toraja (Indonesia)', *Southeast Asian Journal of Social Science*, 21: 62–72.
—— (1998) 'More than an Ethnic Marker: Toraja Art as Identity Negotiator', *American Ethnologist*, 25: 327–51.
Adams, M. (1973) 'Structural Aspects of Village Art', *American Anthropologist*, 75: 265–79.
Azarya, V. (2004) 'Globalization and International Tourism in Developed Countries: Marginality as a Commercial Commodity', *Current Sociology*, 52: 949–67.
Bateson, G. (1973) 'Style, Grace and Information in Primitive Art', in A. Forge (ed.), *Primitive Art and Society*, London: Oxford University Press.
van Beek, W.E. (2003) 'African Tourist Encounters: Effects of Tourism on Two West African Societies', *Africa*, 73: 251–89.
Causey, A. (1999a) 'The *Singasinga* Table Lamp and the Toba Batak Art of Conflation', *Journal of American Folklore*, 112: 424–36.
—— (1999b) 'Stealing a Good Idea: Innovation and Competition among Toba Batak Woodcarvers', *Museum Anthropology*, 23: 33–46.
—— (2000) 'The Folder in the Drawer of the Sky Blue Lemari: A Toba Batak Carver's Secret', *Crossroads*, 14: 1–34.
Chibnik, M. (2003) *Crafting Tradition: The Making and Marketing of Oaxacan Wood Carvings*, Austin, TX: University of Texas Press.
Dalton, R. (2006) 'Guinea Experts Cry Foul on Tribal Exhibits', *Nature*, 440: 722–3.
Douglas, N. (1996) *They Came For Savages: 100 Years of Tourism in Melanesia*, Lismore: Southern Cross University Press.
Dougood, R.C. (2005) ' "Ol i kam long hul bilong Wotñana" (They Come from the Hole of Wotñana): How a Papua New Guinean Artefact Became Traditional', in T. Otto and P. Pedersen (eds), *Tradition and Agency: Tracing Cultural Continuity and Invention*, Aarhus: Aarhus University Press.
Errington, F. and Gewertz, D. (1989) 'Tourism and Anthropology in a Post-Modern World', *Oceania*, 60: 37–54.
Fischer, J.L. (1961) 'Art Styles as Cultural Cognitive Maps', *American Anthropologist*, 63: 79–93.

Foster, R.J. (2002) *Materializing the Nation: Commodities, Consumption and Media in Papua New Guinea*, Bloomington, IN: Indiana University Press.

Gewertz, D.B. and Errington, F. (1991) *Twisted Histories, Altered Contexts: Representing the Chambri in a World System*, Cambridge: Cambridge University Press.

—— (1999) *Emerging Class in Papua New Guinea: The Telling of Difference*, Cambridge: Cambridge University Press.

Harrison, S. (1993) *The Mask of War: Violence, Ritual, and the Self in Melanesia*, Manchester: Manchester University Press.

King, C.R. (2000) 'The (Mis)Use of Cannibalism in Contemporary Cultural Critique', *Diacritics*, 30: 106–23.

Levantis, T. (1998) 'Tourism in Papua New Guinea: A Comparative Perspective', *Pacific Economic Bulletin*, 13: 98–105.

MacCannell, D. (1992) *Empty Meeting Grounds: The Tourist Papers*, London: Routledge.

May, R. (1977) *The Artifact Industry: Maximizing Returns to Producers*, Discussion Paper No. 8, Boroko, PNG: Institute for Applied Economic and Social Research.

Mead, M. (1938) *The Mountain Arapesh II: Arts and Supernaturalism*, Garden City, NJ: The Natural History Press.

Ness, S.A. (2005) 'Tourism–Terrorism: The Landscaping of Consumption and the Darker Side of Place', *American Ethnologist*, 32: 118–40.

Otto, T. and Verloop, R.J. (1996) 'The Asaro Mudmen: Local Property, Public Culture?', *The Contemporary Pacific*, 8: 349–86.

Pratt, M.L. (1992) *Imperial Eyes: Travel Writing and Transculturation*, London: Routledge.

Root, D. (1996) *Cannibal Culture: Art, Appropriation, & the Commodification of Difference*, Boulder, CO: Westview Press.

Senft, G. (1999) 'The Presentation of Self in Touristic Encounters: A Case Study from the Trobriand Islands', *Anthropos*, 94: 21–33.

Silverman, E.K. (1999) 'Tourist Art as the Crafting of Identity in the Sepik River (Papua New Guinea)', in R.B. Phillips and C.B. Steiner (eds), *Unpacking Culture: Art and Commodity in Colonial and Postcolonial Worlds*, Berkeley, CA: University of California Press.

—— (2000) 'Tourism in the Sepik River of Papua New Guinea: Favoring the Local over the Global', *Pacific Tourism Review*, 4: 105–19.

—— (2001a) *Masculinity, Motherhood, and Mockery: Psychoanalyzing Culture and the Iatmul Naven Rite in New Guinea*, Ann Arbor, MI: University of Michigan Press.

—— (2001b) 'From Totemic Space to Cyberspace: Transformations in Sepik River and Aboriginal Australian Myth, Knowledge and Art', in J. Weiner and A. Rumsey (eds), *Emplaced Myth: Space, Narrative and Knowledge in Aboriginal Australia and Papua New Guinea Societies*, Honolulu, HI: University of Hawai'i Press.

—— (2003) 'High Art as Tourist Art, Tourist Art as High Art: Comparing the New Guinea Sculpture Garden at Stanford University and Sepik River Tourist Art', *International Journal of Anthropology*, 18: 219–30.

—— (2004) 'Cannibalizing, Commodifying, and Creating Culture: Power and Creativity in Sepik River Tourism', in V. Lockwood (ed.), *Globalization and Culture Change in the Pacific Islands*, Upper Saddle River, NJ: Prentice-Hall.

—— (2005) 'Sepik River Selves in a Changing Modernity: From Sahlins to Psychodynamics', in J. Robbins and H. Wardlow (eds), *The Making of Global and Local Modernities in Melanesia: Humiliation, Transformation and the Nature of Cultural Change*, Burlington, VT: Ashgate.

Smith, M.F. (1994) *Hard Times on Kairiru Island: Poverty, Development, and Morality in a Papua New Guinean Village*, Honolulu, HI: University of Hawai'i Press.

—— (2002) *Village on the Edge: Changing Times in Papua New Guinea*, Honolulu, HI: University of Hawai'i Press.

Steiner, C.B. (1999) 'Authenticity, Repetition, and the Aesthetics of Seriality: The Work of Tourist Art in the Age of Mechanical Reproduction', in R. Phillips and C. Steiner (eds), *Unpacking Culture: Art and Commodity in Colonial and Postcolonial Worlds*, Berkeley, CA: University of California Press.

Weiner, J.F. (1995) 'Anthropologists, Historians and the Secret of Social Knowledge', *Anthropology Today*, 11: 3–7.

West, P. and Carrier, J.G. (2004) 'Ecotourism and Authenticity', *Current Anthropology*, 45: 483–98.

4 'Everything is truthful here'

Custom village tourism in Tanna, Vanuatu

Prue Robinson and John Connell

Tourism is of enormous importance in the Pacific island state of Vanuatu, as the most important sector of the economy. Much of it based on small-scale resort tourism centred on the capital, Port Vila, on the main island of Efate, but promotional literature, guidebooks and the in-flight video on all incoming Air Vanuatu flights emphasise the presence of more traditional cultures in other islands and especially those in 'custom villages' on the island of Tanna. Almost all tourists who visit Tanna intend to see Yasur volcano and most also visit a 'custom village'. Although expectations vary, they usually expect to witness a distinctive cultural way of life, which has been portrayed as unique and authentic. This chapter examines how tourist needs and expectations are mediated by tourism operators, how villagers portray different versions of 'tradition' and 'authenticity' and how attitudes of tourists to the performance of visual culture in the villages vary. It traces the constant fluctuation between tourists', entrepreneurs' and villagers' objectives and expectations as each group seeks interrelated objectives.

Custom villages in Tanna perform 'traditional' dances, and demonstrate other facets of 'traditional' knowledge, based upon *kastom*. The Bislama (Vanuatu pidgin English) word *kastom* is widely used to refer to a larger meaning in the Vanuatu context as 'not merely an odd collection of dances and rituals but a whole way of life which dictates almost all of one's actions' (MacClancy 1981: 20). In a contemporary sense, it tends to be used by the people of Vanuatu (ni-Vanuatu) to 'characterise their own knowledge and practice in distinction to everything they identify as having come from another place' (Bolton 2003: xiii). Custom villages are widely advertised as having these distinctive *kastom* attributes, and so provide authentic representations of ni-Vanuatu culture; hence tourists visit them to have an authentic experience of that culture. However throughout Vanuatu, including Tanna, what is claimed as *kastom* embodies epistemological and practical changes that have occurred since contact.

The problematic concept of 'authenticity' has been addressed from a diversity of perspectives. Even at its simplest, the tourist quest for authenticity is the search for difference, exoticism, the 'other' and 'real' cultures and context. This involves both a search 'grounded in the belief that what we have lost can be found in Others more "primitive" and therefore more natural than ourselves' (Shepherd

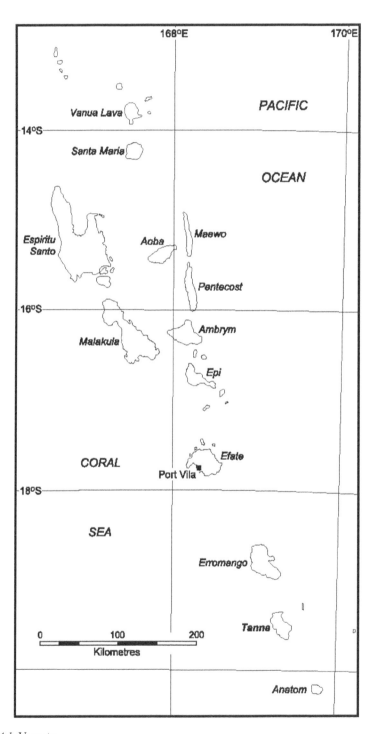

168°E   170°E

PACIFIC

Vanua Lava

14°S

OCEAN

Santa Maria

Espiritu
Santo

Maewo

Aoba

Pentecost

16°S

Ambrym

Malakula

Epi

Efate

Port Vila

18°S

CORAL

SEA

Erromango

0   100   200

Tanne

Kilometres

Anatom

Map 4.1 Vanuatu.

2002: 192; Kontogeorgopoulos 2003), and some form of escape both from the monotony and blandness of everyday life (MacCannell 1976) and from alienation within a seemingly postmodern, globalised and homogeneous world (Wang 1999; Grunewald 2002; Olsen 2002; Reisinger and Steiner 2006). Some tourists are said to be motivated by an 'anthropological desire' to find out more about remote and different communities (Smith 2003: 18) and then to 'revel in the otherness of destinations, people and activities because they offer the illusion or fantasy of otherness, of difference and counterpoint to the everyday' (Craik 1997: 114). At the very least authenticity should 'invoke the distinctiveness of local interactions' (Chaney 2002: 205). In short, tourists searching for authenticity are seeking both an 'escape', however temporary, from modernity, and some degree of discovery in an unfamiliar context. However it is evident that in very many contexts tourists are not necessarily in search of immersion in 'authenticity' but are often primarily in search of difference, pleasure, distraction or entertainment, without any discernible ideological objectives (Connell 2007). Custom villages may simply be one of the 'must-sees' in Vanuatu.

Moreover it was long ago pointed out that authenticity is an illusion that cannot actually exist (Lévi-Strauss, cited by Shepherd 2002: 184) but is a 'subjective attribute that circulates as an arbitrating mark' (Smith 2003: 22; Shepherd 2002: 190), hence attractions may be viewed as authentic even when entirely staged (Gibson and Connell 2005). This is particularly so where tourists have limited knowledge of local cultures. Hence past 'traditional' events and customs may be enhanced, truncated or otherwise changed to enhance the experience of exoticism and otherness (Silver 1993). The complexity and purpose of such changes, and reactions to them, are examined here.

Vanuatu

Vanuatu is a small island state of more than 80 populated islands (Map 4.1) and about 190,000 people, almost all of whom are Melanesian (ni-Vanuatu). Traders and missionaries, from the first quarter of the nineteenth century, brought significant external contact. Vanuatu, then the New Hebrides, was a joint French and British colony from 1906 until independence in 1980. Most of the population remains rural and engaged in subsistence agriculture; there is limited infrastructure (such as sealed roads, piped water and electricity) in most outer islands and there has been considerable migration from distant islands to Port Vila. Vanuatu is classified by the United Nations as a least developed country.

Tourism in Vanuatu, which currently markets itself as 'another time, another pace', first became significant in the 1970s, when annual tourist numbers passed 5000, after the first resort had been completed in 1969. Most tourist facilities are around Port Vila. Although annual tourist numbers passed 50,000 in the 1990s there have been few attempts to spread tourism to other islands, where there are limited facilities, other than through cruise ships. Most tourists are from Australia, followed by New Zealand and New Caledonia. Analysis of guest book entries showed that this was exactly the same in Tanna.

Tanna has a population of about 21,000, spread over six language groups and more than 100 villages, with no real urban centre. It has one small hotel, a handful of tourist bungalows, no sealed roads, a couple of dozen vehicles and access by small plane. It is a relatively poor island, cash incomes are small, infrastructure is limited and there has been significant migration from Tanna to Port Vila. The number of tourists visiting Tanna is less than 10 per cent of all tourists; few stay more than a couple of days. In the 1980s the Yasur volcano brought a handful of visitors, and the presence of the John Frum cargo cult (Brunton 1981) in several villages also brought some tourists; at least one such village had been encouraged to 'revert to the primitive' by no less than a *National Geographic* stringer in order to encourage tourism (Douglas 1996: 197; Miles 1998: 77). During the 1990s 'custom village' tourism began to grow in importance, initially centred solely on the village of Yakel (Connell 2007). The first small hotel was built in the 1960s, but even by the mid-1990s there were few places to stay. The island can accommodate no more than about 100 tourists at any one time, but that limit is never reached.

Vanuatu largely markets itself as a country with a distinct Melanesian culture, alongside both Anglophone and Francophone culture, and a place of resort-style hotels, amidst interesting landscapes and 'exotic' villages. Guidebooks take a similar perspective. The Moon Handbook states 'No other country harbors as many local variations. The glamorous duty-free shops, casinos and gourmet restaurants of the cosmopolitan capital, Port Vila, contrast sharply with unchanging traditional villages just over the horizon' (Stanley 2004: 879). The Lonely Planet guide introduces Vanuatu with: 'It's a rare traveller who doesn't like Vanuatu, as there's something to suit most tastes: from honeymooners looking for a romantic relaxed time away to adventurers keen to visit remote islands where villagers live as they have done for centuries' (Bennett and Harewood 2003: 9). Within Vanuatu various brochures emphasise the attractions of different islands and those for Tanna always mention custom villages. The most common of these, the Jasons *Vanuatu Visitor Guide*, in its 2006 edition, states: 'Tanna's 20,000 inhabitants have largely retained their original custom and culture (kastom), preferring to turn their back on modern ways'. A nationally produced brochure from the early 2000s, advertising Tanna and one of the local bungalows, states:

> Tanna is a world of its own quite different from the rest of Vanuatu. The traditions of Tanna are ancient and very strong and put the Tanna people in a category apart. Most villages in Tanna still live according to the ancient customs and traditions with no other clothes than grass skirts and nambas (penis sheaths).

Advertisements also emphasise that Tannese customs can be viewed at special performances in Port Vila given by people from 'a real kastom village' (Figure 4.1). In such contexts 'culture' is defined essentially as a visible and aural phenomenon, centred on distinctive dress and dance styles.

Tanna is usually singled out as having an accessible volcano, the John Frum cargo cult and also its 'custom villages'. Air Vanuatu's standard in-flight video,

Figure 4.1 Advertising *kastom*.

produced by the Vanuatu National Tourism Office, mentions 'the magic island of Tanna' where villagers have 'their own lifestyle completely uninfluenced by western eyes'. The Moon Handbook introduces Tanna as being 'renowned for its active volcano, potent kava, coffee plantations, custom villages, cargo cultists, exciting festivals, strong traditions, magnificent wild horses, long black beaches, gigantic banyan trees, two-meter-long yams and day-tripping packaged tourists' (Stanley 2004: 924). Lonely Planet is even more effusive: 'lush undisturbed rainforests, heady night-perfumed flowers, coffee plantations, plains where wild horses gallop, mighty mountains, hot springs, waterfalls and presiding over it all fuming, furious Mt. Yasur . . . Christianity, cargo cult (where believers act like Europeans so that wealth "cargo" will come their way) and *kastom* (traditional culture) are important and all natural phenomena have a fourth dimension of spirituality and mystique' (Bennett and Harewood 2003: 89). Other guidebooks take similar perspectives, while the 'tourist intermediaries' of newspaper and magazine articles embellish such themes both graphically (Skrivankova 2002) and visually, in particularly contrived form.

This chapter examines the five custom villages on Tanna that were engaged in the tourist industry in 2003 and 2004. Villagers were interviewed over a period of two months in 2004 following a preliminary study in 2003 and a brief return visit was made in 2007 (Connell 2007). All the tourist operators were interviewed, most on several occasions, and about 70 tourists were interviewed over the two periods. One of these custom villages was Yakel, long renowned as the principal custom village in Tanna, through having seemingly retained the old ways more successfully than elsewhere. In several guidebooks Yakel takes a particularly prominent place, whereas other custom villages are not separately identified or have a more fleeting presence in a single edition. Photographs from Yakel are the principal means of marketing Tanna. The four other villages have more recently become 'custom villages' and at least two are relatively remote, even for the few tourists who visit Tanna.

Tourism operators

Almost all visits to custom villages are arranged within Tanna by four or five operators resident there, both Melanesian (ni-Vanuatu) and expatriate, usually in loose association with various bungalow 'resorts' where tourists stay and which are usually owned by the tour operators. The obvious objective of these entrepreneurs and middlemen is to make profits. This depends on tourists gaining satisfaction and their costs not being too high. Consequently their choice of custom village depends on whether there is a 'good performance', seemingly exuding both spontaneity and authenticity, and on the payment requested by the villagers. Pragmatically performances, and thus authenticity, were judged by operators (and by tourists) in terms of enthusiasm, spontaneity and traditional dress, that is by the 'performer's commitment to the dance' (Daniel 1996: 785). Visual and aural sensations dominate. Villages are thus put under some degree of pressure to perform 'adequately', and to some extent are played off against each other, though social ties between some entrepreneurs and some villages limit this pressure.

Yakel has something of a comparative advantage, through its adherence to 'old custom' and opposition to education, government and religion. Pressure has tended to be put on other villages to be rather more like Yakel in their enthusiasm for old custom. One operator thus stated:

> Yakel is a *nambas* village, which means women don't wear their tops and men don't wear their bottoms; other villages didn't want to do that. I said people wanted to see real custom but they said that's old custom, we are now evolved and we don't want white men to see us like that, we are embarrassed. I said you should be proud like the Indians, the Aboriginals and the Maoris when they find their roots in custom dress. But they are just starting to acquire pride in being dressed as western people and so on. They refuse to do *nambas*.

Consequently Yakel has been the main destination of custom village tours, but this has created a problem for operators. Yakel is seen to be more likely to demand higher fees, put on a more truncated performance and involve fewer performers, and thus less spectacle. One operator concluded:

> Yakel custom village gets a bit jaded after a while so we give them a break because they are not performers; we want spontaneity, we want original real traditions. I told Yakel they needed a break because they were looking sick of it and you can see it.

Another phrased it rather similarly:

> In the last year they have really stopped showing people through their villages and only five or six people turn up for dancing and they only wanted to sell their handicrafts and to charge $20 – I don't think there is any value in that at all.

But other villages could sometimes lose their enthusiasm: 'we spread the money around the custom villages until they become complacent'. Both European and Melanesian tour operators shared similar views: 'they must dance and act properly otherwise the tourists are not happy and they complain'. However, many visits to villages are arranged at very short notice (since communications are poor), other village activities must be put aside and quickly summoning large numbers of villagers is not always possible.

Yeniar (Iweniar) village particularly gained from the occasional exclusion of Yakel, its location closest to the two main sources of tourists and its social ties with bungalow owners. Another village also gained from tensions elsewhere: 'Before I used Imayo village but after some mismanagement and complaints, I sent people to this place and said "If you wear *nambas*" and they said "OK".'

Ultimately operators recognised that it was the villagers' right to choose whether or not and how to participate, thus:

> Yakel is a good example of that; it is up to their community to work out where they want to place themselves. They were the first here but I guess money corrupts. It's not corrupting their culture I suppose, but it's human nature.

The number of performers, the colour and exoticism of their dress (or lack of it) and the drama of the performance are perceived as crucial to the portrayal of exoticism, otherness and mystery. Limited incomes in villages encourage regular participation; one operator, asked if villagers would tire and become stale from regular dances, said: 'No they won't get tired, because that's their only money, from tourism; if they are tired, no money.' Even more bluntly, as another put it: 'if they don't have tourists, they don't have money.' Villages were also expected to display particular characteristics:

> Big change. The village is kept clean; before, when tourists didn't go, there were pigs and pig shit everywhere, they had no flowers. Now that tourists come, they keep the animals away, keep the village clean, plant flowers and so they get the tourists' interest.

Operators maintained surveillance over, and encouraged the regulation of, space.

Figure 4.2 Yakel village.

Operators naturally tended to enthuse and talk up their 'product'; one stated that what you see is:

> still totally unique and a natural part of their life, not like Hawai'i where you see a Polynesian dance put on for totally commercial reasons; here if you go to a custom village what you see is their culture, their daily culture. That's just how they live, that's the best part of it.

Another argued that, unlike in other parts of the island Pacific, 'Vanuatu is unique to have culture and all the physical attributes of that culture'. But another operator later confided that 'the authentic villages are the ones you can't see'. Though operators were well aware of their role as 'brokers in authenticity' (Adams 1984) they were ultimately never quite sure what was the best approach to marketing and thus the best performance to encourage from villagers, especially at Yakel, where the performance had scarcely changed over time:

> Yakel – that's how they used to be. You must question a little bit on how they are today. You drive there and there is a brand new 4WD sitting out front. But Yeniar – how they practise their custom today is very interesting and I am finding that tourists prefer to see how they live today and how they practise it.

Figure 4.3 Yeniar village.

Others recognised paradox:

> Yakel is authentic, [but] as soon as they accept money are they authentic? Well it's a bit of an interesting paradox because Yakel is the perfect example – the most successful kastom village there is. First they rejected the white man and the missionaries, then they opened their village and people paid to come and see them so they just rejoined Western society.

Moreover, as another operator explained, the paradox extends further: 'kastom villages that are not that strong in kastom are not that expensive and the prices go down. This village has a lot of education so they know to keep their price at 1400 vatu [A$17.50].' Custom is expensive; as another operator noted, 'the more traditional, the higher the tourist entrance fee', though the operators gained about two-thirds of the income generated. Operators were interested in keeping village entry fees down, whilst promoting the most exotic and colourful version of custom.

Villages and villagers

At various times in the past decade, different villages in Tanna have been identified, and identified themselves, as 'custom villages'. Their participation in the tourist industry in this context, the only significant way of participating in and gaining an income from the industry in Tanna, has been a function of various factors, including religious beliefs, previous experience of tourism, attitudes to culture and need for income.

Missions were powerful in Tanna though 'ni-Vanuatu did not suddenly lose all their custom on conversion to Christianity [but] people merged the two, believing in both the power of God and spirits, ghosts, sacred stones and sorcery' (Mac-Clancy 1981: 72; Bonnemaison 1994: 111–12). Many newer Christian churches, such as the Assemblies of God, were most hostile to *kastom* in its visible forms. A member of the Living Waters church stated '*Kastom* and church; you can't ride two horses.' In two of the five villages only some villagers and, in one, none of the women participated in custom performances since particular churches had banned such activities. Elsewhere it was stated that commitment to Christianity required the abandonment of *kastom*: 'If you do *kastom* dances you can't go back to the church and be listened to because you have fallen down in the eyes of God and will be judged.'

Tanna villagers no longer retain *nambas* and grass skirts, what is locally referred to as 'old *kastom*', but most villages have taken up 'new custom', or 'custom for today', that includes cloth (calico) skirts and trousers. As one villager observed, 'Here we do *kastom* for today, because the missionaries came over and gave us calico; we dance with calico, we use many things from overseas like people use for Christmas, such as tinsel; we use it for culture.' 'New custom' villages include those aligned to the John Frum cult, but these are not involved in the tourist industry as 'custom villages', though they had a more prominent role in earlier years (de Burlo 1986). Yakel is distinguished through having apparently withstood modern education, religion and 'modernity' in general (if not tourism). In every village custom is understood and practised in different ways (Robinson 2004: 36–63; see also Miles 1998; Bolton 2003).

There is clear recognition of change in the 'new custom' villages: 'before people used *nambas*, people killed each other and ate each other, fought, but now we have changed our tradition', and 'Cloth is from white people and not *kastom*, but neither are tables, chairs, pots and some food but we use them because without them life would be hard.' In many villages there is however some reluctance to engage fully in the modern world. Relatively few people go to church, and few children go to high school (partly for economic reasons). A general perspective on education is that if everyone goes to school 'our culture will lose out; we must leave some people to stay back to take part in our culture.' Similar perspectives relate to other aspects of change. However villagers consciously hold on to certain facets of custom, well aware that it is a means to make money in a struggling island economy, and so in turn custom becomes valued as a drawcard for tourism.

Villages and villagers are able to negotiate their involvement with the tourist industry. The principal reason for participation is income generation, but pride in past traditions is important. Income is welcomed where remittances, some wage employment and limited marketing of crops are the only, partial alternatives. Income varies substantially between villages, which receive between 500 vatu (A$6.25) and 1500 vatu ($19) per visitor, amounting to about $500 per week for the most successful village, Yakel, and less than half that in the eastern villages. Yakel was estimated by one operator in 2003 to earn about $20,000 a year, which corresponds with recent estimates. In each case, but especially in Yakel, tourist incomes are superior to those from other sources. On the eastern side of Tanna income-earning opportunities are particularly scarce. As one Tabau village leader suggested, 'we don't sell our custom, we just need some help', here to construct a water pipe that would benefit the whole village and not merely those engaged in tourist performances. Village leaders also intended to use the performances as a means of playing a larger part in the tourist industry but, without the capital, marketing and transport facilities to do so, they depended on the operators for business.

Some villages had stopped participating in the industry, because of disputes over the use of the income, disputes within the village over who should be the beneficiaries (especially in villages where only a fraction of people participated) and disputes between villagers and operators over the price. Between 2004 and 2007 Yeniar dropped out of custom village tourism, after conflict with a neighbouring village over income distribution, but several villagers migrated to the village of Ipai, which became a new custom village. More villages have sought to enter the industry but limited tourist numbers have made this largely impossible.

At Tabau village, where tourism was quite recent, the rationale for participation was also expressed in a cultural sense. The village chief argued that the impact of both church and education had been considerable: 'school kills *kastom*, but keeping *kastom* and knowledge together will keep understanding about how our grandparents lived, so generation after generation will remember *nambas* and grass skirts, true *kastom* and not the less distant past.' Tabau has thus reinstated wearing *nambas* in performance, partly as mimicry of Yakel, partly in homage to the past and partly as a guarantee that tour operators will include Tabau on their schedules.

In each village performances for tourists are unlikely to occur more than twice a week, because of small tourist numbers, and when tourists arrive dances are performed irrespective of their numbers. The performance in custom villages usually lasts no more than an hour, and combines some elements of both entertainment and education. At Yeniar, for example, the visit commenced with a village tour that showed various facets of 'village life', from the construction of a leaf stretcher to coconut cutting, making *laplap* (a characteristic 'pudding' based on grated yams and coconuts), fire lighting, fruit and vegetable sampling, the throwing of a stick to determine the site for yam planting, village houses, and a game. The chief played a panpipe accompanied by singing. Tourists are shown how to use bows and arrows. Three different dances were performed followed by opportunities to purchase handicrafts and a range of artifacts. Other villages had less comprehensive variants of this. One Yakel dance showed how the missionaries were expelled and old *kastom* re-established. The two more recent village entrants into the custom village scene mainly performed dances developed for tourists with some doing songs in English without any traditional meaning; two dances were 'where we come hand in hand together to improve Vanuatu, the next is in English saying fly over the water to come and see the volcano and say goodbye to our village.' Another village has a dance called 'penalty' that involves a mime of football. Customs, such as rituals for the planting of crops, are explained. Villagers usually participate enthusiastically in dances, though numbers may vary from five to fifty, and do not have to be cajoled into performing. Performances are much more informal, and less obviously staged, than in many 'folk village' contexts in other parts of the world (for example Rees 1998; Johnson and Underiner 2001; Zeppel 2002; Dyer *et al.* 2003; Xie 2003).

Nonetheless, as in many other tourist contexts, performances are modified (usually by being shortened) and recently devised songs, dances and games are incorporated, some of which directly involve tourist participation. At one village dances had been taught to younger adult villagers by the older residents though many had little knowledge of their significance or even name, as dances were essentially returning as empty practice, decontextualised from local society (Hobsbawm and Ranger 1983). Nonetheless dances have been retained in custom villages, whereas most villages not involved in tourism had forgotten them, and those who performed them clearly enjoyed the experience and the interaction with tourists (and the monetary reward). Many villagers themselves argued that 'doing *kastom* dances for tourists makes the *kastom* strong'. Others saw little distinction between important elements of custom and change: 'the future will be the same as Fiji and Hawai'i if the thinking of young people doesn't maintain the values of *kastom* and church.'

Modern origins may however be denied. One villager, when asked about face paint, purchased from local stores, stated that she got it from rocks in the creek. In Tabau village, where villagers were again willing to wear *nambas* for the performance, the chief stated: 'Everything is truthful here; when tourists come here we take off our trousers and wear *nambas*', though this was intended as a rebuttal of Yakel, where villagers take off their *nambas* when tourists are not there but claim otherwise. Nonetheless in one village it was pointed out that tourists

> don't like pretend, they like real culture like Yakel. But because Yakel is too far away we wanted to make something that demonstrates what Yakel does. Tourists think it is not real but when they see it they really like it.

Villagers themselves therefore accept the supposed primacy of Yakel in the authenticity stakes.

Creativity and change thus prevent routine and consequent 'performance death' (Daniel 1996: 790) and enable a degree of spontaneity, but by new forms of incorporation and thus divergence from *kastom*. However briefly transferring performances and objects from the cultural sphere to the touristic sphere does not necessarily mean the loss of local control and meaning. Despite the desire for income and dependence on operators for visitors, villagers retained significant autonomy and authority since without them custom village tourism would be impossible. At the very least, villagers are far from passive participants in the tourist industry.

Tourists

Tourists visited Tanna primarily to see Yasur volcano but usually also to visit a custom village; if they had not initially been interested in seeing one, or knew little about them, they were encouraged by local operators. Their motivations for visiting a custom village varied considerably and were often not particularly subtle: 'We just want to see simplicity – the simple life is what we came to see'; 'I wanted to see a more traditional way of life, I don't think Le Lagon [resort hotel] in Vila exposed us to traditional food, dance and way of life'; 'We wanted to see how people manage without electricity and running water and how they live with the land'; 'to see culture not corrupted by the twenty-first century' and 'pure culture'. One woman wanted to see 'how families live and how they interact; I'm very interested in anthropology to see how different people live.' Otherwise tourists never mentioned 'professional' interests or any particular detailed knowledge of, or desire to learn more about, local culture and history. Indeed their motives were often quite different and more superficial than those who sought substantial engagement with 'other' cultures. Nevertheless, most certainly expected to see what was behind any performance: 'we want to see what their lives are really like and not cultured history.' Most had little idea of what such expectations might actually mean. Our 'expectations were of primitive Africa but we had limited knowledge.' An unknown Africa was a repeated trope for authenticity. Given the lack of knowledge of ni-Vanuatu culture, authenticity could only be perceived in a relative and spatial context.

Tourists were sometimes offered visual culture/authenticity in the simplest possible terms by operators who were not unduly subtle; one middle-aged female tourist was told by an operator that the attraction of a particular village was that she would see 'lots of tits and bums'. Other tourists had few illusions that what they would see would be a performance: 'we know it's put on for the tourists' show; we know they will put on other clothes but it gives some idea of their tradition and how they lived in the past.' Alternatively 'it takes away a little bit knowing they are not always wearing the grass skirts; I feel like a tourist.' There were no evident distinctions of gender or age in people's expectations, but previous experience had some bearing, largely in defusing grandiose expectations of difference.

The most positive experiences for tourists centred on aspects of cultural pride, uniqueness and naturalness. It was 'extremely unique, never seen anything like it; just totally different from elsewhere as it is not for commercial reasons, seems like they are doing it for more personal reasons.' Most found the villages to be different from their expectations: 'We loved the village; it was not like a neat town in Africa, it was more scattered and unbelievably clean; we expected it would reek.' Another stated that 'this village is better than those in Africa – less stage-managed; it is a normal village and not set up for tourists. It is not a ready-made village as seen in Swaziland.' Or, in comparison with Port Vila, 'authenticity is very important as I noticed it was lacking in Vila, you have a sense of being with a throng of tourists – it is very superficial and gives less meaning to my life.' For some the experience was more 'authentic' than anticipated and some found Yakel especially somewhat challenging and confronting: 'Not for the kids, a bit scary and uncomfortable' and 'we have come to experience Vanuatu culture, but did not want to see naked women – not the reason we came.' One operator stated that they took 'Europeans, people who have travelled and those seeking real authenticity' to Yakel but families and less well-travelled tourists 'would find Yakel too confronting, and since we want them to enjoy the experience we take them elsewhere.'

For some tourists it was 'not put on or artificial' whereas just as many others 'would have preferred to see something not Westernized' or 'it sounded like the traditional culture was still pretty much alive, so I'm a little bit disappointed.' Many recognised and accepted that they were involved in a performance: 'It's pretty much for the tourists. I have done two expeditions to Irian Jaya [West Papua] and have seen the custom dance there; it's a big difference.' Another placed the Papua New Guinea highlands in the same context. For others authenticity had a temporal rather than spatial dimension: 'To achieve authenticity you would need to stay for six months not just two-day trips.' Another believed that staying overnight in a village would have secured authenticity. Many sought to move further behind the performance: 'being here is just beautiful; I want to be with the people themselves, in their day-to-day lives', though this could verge on the patronising: 'they have a great deal of pride and dignity and show they are quite capable to be independent, even though they are aware they could improve their lot.' The major disappointment for some was the inability to go beyond the performance: 'I would have enjoyed a tour of the village, to see inside houses, how they live, the facilities or lack of facilities.'

One further group were not sure what they were seeing, or how this fitted into notions of local culture: 'I would have preferred them to wear *nambas*; I don't want them to wear Western things just because we are here.' For them and indeed for most tourists there was no real expectation or knowledge of what was involved. The performance too was shrouded in some degree of mystery. Even a thorough reading of the more detailed guidebooks and tourism brochures provides minimal information on Tannese society, past or present, which is both varied regionally and complex (Bonnemaison 1994), hence even superficial prior knowledge is rare and perceptions are strongly influenced by operators, other tourists and interme- diaries, and the actual performances. Moreover the primary motivation of most tourists in Tanna is to see Yasur; hence, for many, a visit to a custom village is either secondary or unanticipated, so many had few if any preconceptions. Few required or expected a detailed understanding and analysis of the past – they were not anthropologists or social historians. Since tourists had no knowledge of Tan- nese culture they were able to believe in various versions of 'tradition' – easily duped and even willing to be duped: 'the more exotic it looks, the more you think it different; the more different and the more exciting it is.'

Not only had tourists diverse, and vague, expectations of what they might see, they had equally diverse responses to what they had seen, centred on very dif- ferent notions of (in)authenticity, from 'nothing is authentic now' to 'that was more authentic than we had ever seen'. Only their previous experiences of what they saw as similar contexts differentiated them. Relativity was critical. Those who had the most definite perception of authenticity – usually finding it absent in Tanna – were those who had travelled most, usually either in Africa or elsewhere in Melanesia. Achieving a sense of witnessing authenticity was dependent on per- sonal interpretations, expectations and prior knowledge of local culture.

Within the village

All the tourists in custom villages expected to see a 'traditional' place and per- formance, perhaps gain status from so doing, and hence convince themselves (and later others) that this is what they had seen, and thus acquire some degree of cultural capital. Consequently there was some 'projection of tourists' own beliefs, expectations, preferences, stereotyped images and consciousness onto toured objects' (Wang 1999: 355). Less commonly tourists adopted a 'questioning gaze' (Bruner 2005: 95–100) that challenged any notions of accuracy and authenticity. In short, tourists thus often saw what they wanted to see, but they also usually observed a 'back stage' somewhat different from the staged performance, where notions of tradition seemed to be absent (Connell 2007). Many tourists observed a 'back stage' where the everyday life of villagers displayed 'modern' components not evident in the performances. Shoes, clothes, bicycles and bibles were sup- posedly tucked away but sometimes visible. One was disappointed to see clothes hanging on a line in one village: 'they could have hidden their clothes' to preserve the illusion rather than the 'reality'.

Other tourists were sometimes enthusiastic about having penetrated that back stage: 'I doubt the authenticity of the grass skirts as we saw a man running without one; viewing things from the back of the truck is authentic.' Or 'I don't want to see culture if it's deliberately put on; I love watching an adapting thing – that's interesting.' As some tourists found at Yakel particularly, actually seeing beyond the 'front stage', and thus being able to decry what they perceived as artificiality, was also a means of claiming status and even pleasure (Connell 2007). Inadvertently, they were acquiring cultural capital both through their presence at esoteric tourist sites and sights in remote places, but also through their ability to see beyond them. Viewing elements of the back stage and experiences elsewhere ensured that most recognised the inauthenticity of the contrived performances (cf. Shepherd 2002) but nevertheless found this entertaining and were content to believe that, whether authentic or not, they had seen something unique and different.

Paradoxes thus appeared. Yakel, the village most opposed to modernity (specifically in the form of education and religion) has selectively benefited the most from one particular facet of modernity – tourism – that it has not challenged. Yakel villagers have several four-wheel-drive vehicles, alongside significant investments in tourist bungalows nearby, though the village itself is without electricity or piped water. One of the authors observed a Yakel woman hastily getting rid of her blouse as outsiders approached (a remarkable converse of the situation in other remote ni-Vanuatu villages). Authenticity pays.

Just as tourism and visual culture are inherently intertwined, so too are the relations between operators, local people and tourists, and there is some degree of collusion and collaboration between all of them, as tourism is commodified in a distinctive form. Indeed there is a form of collusion and collaboration between the need for income and the display of custom. In the end tourists tended to validate authenticity against the images they had seen in brochures or on video: the 'real' was validated by the 'image'. Villagers were least concerned with the 'orientalist' perception of the past and authenticity that the operators required of them and that some tourists wished to see, although, like the tour operators and for the same reasons, they were anxious to give value for money. The tourists sought value for money and seeming authenticity was usually part of that, though a minority did seek the normal, everyday life of the back stage. Both villagers and tourist operators have effectively become cultural entrepreneurs mediating contemporary village life and custom performances, and carefully balancing front stages and back stages, to satisfy both tourists and villagers, as a mean of securing continued tourism.

Conclusion

In every custom village performances are staged for tourists. The performance is intended to demonstrate living tradition, or what is said to be *kastom*, but almost all tourists recognise this as something of a staged performance, or 'pseudo-event' (Boorstin 1961) that may have only tangential bearing on contemporary village life, which can sometimes be seen behind the 'stage'. Ironically the staged event is inauthentic but in some respects 'traditional', whereas the village life that is fleetingly observed out the back, but intended to be hidden from view, is authentic – a small part of lived experience – but modern and therefore excluded from the show. For a few tourists their brief glimpses of the back region, and their few moments of participation in it, were where they experienced moments of authenticity. In Yakel especially the ability to deny and decry the performance as false or inauthentic and celebrate the briefly glimpsed 'real world' gave the greatest pleasure, both because it seemed 'real' and because it was intended to be hidden (Connell 2007).

However, despite constant assumptions that tourists in rural areas of developing countries are often seeking authenticity, in many cases – and certainly here – they were not particularly seeking authenticity, as opposed to an enjoyable and unique experience, and for the most part were content to enjoy a performance and an experience, however brief, of village life so different from their own worlds and lives. A large part of that enjoyment came from witnessing the performance in the midst of a remote Tanna village, in a largely natural setting and with other villagers, rather than in a hotel (cf. Bruner 1996: 175). Although a few tourists in Tanna were in 'experiential mode' (Cohen 1988), they were a minority. Some tourists perceived a lack of authenticity, relative to other places or even to the brochures, while others discovered a 'back stage' with pleasure, though still others would have welcomed more insight into contemporary life. Ultimately the need for, perception of, and demands on authenticity were very varied.

In Tanna, not only has there been some degree of cultural revival, after certain activities have lapsed, but there has also been 'invented tradition' in which certain values and norms are promoted and displayed that imply continuity with the past but are largely fictitious (Hobsbawm and Ranger 1983). In a sense much 'new custom' can be seen as both invented tradition and 'emergent authenticity' (Cohen 1988). In this context culture is a visual and aural phenomenon, combining animated dances, face painting, hints of historic warfare and some songs sung in English. Processes of cultural revival, invention (as in the 'football' dance) and commodification occur simultaneously, and 'new custom' can incorporate all manner of changes, to the extent that 'authenticity' – and therefore invented tradition – have limited indigenous meaning, where continual creativity, diffusion and change are central to ethnicity and identity (Tilley 1997), and culture has always evolved and changed.

Villagers have entered the industry, and constituted 'custom villages', primarily for income generation, but restoration of part of the past is central to participation. Like tourists they too have no clear concept of authenticity (cf. Cohen 1988), but have discovered what aspects of the past can be adequately enacted to the general satisfaction of most tourists. Authenticity is thus socially constructed, flexible and negotiable, and villagers see their performances as not being problematically 'reconstituted through a trope of primitive Others' (Yea 2002: 189) but selectively drawn from a vibrant and culturally meaningful past. Indeed, as also for the Chambri and Trobriand islanders of Papua New Guinea (Gewertz and Errington 1991; Senft 1999), real meaning has largely been emptied from relations with tourists, but remains central to elements of core identity such as kinship relations and exchange relationships (Bonnemaison 1994) beyond the 'empty meeting grounds' (MacCannell 1992) where brief tourist encounters take place.

Villagers performed an identity with which they were comfortable, whether invented, borrowed, preserved or restored, and that, ironically, through a version of tradition that enabled income generation, moved them further from it. Each had their own, fluctuating, view of what *kastom* was and is, and how it should be presented – whether 'new' or 'old' – and by whom. Participation and the form of participation emphasised conscious choice, local agency and self-reflexivity (Tilley 1997). Villagers were aware of external images of themselves, and might actively counter these images and construct alternative visions of their history and culture. As in the parallel context of 'world music', performances displayed a 'strategic inauthenticity' in emphasising natural places and people, though the lives of the performers had become more 'modern' (Taylor 1997) and, as in other contexts, such as that of the Asaro village mudmen in Papua New Guinea (Otto and Verloop 1996), had evolved to meet national needs in the promotion of tourism and the needs and expectations of tourists.

Tourists are not (usually) social scientists and have no detailed knowledge of ni-Vanuatu cultures. Here, as elsewhere, tourists accepted 'pseudo-events' partly because they could not distinguish inauthenticity (Cohen 1988: 383; Wang 1999: 355) but partly because they preferred, wanted and needed to have seen 'authenticity' either as performance or lived experience. Indeed, like tourists at performances in Bali, 'the basic metaphor of tourism is theater, and the tourists enter into a willing suspension of disbelief' (Bruner 1996: 176) except for those who claimed the alternative authenticity of the back stage, or decried the performance as inauthentic as they have done in similar contexts elsewhere in Vanuatu (Tilley 1997: 79). Nonetheless, for the most part, 'gullible tourists are only transitory "guests" easily manipulated by the wealth of experience acquired by their welcoming "hosts"' (Tilley 1997: 75). Cohen has noted that 'the modern tourist-pilgrim is damned to inauthenticity' (Cohen 1988: 373) but also conditioned to it and, as Britton has pointed out, 'tourists are by and large conditioned to look for the qualities associated with a cultural model, staged performance or life-style representation, rather than its authenticity or the reality of the social life it is part of' (1991: 455). Thus 'the past-in-the-present is now a look, not a text' (Morris 1988: 2) for casual visitors – necessarily superficial and which even casual visitors assume is not the 'real thing' and from which little therefore can be learned. It is what one does in Tanna.

The tourism industry, here including villagers, entrepreneurs and tourists, is both consumer and producer, but also destroyer, of cultural diversity. One tourist inadvertently recognised the paradox that 'the biggest catalyst for change will be more exposure, many people tramping through, but the money will preserve the culture.' Although one argued that 'it is important to keep it as natural as they can' it was also recognised that 'without tourism there would be no engagement with the outside economy and that is detrimental as it can't stay as a living museum'. However the custom villages were only part-time living museums and links to past cultures.

Through their ability to be flexible about *kastom*, even in the face of low tourist numbers and scheming operators, villagers were able to retain considerable agency. Power relationships constantly changed according to the shifting identities and aspirations of villagers, operators (brokers) and tourists, all of who had some degree of power and agency (see Cheong and Miller 2000). Through villagers' control of *kastom* and their resistance to unwelcome pressures they have escaped what has elsewhere been described as the constant 'tension between the lived and the aesthetic' (Clifford 1988: 247; Kirtsoglou and Theodossopoulos 2004). Although income is central, Tanna villagers are pleased and proud to be, however briefly, guides to the richness and uniqueness of their culture. Whereas the local tourist industry has constantly fluctuated and changed according to diverse perceptions, disputes and the complexity of three-way interactions (alongside diverse attitudes within individual villages), ultimately almost everyone collaborates in the commodification of cultural difference in this highly ambiguous 'contact zone', where 'traditions are constantly salvaged, created and marketed in a productive game of identities' (Clifford 2000: 101). Indeed, as Clifford goes on to point out, there are both 'inventions of tradition and traditions of invention' (2000: 102). Collusion and collaboration, illusion and authenticity, all contribute to development at these grassroots, as culture is consumed, produced and dissolved in the service of social and economic development.

References

Adams, K. (1984) 'Come to Tana Toraja, Land of the Heavenly Kings: Travel Agents as Brokers of Ethnicity', *Annals of Tourism Research*, 11: 469–85.

Bennett, M. and Harewood, J. (2003) *Vanuatu*, 4th edition, Melbourne: Lonely Planet.

Bolton, L. (2003) *Unfolding the Moon. Enacting Women's Custom in Vanuatu*, Honolulu, HI: University of Hawai'i Press.

Bonnemaison, J. (1994) *The Tree and the Canoe. History and Ethnography of Tanna*, Honolulu, HI: University of Hawai'i Press.

Boorstin, D. (1961) *The Image. A Guide to Pseudo-Events in America*, New York: Harper and Row.

Britton, S. (1991) 'Tourism, Capital and Place: Towards a Critical Geography of Tourism', *Environment and Planning D. Society and Space*, 9: 451–78.

Bruner, E. (1996) 'Tourism in the Balinese Borderzone', in S. Lavie and T. Swedenburg (eds), *Displacement, Diaspora and Geographies of Identity*, Durham, NC: Duke University Press.

—— (2005) *Culture on Tour: Ethnographies of Travel*, Chicago: University of Chicago Press.

Brunton, R. (1981) 'The Origins of the John Frum Movement: A Sociological Explanation', in M. Allen (ed.), *Vanuatu: Politics, Economics and Ritual in Island Melanesia*, Sydney: Academic Press.

de Burlo, C. (1986) 'Cultural Resistance and Ethnic Tourism in South Pentecost, Vanuatu', in R. Butler and T. Hinch (eds), *Tourism and Indigenous Peoples*, London: International Tourism Business.

Chaney, D. (2002) 'The Powers of Metaphors in Tourism Theory', in S. Coleman and M. Crang (eds), *Tourism: Between Place and Performance*, New York: Berghahn.

Cheong, S. and Miller, M. (2000) 'Power and Tourism: A Foucauldian Observation', *Annals of Tourism Research*, 27: 371–90.

Clifford, J. (1988) *The Predicament of Culture*, Cambridge, MA: Harvard University Press.
—— (2000) 'Taking Identity Politics Seriously: The Contradictory, Stony Ground . . .' in P. Gilroy, L. Grossberg and A. McRobbie (eds), *Without Guarantees: Essays in Honour of Stuart Hall*, London: Verso.
Cohen, E. (1988) 'Authenticity and Commoditization in Tourism', *Annals of Tourism Research*, 15: 371–86.
Connell, J. (2007) 'The Continuity of Custom? Tourist Perceptions of Authenticity in Yakel Village, Tanna, Vanuatu', *Journal of Tourism and Cultural Change*, 5: 71–86.
Craik, J. (1997) 'The Cultures of Tourism', in C. Rojek and J. Urry (eds), *Touring Cultures: Transformations of Travel and Theory*, London: Routledge.
Daniel, Y.P. (1996) 'Tourism Dance Performance: Authenticity and Creativity', *Annals of Tourism Research*, 23: 780–97.
Douglas, N. (1996) *They Came For Savages: 100 Years of Tourism in Melanesia*, Lismore: Southern Cross University Press.
Dyer, P., Aberdeen, L. and Schuler, S. (2003) 'Tourism Impacts on an Australian Indigenous Community: A Djabugay Case Study', *Tourism Management*, 24: 83–95.
Gewertz, D. and Errington, F. (1991) *Twisted Histories, Altered Contexts: Representing the Chambri in a World System*, Cambridge: Cambridge University Press.
Gibson, C. and Connell, J. (2005) *Music and Tourism: On The Road Again*, Clevedon: Channel View Publications.
Grunewald, R. (2002) 'Tourism and Cultural Revival', *Annals of Tourism Research*, 29: 1004–21.
Hobsbawm, E. and Ranger, T. (1983) 'Introduction: Inventing Traditions', in E. Hobsbawm and T. Ranger (eds), *The Invention of Tradition*, Cambridge: Cambridge University Press.
Jasons Vanuatu Visitor Guide (2006), Port Vila: Jasons.
Johnson, K. and Underiner, T. (2001) 'Command Performance: Staging Native Americans at Tillicum Village', in C. Meyer and D. Royer (eds), *Selling the Indian: Commercializing and Appropriating American Indian Cultures*, Tucson, AZ: University of Arizona Press.
Kirtsoglou, E. and Theodossopoulos, D. (2004) 'They are Taking our Culture Away: Tourism and Culture Commodification in the Garifuna Community of Roatan', *Critique of Anthropology*, 24: 135–57.
Kontogeorgopoulos, N. (2003) 'Keeping up with the Joneses: Tourists, Travellers and the Quest for Cultural Ethnicity in Southern Thailand', *Tourist Studies*, 3: 171–203.
MacCannell, D. (1976) *The Tourist: A New Theory of the Leisure Class*, Berkeley, CA: University of California Press.
—— (1992) *Empty Meeting Grounds: The Tourist Papers*, London: Routledge.
MacClancy, J. (1981) *To Kill a Bird with Two Stones: A Short History of Vanuatu*, Port Vila: Vanuatu Cultural Centre Publications.
Miles, W. (1998) *Bridging Mental Boundaries in a Postcolonial Microcosm: Identity and Development in Vanuatu*, Honolulu, HI: University of Hawai'i Press.
Morris, M. (1988) 'At Henry Parkes Motel', *Cultural Studies*, 2: 1–47.
Olsen, K. (2002) 'Authenticity as a Concept in Tourism Research', *Tourist Studies*, 2: 159–82.
Otto, T. and Verloop, R. (1996) 'The Asaro Mudmen: Local Property, Public Culture', *The Contemporary Pacific*, 8: 349–86.

Rees, H. (1998) 'Authenticity and the Foreign Audience for Traditional Music in South-west China', *Journal of Musicological Research*, 17: 135–61.

Reisinger, Y. and Steiner, C. (2006) 'Reconceptualizing Object Authenticity', *Annals of Tourism Research*, 33: 65–86.

Robinson, P. (2004) *Kastom or Customized? Community-Based Tourism in Tanna, Vanuatu*, unpublished honours thesis, University of Sydney.

Senft, G. (1999) 'The Presentation of Self in Touristic Encounters: A Case Study from the Trobriand Islands', *Anthropos*, 94: 21–33.

Shepherd, R. (2002) 'Commodification, Culture and Tourism', *Tourist Studies*, 2: 183–201.

Silver, I. (1993) 'Marketing Authenticity in Third World Countries', *Annals of Tourism Research*, 20: 302–18.

Skrivankova, N. (2002) 'Custom Made', *Sydney Morning Herald*, 29 March.

Smith, M. (2003) *Issues in Cultural Tourism Studies*, London: Routledge.

Stanley, D. (2004) *South Pacific*, 8th edition, Emeryville, CA: Moon Handbooks/Avalon Travel.

Taylor, T. (1997) *Global Pop: World Music, World Markets*, London: Routledge.

Tilley, C. (1997) 'Performing Culture in the Global Village', *Critique of Anthropology*, 17: 67–89.

Wang, N. (1999) 'Rethinking Authenticity in Tourism Experience', *Annals of Tourism Research*, 29: 349–70.

Xie, P. (2003) 'The Bamboo-Beating Dance in Hainan, China', *Journal of Sustainable Tourism*, 11: 5–16.

Yea, S. (2002) 'On and Off the Ethnic Tourism Map in Southeast Asia: The Case of Iban Longhouse Tourism, Sarawak, Malaysia', *Tourism Geographies*, 4: 173–94.

Zeppel, H. (2002) 'Cultural Tourism at the Cowichan Native Village, British Columbia', *Journal of Travel Research*, 41: 92–100.

5 The whole nine villages

Local-level development through mass tourism in Tibetan China

Kelly Dombroski

This chapter examines the potential for local-level development through mass tourism, drawing on a case study of a group of nine villages of ethnic Tibetans who are indigenous to the Jiuzhaigou Biosphere Reserve of Sichuan Province, China. Rather than merely examining the impacts of mass tourism on a village community, it looks at the other side of the equation: the reactions that communities have to tourism, the agency that individuals, groups and the community as a whole exhibit in responding to tourism, and even the role that communities and individuals themselves play in facilitating and promoting mass tourism. Through the responses of the villagers within the Jiuzhaigou Biosphere Reserve and the local government over time, mass tourism has in fact been a positive force for village-level development that is both sustainable and empowering. This chapter introduces the Jiuzhaigou Biosphere Reserve, outlines local understandings of wellbeing and investigates how the development of tourism has contributed to its enhancement. It then moves on to look at the particular conditions of tourism and governance in Jiuzhaigou that enabled this and concludes that, when appropriate regulation is in place, mass tourism may be even more appropriate in some cases than 'alternative' or individualised tourism.

Jiuzhaigou Biosphere Reserve

Jiuzhaigou (literally Nine Village Gully) is a biosphere reserve situated in the north of the Aba Tibetan and Qiang Autonomous Prefecture of Sichuan Province, China. This prefecture has traditionally been inhabited by Tibetans, and Qiang and Han Chinese, with the highest government seats reserved for Tibetans and Qiang. The traditional home of some 1000 Tibetans, Jiuzhaigou Biosphere Reserve is protected as a site of rare natural beauty and scientific significance, mostly because of its karst-formed waterfalls and more than 100 coloured lakes. It encompasses 734 square kilometres (JNNRA/JNSRA 2003), a large reserve by Chinese standards. Just outside the reserve lies the town of Zhangzha, population 6000, which has grown from a few villages to a township supplying labour for the Reserve Administration and servicing the accommodation and entertainment needs of the more than 1 million tourists a year. The reserve is limited to 12,000 visitors per day, who normally stay in one of the 24,000 hotel beds available in the immediate area outside the reserve (ibid.).

Jiuzhaigou was relatively isolated until a government agency began to log the heavily forested valley in 1966. Roads were constructed within what is now the reserve area, with a scientific survey team arriving in 1975. As a result of this survey, which 'discovered' the natural wonders of the area, the provincial government set it aside as a nature reserve. The area became a national-level reserve in 1984, at which time it was officially opened to visitors (ibid.). During the latter part of the 1980s, as the number of tourists began to grow, the villagers both within and just outside the reserve constructed large Tibetan-style guesthouses to host tourists in traditional style. Other small businesses also sprang up within the reserve including horse and yak rides, opportunities to rent traditional Tibetan dress for photo shoots (Figure 5.1) and small restaurants or food stalls. These were all carried out as extra cash-earning opportunities alongside the continued traditional farming activities, which at that time were still permitted within the reserve area.

In 1992 the World Heritage Committee of UNESCO added Jiuzhaigou to the World Heritage List, and in 1998 accepted it into the World Biosphere Network under the Man and Biosphere Programme. This was very much driven by government officials at the regional level, who had high expectations of the volume of tourists and foreign exchange that could be accessed through joining this programme. These hopes were not realised, however, and years went by when huge hotels sat relatively empty: infrastructure was not sufficiently developed in the region for foreign tourists to come in anything but the smallest of numbers (Dombroski 2005). As with many areas in China, it took some time for government officials to recognise the benefits of the steadily increasing volumes of domestic

Figure 5.1 Mass tourists dressed up in 'Tibetan' clothing.

tourism (Aramberri and Xie 2003; Wen 1998; Zhang 1997). By 1997, annual tourist numbers (mostly domestic) had reached 200,000 (Thorsell and Sigaty 1999), by 1999, 580,000 (Lindberg *et al.* 2003), and 2003 saw a staggering 1.2 million tourists pass through the gates (Dombroski 2005). In 2004, all expectations were exceeded with a final figure of 1.91 million tourists, although this growth rate did not continue in 2005, with officials estimating around 2 million tourists (Keke, personal communication December 2005).

The nature of tourism in Jiuzhaigou is undoubtedly classic mass tourism: tourists come in busloads with guides, visit Jiuzhaigou as part of a round trip of the sights in the area, stay at reduced rates in three-star hotels and are mostly middle-aged. The nature of Chinese mass tourists is slightly different from their Western counterparts, with the emphasis being on seeing everything rather than relaxation (with photos at appropriate landmarks as 'proof' of their exploits), travelling with colleagues and business partners more than family members, and generally having a much higher tolerance of crowds, built facilities and 'staged authenticity' (MacCannell 1999) than tourists hailing from the Western world (Sofield and Li 1998). Tourists stay one to two nights in Zhangzha, entertained in the evening by one of the many innovative and contemporary Tibetan song and dance performance troupes. Foreign tourists are mostly overseas Chinese, Taiwanese, Japanese and Korean groups, some American tour groups, and independent tourists from Europe, Israel and North America.

Management of the reserve is undertaken by the Jiuzhaigou Biosphere Reserve/ National Scenic Reserve Administration, which also constitutes the local government of those villages lying within the reserve. A different local government office of an equal level manages nearby Zhangzha and the other non-reserve villages; both offices are subject to the prefecture government. Unlike other local governments, positions in the Reserve Administration are not elected, although there are several local people in management positions. The nine villages within the reserve are grouped into three administrative areas, each of which has an elected representative who speaks on its behalf to the Administration. This representative is responsible for distributing monetary and other compensation to the villagers, as discussed later.

Village wellbeing and tourism

This study was limited to ascertaining the changes in wellbeing for the indigenous people who reside within the Jiuzhaigou Biosphere Reserve, and did not address equally important issues for other local groups (indigenous and non-indigenous) residing nearby the reserve. Although it is common practice in academic literature to advocate the use of indigenous concepts and practices within fieldwork methodologies, especially in terms of understanding indigenous concepts of 'development', 'wellbeing' or 'the good life' (Gegeo 1998), in practice the preference of researchers has often been to define development in terms of basic needs, income, social capital or according to any number of ideas that have surfaced in academic literature over the last 30 years or so. The methodology of this research is firstly to understand the perception of 'the good life' or wellbeing held by the people who reside within the Jiuzhaigou Biosphere Reserve, and to then ascertain how contact with mass tourism had affected this wellbeing. The following sections outlining the local view of wellbeing are based on three months of qualitative fieldwork with people residing within the Jiuzhaigou Biosphere Reserve. Methods used included focus groups, semi-structured interviews and several participatory diagramming exercises, but of most value was an intensive three-week period of full-time participatory observation residing in and sharing work with one particular village community. Interviews and exercises were triangulated by further observation, and initial observations were likewise tested for their validity through questions in further interviews.

Local understandings of 'wellbeing'

Discussions with people in Jiuzhaigou based around the idea of wellbeing invariably centred on health. The view of health that the village members held focused on physical health and wellbeing. Health was much more than an absence of sickness, but incorporated ideas of strength, youth and the ability to work and earn a living, thus: 'Health is important. If you have good health you can work, you can have a good life' (35-year-old truck driver).

Villagers emphasised the ability to support oneself and one's family as key criteria in achieving a good life, rather than just health for its own sake. The ability to earn a livelihood was considered paramount, despite the fact that since 2001 each person in the reserve received a significant income as compensation for both the disallowing of farming (as in all catchment areas of the Yangtze River, as part of the national flood prevention programme) and the closing of guesthouses within the reserve as tighter environmental sustainability policies were introduced. This had not seemed to change their attitudes towards earning a living because of the perceived instability of government policies:

> I have no complaints about the money, except they [the Administration] can change their policies at any time, depending on who is in charge. I worry about this . . . If it [the money] stops, I will go back to farming. Business is more reliable [than government money].
>
> (Retiree, 70 years)

The second major objective in obtaining a good life was increasing one's life choices through education. This was ranked highly by those interviewed, and was also seen to be valuable from the fixation of everyday conversation around the topic – parents exhaustively analysed everything from the quality to the accessibility of the education of their children. Just as the previous generation was able to expand their choices through tourism businesses (they could choose to be farmers, or business people, or both), the current generation's choices were considered to be expanded through education. Young people had varied goals – to be dance teachers, entrepreneurs, translators, administrators, certainly a much wider range than their parents or grandparents. Yet education or study was not something forced upon the young, and the Tibetan family study habits differed markedly from the strict parental control over young people's schooling plans and choice of subjects generally exhibited amongst Han Chinese families in China. This may be because for Jiuzhaigou Tibetans education is meant to increase choice, not give status to the family. Han Chinese cultural values and practices in education have not spread to Tibetans in Aba Prefecture, probably because, despite being subject to the state education system, the teachers are still mostly Tibetan and local.

Third, the people of Jiuzhaigou placed a high value on the family and the cultural and social obligations of the family. Those who were considered the poorest in the village were those without children at home – despite these elderly people receiving extra financial benefits from the Reserve Administration. Families were considered a source of love and belonging, not just to the immediate family, but to the extended family of the village, where many are related if not through blood then through marriage. Belonging to one's village, or belonging to Jiuzhaigou, were positive identity markers for all interviewees and focus groups. The people of Jiuzhaigou maintain a separate language and distinctive customs from other Tibetans in their county, citing a proverb along the lines of 'When you climb over a mountain to another valley there is a different culture.' They also practise a version of the Bon religion, a mix of pre-Buddhist animism and Tibetan Buddhism quite distinct from that practised by Lhasa Tibetans.

Changes in wellbeing

In discussing concepts of wellbeing and the good life with villagers, the present situation was always addressed in comparison with the past. The most common timeframe for discussion was the past 20 years, from the early 1980s when tourism was a mere trickle, to the substantial flows experienced in the last four to five years. Changes in each area of wellbeing for the good life were mostly positive, and mass tourism was considered a beneficial, if somewhat bothersome, phenomenon.

In terms of livelihoods, despite the fact that farming and hosting guests is no longer permitted, villagers received cash that more than compensated for the earnings they gained in either activity. Each person, no matter their age or involvement in these activities previously, received 3000 RMB (US$375) in 2001, 5000 RMB in 2002, 7000 RMB in 2003, and 7500 RMB in 2004. For example, one participant explained how, when running their guesthouse, they had made around 20,000 RMB per year, yet with all the compensation provided to his five-member family they now brought in 35,000 RMB (US$4375) from that source alone. When he added this to his income and his wife's income, they were earning some 71,000 RMB (US$8875) per year. A county government official stated in a personal communication that the average individual income within Jiuzhaigou was 18,000 RMB per year, compared with an average of 1200 RMB per year for the rest of the county – although these figures could not be officially confirmed and this difference was purely due to Jiuzhaigou's tourism-based employment.

This has led to locals exercising increasing discernment over which jobs they will take up. One example among many is that construction must be done by Han Chinese outsiders brought in for that purpose, since locals (both within the reserve and in nearby Zhangzha) see this work as too dangerous and hard for the return. Local people prefer to own their own businesses over all other employment, and do not do physical labour or tour-bus guiding, as these jobs have long, hard hours for less financial return. All the jobs of people who live within the reserve are tourism-related. Women from Panya village stated that, even if they were allowed to return to farming, they most definitely would not: tourism-based employment was considered definitely preferable to the hard physical labour of farming. The villagers take satisfaction in having the means to consider various job opportunities and take only those they consider worthwhile.

In terms of family, cultural and social wellbeing, the extra attention and status accorded to Jiuzhaigou Tibetans through tourism has meant that certain cultural practices have been revived, allowed by the government in order to attract tourism and development. Economic development in China is thought to lead to a final socialist state (Knight 2003), where, with the removal of economic inequalities, a 'natural withering of ethnic differences' would occur (Tsering 1999). Yet these aims and goals of the Chinese Communist Party are held in tension with the requirements for ethnic difference in tourism development. In fact, it has been well documented that tourism promotion has encouraged the revitalisation of traditional ethnic practices in many areas, which in some cases had been neglected, discouraged or even prohibited for years (see for example Doorne *et al.* 2003; Oakes 1993; Sofield and Li 1998). An example of this is the case of the Miao people, who were exhorted by tourists not to change their culture with modernisation, not realising that in fact the Miao people had been permitted to revitalise their traditional practices only very recently for the sake of tourism (Oakes 1993). Tourism has also had the effect of raising the status of minority peoples (and the perceived profitability of their tourism ventures) to the extent where it has been known for Han Chinese 'impostors' to pass themselves off as minority people (Li 2003).

Both these effects have been observed in the case of Jiuzhaigou, where local customs are being revalued and revitalised through economic development and the increased value placed on their culture by themselves, tourists and the state. In particular, locals revel in their increased economic ability to participate in what was previously the domain of rich Tibetans: beautiful traditional clothing and jewellery, forests of prayer flags, the keeping of a special room as a shrine within the house, pilgrimages to Lhasa or other special religious sites within the area, the traditional painting and carving of their houses inside and out. Tourism has encouraged local people to identify not only with their own distinct local cultural roots as Jiuzhaigou natives, but also with Tibetans in general, as increased contact with Han Chinese and other Tibetans through tourism and education has strengthened their identification with Tibetans and other minority peoples.

Yet despite these well-recognised benefits, local women especially worried over the changes taking place in the young people, thus: 'This generation is very different, very Han. Their lifestyle and their thinking are different . . . The children have a Han education. They have become more and more Han' (46-year-old mother of four). The young people themselves, however, believe they are very committed to their families, their communities and their culture: many of them intend to return to live in their villages even after university education. Issues such as speaking Mandarin, listening to Chinese music, carrying mobile phones and Western dress are in fact widespread amongst minority youth in the wider county and prefecture, and are not therefore directly related to tourism. Moreover, the pride that the Jiuzhaigou teenagers exhibit over their home villages, and the amount of opportunities for employment within the home area, mean that assimilation into Chinese culture is not actually a goal for them, as it is for many of their poorer counterparts in the region. The tensions of globalisation, development and education mean that Jiuzhaigou youth have greater access to a wider range of consumer goods and choices, but at the same time have a greater understanding of what it is that sets them apart as a distinct culture regionally, nationally and internationally.

As Lee (2001) argues, minority students can subvert the education system to benefit their own culture, and empowerment is rooted in their commitment to helping their communities and ethnic groups. 'It is important to realise that when members of a minority group adopt education as a mobility strategy, they are not necessarily validating the educational project of making them into compliant and subordinate citizens' (Harrell and Ma 1999: 217). The young people of Jiuzhaigou certainly showed a commitment to their communities, and desired to use their education to assist in the conservation and development of the reserve. Although it cannot be predicted how this will change over time, students get most of their exposure to Han culture through their educational choices, not through exposure to tourism. Those students who did not continue their high school education are firmly absorbed into the village community, where the differences between the locals and tourists only serves to strengthen cultural identity and identification with one's family and community. As in every community, there is some tension between generations over what is important in culture, yet older generations do not attempt to control the younger generations in their efforts to work out their cultural identity.

Although much of the change in the past 20 years could be interpreted through a nostalgic frame that idealises the pre-tourism past and focuses on these current tensions, there was little trace of romantic imaginings of their previous lives amongst the people of Jiuzhaigou themselves. As Rangan points out, it does seem ironic that Western scholars are often so critical of development when 'voices from the margins – so celebrated in discourses of difference and alternative culture – are demanding their rights to greater access to a more generous idea of development' (Rangan, cited in Scheyvens 2002: 7). In spite of the concerns of the older generation over the changes in the younger generation, it was still made manifestly clear by all interviewees and focus groups of all ages that their wellbeing had been enhanced dramatically. Their ability to earn a livelihood and have a healthy life, their access to greater freedom of choice and education, and their increased freedom to express their culture are attributed mostly to tourism – the benefits of tourism economically and the fact that previous repressive policies have been relaxed. The tension between greater social and economic freedom and greater regulation over their traditional lands is resolved for most participants through their understanding of choice and freedom. Generally, the people have a view of choice that embraces the increase in opportunity provided through economic and social development, whereby, although their lives have been increasingly regulated by the reserve policies, the regulations involve minor things in comparison with the increased choice afforded through increased employment and education opportunities. However, although the majority of participants, both male and female, tended to see that tourism had brought increased freedom of opportunity, several male participants felt that the regulation of the reserve was excessive. This is mostly to do with traditional roles: some men miss the time they spent hunting wildlife, a practice now forbidden, whereas women feel that they have been released from some of their more physically demanding traditional roles such as farming and the collection of firewood and water.

Although it may seem enough to look at the villages themselves and the villagers' views on changes in wellbeing through tourism, this is a limited perspective. The following sections will discuss why and how mass tourism has assisted in this increased wellbeing, as well as exploring the issue of sustainability, and the relevance of this case to other areas of mass tourism and village communities.

Special conditions of tourism and governance in Jiuzhaigou

Jiuzhaigou shows not only that it is possible for villages to enhance their wellbeing or experience of 'the good life' through mass tourism, but also that sustainable and empowering community development (in the Western academic sense) can be facilitated by mass tourism. However, this has been aided by the particular conditions of Jiuzhaigou: the characteristics of tourism in the area, the attitudes and skills of local entrepreneurs, and the policies of the local government. These special conditions are outlined below.

Characteristics of tourism in Jiuzhaigou

Many of the characteristics of tourism in Jiuzhaigou enable it to be a successful example of local-level development through mass tourism. First, Jiuzhaigou is a protected area. The protected area is clearly delineated, and regulation is in place to ensure environmental sustainability and to protect the local people's way of life. The area is managed by the Reserve Administration, which is also the local government for the villages within the reserve, reducing much of the compromise that goes on in many sites of mass tourism with negotiations between tourism industry representatives, local governments, businesses and local residents, as for example in Hawai'i (Van Fossen and Lafferty 2001). As Weaver (2000) shows, sustainable mass tourism is possible through regulation of the industry, and in fact is more sustainable long term than 'circumstantial alternative tourism' – tourism that is alternative only because it is small in scale, but is not regulated necessarily to be so. Compared with the situations found in many of the surrounding areas, the particular situation of Jiuzhaigou as a protected area has lent itself both to increased environmental sustainability and to the prioritising of local development over and above outside-owned development. The Administration is able to do this by regulating the environmental conditions of the reserve, as well as restricting the number of businesses within the reserve to just one member of each reserve household, and immigration is achieved only through marriage to an indigenous reserve resident.

Second, the immense popularity of Jiuzhaigou as a tourism site has meant that there are no financial limitations to implementing their sustainability plan, as is often the case in China where financial difficulties lead to 'paper parks' (Jim and Xu 2003) – protected on paper but not in practice. The amount of money made by the reserve allows for compensation to be paid to locals, and for local-level developments such as transport to school and decent roads, electricity and piped water. The Administration regularly acquires the services of international consultants and pays its staff fairly well, hence reducing the likelihood of people abusing their positions. This has so far been to the benefit of local development and sustainability while contributing to local wellbeing. The rapid increase of tourist numbers over the last three years is of some concern, although the Administration is reviewing current carrying capacity policy. The reserve has continued to strengthen its environmental protection policies over time, as tourist numbers have risen, so these have yet to pose a problem *within the reserve*. Of course, the brunt of the environmental effects will be felt outside the reserve, where the hotels are situated. The county government no longer issues permits for the building of hotels, but further research would be required to assess the extent of environmental damage outside the reserve. The reserve makes quite enough money to be self-sustaining, and it would be self-defeating to attempt to increase tourism further from their perspective; however, there may be pressure from businesses in the surrounding area to continue growing for their benefit.

Third, the fact that Jiuzhaigou is home to an indigenous minority means that there is some value in allowing local residents to remain on their lands in the reserve. In many tourism sites in China, locals are unceremoniously moved off the site to make way for tourists, with little or no compensation, as at the Nanshan cultural tourism zone in Hainan (Li 2004). Jiuzhaigou is situated within the Aba Tibetan and Qiang Autonomous Prefecture, a region within Sichuan province that is self-governed by Tibetans and Qiang. In addition to the natural sympathies of self-government, there are actually Chinese government requirements in autonomous areas that 'authorities responsible for development [must] accept *and respect* the right of the community to make a decision for or against its involvement and the use of its resources in tourism development; and then to have the power to carry out that decision' (Sofield and Li 2003: 90, their emphasis).

Prefectures are also a small enough unit not to be a threat in terms of sovereignty movements, unlike larger autonomous areas such as Tibet and Xinjiang, hence leniency is not seen to be encouraging a split, but rather encouraging the prefecture inhabitants to view the central government in a mostly positive light.

Fourth, the local people are not the main focus of the tourists in Jiuzhaigou. Rather, it is the fame of the Jiuzhaigou scenery that draws them: as the saying goes, 'After you have visited Huangshan, you will never want to see another mountain; after you have visited Jiuzhaigou, you will never want to see any other water.' This has meant that tourist routes follow scenic routes (which are not close to villages), and tourism has not therefore been intrusive within local village life. The exception is one village that has recreated itself as the 'official minority village' where tourists may wander and peer at people's houses on the ground levels – a trade-off for being able to run businesses from home rather than travelling to the business centre or clothes-renting sites (Figure 5.2). Most people from other villages bus to work at the popular scenic sites or the business centre. However, this is not the reason why people travel hundreds of kilometres to visit Jiuzhaigou: if they were interested in viewing minority cultures or Tibetan culture the 'place' to go would be Yunnan or Tibet. Tourism in China is very much based on the status gained by having visited 'key' heritage sites (Sofield and Li 1998). There are certain places people 'must' visit in Sichuan for various reasons: the biggest Buddha in the world at Leshan, the most amazing waterfalls in Huanglong, and the most beautiful water in all of China at Jiuzhaigou. Tourists stick to the prescribed route, and make sure they 'prove' their visit by photos at appropriate recognisable landmarks. Villagers feel that they are fortunate that Jiuzhaigou's reputation is for its beautiful water, and they do not have to experience more intrusive forms of tourism. Other nearby areas in Sichuan do not have this kind of resource to attract tourists, and must use themselves, their homes and their bodies if they are to do so. These areas are able to attract Western, independent tourists as 'authentic' sites (proclaimed as such by the Lonely Planet guidebook), but only those Chinese tourists who have individual transport.

Fifth, and following on from this, because of these unique natural and cultural features, Jiuzhaigou has been signed up to several international agreements that help regulate the way the reserve is managed – in favour of both sustainability and

Figure 5.2 Official minority village special entrance.

local people's empowerment. Although the reasons for signing up to these pro-grammes may be more along the lines of achieving international recognition and hence status and more tourists (Nyiri and Breidenbach 2005), Jiuzhaigou must still fulfil the environmental and development criteria of these programmes. The Man and Biosphere Programme in particular emphasises people as an integral part of the biosphere rather than separated from it, as is common in conserva-tion practice in many parts of the world, China included. The Green Globe 21 Programme is an Australian-origin ecotourism standard that requires that fairly stringent environmental standards and community development issues be taken into account, including extensive consultation with locals and the education of visiting tourists (Green Globe 21 International Ecotourism 2003). Regular reas-sessments by Green Globe and World Heritage consultants (Dombroski 2005) require Jiuzhaigou to keep up with international research and best practice on sustainability and development issues in tourism.

Sixth, and most importantly for this discussion, Jiuzhaigou is a site of domestic mass tourism. This is significant in that there is little need to cater for Western-style individualistic off-the-beaten-track walks and niche market or luxury accommo-dation, as domestic mass tourism in China emphasises value for money over high quality. Discussions with tourists in Jiuzhaigou confirmed what can be observed at tourism sites all over China: Chinese tourists are accustomed to being managed by tour guides, and prefer to pursue a kind of 'snapshot tourism' in which they pose beside the pre-planned scenic stop-sites of the tour site. This cultural prefer-ence for guided and interpreted tours means that regulations concerning tourists are easily implemented through guide education and the planning of walkways around appropriate sites. For instance, of the current nine villages, only three are on the tourist route, and only one actively attracts tourists as a self-denoted 'official' site. The other villages are rarely visited – villagers in Panyazhai, an 8-kilometre walk uphill from the main route, could remember only one group of tourists coming to visit: independent Western tourists who were illegally hiding out in the park for the night.

Obviously, mass tourism requires a lot of preparation and planning in the form of creating and maintaining tourist routes, ensuring adequate toilet, transport and food facilities and the minimising obvious environmental strains that come with these requirements. But, as the people of Jiuzhaigou explained, tourists here are for the most part well-behaved and keep to the 'rules'. It is the independ-ent tourists who are often unsatisfied with the pre-prepared nature of tourism in Jiuzhaigou, and who attempt to breach environmental, safety and privacy regula-tions by staying in villages during the night or wandering off the official footpaths into forested areas meant to be reserved for wildlife. Thousands of mass tourists keeping to the rules are likely to do less damage and cause less disruption to the reserve and its inhabitants than 100 independent tourists wanting to do their own thing no matter whether it is intrusive or unsustainable.

Villagers say that most Western independent tourists also reject the typical buyer–seller arrangement of tourism businesses, a characteristic common in what Poon (1993) calls the 'new tourist': the post-Fordist consumer who seeks authenticity and reality in their holiday experience. These independent tourists are often seen choosing to ignore the relationship offered by the villagers (a commercial one), and seeking to have a 'real' experience: asking to stay in local homes, eat villagers' food or take their pictures. For villagers who could see up to 4000 people pass through their work site every day, a 'real' relationship with a tourist is absurd – the 'traveller' is still a tourist, primarily interested in 'consuming' what the place has to offer. Even when villagers charge for their hospitality, the reality is that after a full day's paid work and all the unpaid chores that must be done in a village situation with few modern appliances, hosting independent tourists (who normally do not speak Chinese) is more work than it is worth. Independent, foreign tourists are not seen as particularly helpful to local wellbeing and not even very interesting since so many pass through. Mass tourists are predictable, reliable customers who accept the commercial nature of the host–guest relationship, mostly in a friendly and cheerful manner.

Attitudes and skills of local entrepreneurs

Not only have the circumstances of tourism ended up being favourable for village wellbeing and local-level development, but the attitudes and skills of the villagers themselves have gone a long way towards enabling 'the good life' to be further realised. Two attributes of local residents have contributed to this: they are active in seeking opportunities to benefit from tourism in the area, and they accept the need for regulation over tourism and the reserve in order to preserve their wellbeing and sustainability.

The residents of Jiuzhaigou have managed to access increased wellbeing through mass tourism mostly through small businesses: previously guesthouses, yak, horse, and costume renting; now mostly family souvenir stalls and village costume-renting businesses where Tibetan costume dress is rented to tourists who wish to take a photo near the lakes. Even when later government support was introduced (see the earlier section on livelihoods), its purpose was to compensate for the loss of community income when the Administration closed down the guesthouses and animal rides for reasons of environmental protection. All villages have applied themselves to various forms of small business enterprises over the years since tourism began in the valley. Within villages, there has been a concerted effort through internal regulation to ensure that there is equitable access to tourism opportunities by all. For example, the costume renters of Panya had worked individually to gain custom and rent their own sets of clothes, but this had led to increased conflict within the small village. Eventually it was decided to join together in renting out the costumes, sharing the day's earnings equally amongst all workers. Although some tourists complained of a lack of variety and a lack of willingness of the clothes renters to bargain (there is no incentive to drop prices if there is no competition), this benefits local people since their relationships remain harmonious, prices are kept at a high enough level to ensure a decent income for all, and the conditions of work are much more relaxed and enjoyable. Likewise in another village, prior to the closing of guesthouses in 2001, locals agreed to limit each house to 40 beds in order to ensure that anyone who wanted to access tourists could do so.

Such examples show how the attitudes of local people have ensured that 'the good life' is realised and maintained – where wellbeing involves more than just the overtly mentioned aspects of livelihoods, family and cultural wellbeing and education, but also local harmony. Although local people are quick to take advantage of opportunities afforded them through mass tourism, they also understand the necessity of self-regulation in order for everyone to benefit. This valuing of local harmony is critical to local wellbeing in Jiuzhaigou, and may even explain the preference of local people for mass tourists who keep to the 'rules' rather than independent tourists who do not act in harmony with the social context.

Policies of local government

In Jiuzhaigou, not only are the characteristics of tourism and the local population partial to regulation for development, but the administration has the skills and authority to direct this regulation. Although in the 1960s the local government prioritised logging and land reform, more recent administrations have made considerable efforts to ensure local people's wellbeing in the form of job provision, compensation, the building of local facilities and the instigation of the shareholder scheme for the main restaurant. A recent document has a whole section dedicated to community development, a policy prioritised 'to ensure a balance between the integrated development of Jiuzhaigou [as a reserve and tourist site] and the needs of the people' (JNNRA/JNSRA 2003: 42). The relationship between the Administration and the local people is managed by the Residents' Administration Department, whose manager is from within the reserve. It was his responsibility to ensure that each valley resident had a livelihood, while encouraging residents to see the reserve as their own (Dombroski 2005). Despite these aims, the Administration saw this department more as a unit to organise and distribute material development benefits and to persuade residents of the benefits of tourism than as one that should empower residents to choose for themselves, a fairly typical stance of governments elsewhere in similar contexts (Sofield 2003).

Although Administration documents insist that local participation in management is extremely important, in practice this has not been achieved except for the three local representatives, and the various members of the valley who have managed to get jobs within the Administration itself. However, because of these local people in Administration jobs, the situation in Jiuzhaigou would be what the tourism literature terms 'resident-responsive tourism' (Sofield 2003) – where the needs of residents are considered important, but empowerment is not necessarily the end goal. This could well be a point for critique, but Western writers in general need to be careful of criticising the lack of empowerment in China on account of state control when in fact many host communities in Western countries lack empowerment on account of the hegemonic control of large businesses.

In general, the Administration of Jiuzhaigou has been aware of local needs (see JNNRA/JNSRA 2003) and in comparison with many other local governments in

China and worldwide has done well in negotiating the provision of these. The relationships that have developed between local residents and the Reserve Administration – through family ties, business ties, ethnic ties and proximity – have meant that many individuals within the Administration seek to assist local people, despite government policy not requiring the Administration to do so. This has definitely contributed to the quality of life that the local people have been able to experience over the last 20 years of tourism.

Although Jiuzhaigou does not provide official democratic representation and organised participation of local people, it could be argued that to consider only official representation as valid is a cultural value in itself. Many Western cultures emphasise tasks and order over relationships and fluidity. Both Chinese culture and Tibetan culture tend to be relationship-based cultures rather than task-oriented ones, and having a system of unofficial relationship-based 'representation' may be more culturally sustainable and reliable than corruptible 'official' representation. Relationships with the Administration staff are deliberately cultivated by local people, who seek to provide the necessary relationships and educational requirements for villagers to gain employment. This informal system of representation is not random but deliberate, at least on the part of the villagers.

Conclusion

The descriptor 'mass tourism' covers a large range of tourism sites and a large range of tourist types, and the effects cannot be predicted for all situations: mass tourism is a complex force that affects different groups differently. Jiuzhaigou shows how, because of several specific factors, mass tourism can bring significant benefits to people at the village level, and has been a beneficial force in the wellbeing of the nine villages. Community wellbeing has been assisted by the government, by the villages' internal regulation, and by the tourists' being accustomed to a high level of organisation and control. Most importantly, the villagers themselves describe their situation as vastly improved since tourism began, and are for the most part happy with the way things have been managed.

Weaver (2000, 2001) posits that mass tourism may be sustainable as it may limit tourists to specially prepared areas within a small part of a spectacular environment, providing funds for the upkeep of the rest of the reserve. This has been the case in Jiuzhaigou, where the ability to regulate the tourists has been a boon to the management planning of the local government. However, there is a danger in this financial blessing: the tourist becomes the customer to be pleased at all costs, and it is possible that sustainability and local development may start to become sideshows to the central goal of attracting tourists and finance. This is a temptation to which the Administration of Jiuzhaigou has not been immune, as the recent expansion of the marketing department testifies. However, with appropriate policy and the recognition of the special characteristics of Jiuzhaigou that have resulted in its success as a tourism operation, a nature reserve, and a deliberate community development initiative, the Administration has the ability to continue to regulate for its future success. This is more likely to be the case if local people are able to continue to influence the decision-making of the Administration, both formally through elected representatives, and informally through relationship-building and educating themselves and their children to the level required to gain employment in the reserve management.

Jiuzhaigou shows that mass tourism must be analysed on a case-by-case basis and may, in some cases, be a positive presence for indigenous minorities. Many other factors come into play in the changes that are seen in rural minority villages in the west of China and elsewhere. Mass tourism is but one factor, which co-exists and overlaps with other forces such as the globalisation of capitalism, education and the expansion of the mass media. Indigenous rural villages will not be isolated forever, and it would be hypocritical even to require them to be if this is not their desire. The discourse of the noble savage is long outdated, and romantic 'othering' of indigenous people and rural villagers is not an acceptable reason for rejecting all mass tourism developments in areas such as these, especially when examples such as Jiuzhaigou show that, in some cases, local people may consider mass tourism more beneficial to community development than more small-scale but intrusive and individualistic tourism types.

References

Aramberri, J. and Xie, Y. (2003) 'Off the Beaten Theoretical Track: Domestic Tourism in China', *Tourism Recreation Research*, 28: 87–92.

Dombroski, K. (2005) 'Exploring the Potential of Mass Tourism in the Facilitation of Community Development: A Case Study of Jiuzhaigou Biosphere Reserve, Western China', Development Studies Programme, Massey University, Palmerston North.

Doorne, S., Ateljevic, I. and Bai, Z. (2003) 'Representing Identities through Tourism: Encounters of Ethnic Minorities in Dali, Yunnan Province, People's Republic of China', *International Journal of Tourism Research*, 5: 1–11.

Gegeo, D. (1998) 'Indigenous Knowledge and Empowerment: Rural Development Examined from Within' *The Contemporary Pacific*, 10: 289–315.

Green Globe 21 International Ecotourism (2003) Green Globe 21 Company Standard for Travel and Tourism, Version 1.2. Online. Available <www.greenglobe21.com> (accessed 9 July 2004).

Harrell, S. and Ma, E.Z. (1999) 'Folk Theories of Success: Where Han Aren't Always the Best', in G. Postiglione (ed.), *China's National Minority Education: Culture, Schooling and Development*, London: Falmer Press.

Jim, C.Y. and Xu, S.S. (2003) 'Getting out of the Woods: Quandaries of Protected Area Management in China', *Mountain Research and Development*, 23: 222–226.

JNNRA/JNSRA (2003) *Jiuzhaigou – The Heritage for All Mankind: White Paper on Conservation and Development of Jiuzhaigou*, Chengdu: CIP.

Knight, N. (2003) 'Imagining Globalisation: The World and Nation in Chinese Communist Party Ideology', *Journal of Contemporary Asia*, 33: 318–337.

Lee, M.J.B. (2001) *Ethnicity, Education and Empowerment: How Minority Students in Southwest China Construct Identities*, Aldershot: Ashgate.

Li, J. (2003) 'Playing upon Fantasy: Women, Ethnic Tourism and the Politics of Identity Construction in Contemporary Xishuang Banna, China', *Tourism Recreation Research*, 28: 51–65.

Li, Y. (2004) 'Exploring Community Tourism in China: The Case of Nanshan Cultural Tourism Zone', *Journal of Sustainable Tourism*, 12: 175–193.

Lindberg, K., Tisdell, C. and Xue, D. (2003) 'Ecotourism in China's Nature Reserves', in A. Lew, L. Yu, J. Ap and G. Zhang (eds), *Tourism in China*, New York: Haworth Press.

MacCannell, D. (1999) *The Tourist: A New Theory of the Leisure Class*, Berkeley, CA: University of California Press.

Nyiri, P. and Breidenbach, J. (2005) 'Our Common Heritage: New Tourist Nations, Post-Socialist Pedagogy, and the Globalization of Nature', unpublished lecture at the University of Hong Kong, June 2005.

Oakes, T.S. (1993) 'The Cultural Space of Modernity: Ethnic Tourism and Place Identity in China', *Environment and Planning D: Society and Space*, 11: 47–66.

Poon, A. (1993) *Tourism, Technology and Competitive Strategies*, Wallingford: CAB International.

Scheyvens, R. (2002) *Tourism for Development*, Harlow: Pearson Education.

Sofield, T. (2003) *Empowerment for Sustainable Development*, Amsterdam: Pergamon.

Sofield, T. and Li, F. (1998) 'Tourism Development and Cultural Policies in China', *Annals of Tourism Research*, 25: 362–392.

—— (2003) 'Processes in Formulating an Ecotourism Policy for Nature Reserves in Yunnan Province, China', in D.A. Fennell and R.K. Dowling (eds), *Ecotourism Policy and Planning*, Wallingford: CAB International.

Thorsell, J. and Sigaty, T. (1999) 'Human Populations in World Natural Heritage Sites: A Global Inventory', *World Natural Heritage and the Local Community*, Paris: UNESCO.

Tsering, S. (1999) *The Dragon in the Land of Snows: A History of Modern Tibet Since 1947*, New York: Columbia University Press.

Van Fossen, A. and Lafferty, G. (2001) 'Contrasting Models of Landuse Regulation: Community, Government and Tourism', *Community Development Journal*, 36: 198–212.

Weaver, D.B. (2000) 'A Broad Context Model of Destination Development Scenarios', *Tourism Management*, 21: 217–224.

—— (2001) 'Ecotourism as Mass Tourism: Contradiction or Reality?', *Cornell Hotel and Restaurant Administration Quarterly*, 42: 104–112.

Wen, J. (1998) 'Evaluation of Tourism and Tourist Resources in China: Existing Methods and their Limitations', *International Journal of Social Economics*, 25: 467–485.

Zhang, W. (1997) 'China's Domestic Tourism: Impetus, Development and Trends', *Tourism Management*, 18: 565–571.

6 Weapons of the workers

Employees in the Fiji hotel scene

Yoko Kanemasu

Introduction: touristic representation and ideology

'Fiji, the Way the World Should Be' is a popular phrase that has, for over 25 years, represented the Fiji Islands in the international tourist market (Kanemasu 2005). The Pacific island nation prides itself on this slogan, which is typically combined with images of palm-fringed beaches, blue waters, hibiscus flowers and, perhaps most importantly, smiling faces of the people. The people of Fiji, especially indigenous Fijians (henceforth Fijians), are often designated in the world of tourism as 'the world's friendliest people' (Fiji Visitors Bureau 2003), thus widely associated with such qualities as amiability, hospitality and affability. Fijians constitute 50.8 per cent of the nation's population of about 830,000, whose other main ethnic communities include Indo-Fijians (largely the descendants of the indentured Indian labourers brought to Fiji during the colonial rule) (43.7 per cent) and Rotumans (1.2 per cent).

Curiously, the islands were first introduced to the Western world in the late eighteenth century, through the tales of early missionaries, explorers, traders and others, as the 'Cannibal Isles of Fiji', inhabited by hostile and violent cannibals. The antagonistic imagery, embedded in the open conflicts and confrontation that characterised the 'contact period' and the subsequent years of precolonial Western domination of the islands, underwent gradual transformation and eventually gave way to rather 'Polynesianised' imagery of amiable savages, not dissimilar to Tahiti's 'Noble Savages', in parallel with increasing Western commercial and military control of the islands, Christianisation of the islanders, and eventually the 'voluntary' cession of the islands to the British Crown in 1874. If the earlier imagery represented Fijians as the ultimate Ignoble Savages, the later imagery, established in the late nineteenth and early twentieth centuries by travel writers and others, was that of 'Good Savages', conspicuously laudatory and marked by soft primitivism (Kanemasu 2005).

This imagery was further modified and consolidated in the following decades by tourism. The industry emerged in the late 1960s and early 1970s with an influx of foreign capital, which established institutionalised mass tourism as its core, and has since grown into the largest foreign exchange earner in the nation's postcolonial economy (see Varley 1978; Britton 1983; Plange 1996; Korth 2000). It has depended significantly on soft primitivistic marketing centred on the imagery of amiable Fijians, within a tropical idyll of the 'sun, sand and sea' (see Figure 6.1). Representations of ignoble hard savagery of the past, in the form of souvenir war clubs and cannibal forks, and war dance performances, have continued to accentuate, rather than conflict with, the soft primitivistic emphasis. Notably, in the process of touristic incorporation, there has been an increasing association of the imagery with what many Fijians regard as their cultural essence, especially the virtues of hospitality and generosity. Simultaneously, the imagery has been actively promoted by the industry on the basis of its perceived connection to the treasured indigenous cultural heritage. Furthermore, by virtue of its central presence, the imagery has dictated to Fijians a distinctive role in the industry, of providing necessary labour and at the same time embodying the famed amiability and thereby supplying its crucial competitive advantage. Direct service work combined with display of amiability, upon which the success of the industry depends, remains the designated role of Fijians. In a nation of diverse cultures, tourism has thus given somewhat peculiar prominence to Fijians, in contrast with Indo-Fijians and others who are generally absent from the centre stage. The smiles of Fijians have eventually come to command the status of not only a successful advertising image but indeed a symbol of the nation, proudly displayed to the world. Today, the imagery is so prominent that it seems to have become, as an industry representative once said, the country's 'icon' (*Fiji Times* 18 June 2000).

This celebrated imagery may be seen as more than a successful tourism marketing tool or a proud national symbol. Without denying that it does embody

Figure 6.1 Isles of Smiles.

the national pride and aspirations of many (especially indigenous) people of Fiji, one may argue that the imagery has also played an ideological role in colonial and post-colonial Fiji. The term 'ideology' carries various meanings. It is used here to refer to ideas and meanings implicated in the sustenance of domination (Thompson 1984: 4; McLellan 1995: 83; Larrain 1996: 57). In contrast to the coercive subjugation of Fijians in the earlier years, colonial rule depended rather less on direct use of force and more on persuasive measures, which culminated in what may be described as hegemonic rule in the sense discussed by Gramsci (1971). As rule by coercion shifted to 'indirect rule' by the colonial regime in alliance with the eastern Fijian, chiefly elite establishment (along with non-indigenous elite business interests and foreign/multinational corporate interests), the imagery assumed characteristics of a persuasive ideology to elicit and sustain the consent of Fijians to the rule. Indirect rule rested on politicisation of ethnicity and a patronage system championing 'Fijian' interests vis-à-vis those of Indian immigrants, which shaped the internal power structure throughout the colonial era and laid the foundation of post-independence national politics (Durutalo 1985, 1986; Lawson 1990, 1991; Norton 1990). The imagery may be seen as an ideological dimension of this hegemonic structure. In particular, its gratifying and complimentary tone, often accentuated by contrastingly stigmatising imagery of Indo-Fijians established in colonial literature and elsewhere, facilitated polarised ethnic stereotyping and provided normative endorsement and further encouragement of Fijian allegiance to the colonial, and later, post-colonial order. Although the sustenance of such an order cannot be attributed to hegemonic processes alone (see e.g. Halapua 2003), its durability can be said to rest significantly on the consent of the indigenous population. The imagery of the amiable Fijian, along with other hegemonic ideologies (Halapua 2003; Durutalo 1997), has contributed to the production of this consent.

In this context, the touristic imagery of Fijians, although evidently gratifying and privileging, has served as an ideological means of their own subordination. First, the imagery, if without apparent pejorative tones, has continued to define Fijians as 'savages' whose supposedly simple existence closer to the state of nature appeals to contemporary Western romanticism but also endorses their social evolutionary inferiority. The imagery is also marked by a latent emphasis on self-subordination. Fijians are celebrated as a people who represent 'the way the world should be' yet are also often portrayed as much less enviable than that – as willing subordinates, eagerly smiling and anxious to please Western visitors. Amiability, in other words, is inseparably fused with self-subordination, a double-edged notion containing elements of both nobility and ignobility.

Second, the imagery, as an often eagerly embraced self-definition, has served to induce the assent of Fijians to the existing structure of tourism as well as their place in it. The industry, since its early years, has been dominated by foreign, multinational and local European interests, with limited Indo-Fijian and marginal indigenous Fijian participation (Fong 1973; Britton 1983; Korth 2000). For the masses of Fijians, this means that their avenues for sharing the economic benefits of the industry are more or less restricted to hotel wage employment and land leases. The income accruing from land leases is subjected to inequitable distribution and is not an adequate means of participation for most landowners (Durutalo 1997). Similarly, although numerically Fijians are the largest beneficiaries of hotel employment, its economic benefit is rather marginal given the institutionalised practice of allocating jobs along ethnic lines (Britton 1983). The vast majority of Fijian workers are placed in lower-paid direct service positions virtually at the bottom of the industry's job hierarchy. Although this may be partly attributed to the relative lack of formal qualifications and technical skills among them, their role of acting out the celebrated imagery – the industry's primary competitive advantage – plays a major part (Samy 1980; Britton 1983). In establishing and sustaining this structure of the industry, the imagery has suggested that it is 'natural' and perhaps even honourable for Fijians to be placed in the 'frontline' roles and display their prized amiability to visitors: it allows a matter-of-course definition of Fijians as ideal direct service workers, and its laudatory connotation renders their role desirable rather than inevitable. Placing Fijians in direct service jobs with low rates of pay becomes less a matter of economic subordination and takes on an aspect of celebration of Fijian virtues. The imagery, elevated to the status of a treasured national pride, has thus served the sustenance of the existing structure of Fiji's tourism by fostering the assent of Fijians to the position it assigns to them. Although the industry has faced occasional expressions of local discontent in recent years (Kanemasu 2005), the imagery, together with various public relations activities and official measures, counters the development of potentially oppositional actions and ideas by defining amiability as a prized virtue and attaching positive normative value to the touristic role.

Fijian endorsement of this imagery seems considerable today, with what looks like nationwide celebration of it. However, it requires more than a hurried look at documented history to adequately assess the responses of the local population, especially of those communities whose ideas and viewpoints do not always find their way into the public forum. The effects of colonial/touristic representation on those who are represented have been debated vigorously in the past from various perspectives ranging from neo-Marxism, discourse analysis and post-colonial theory to subaltern studies. Whereas many, often following the example of Said's *Orientalism* (1979), have expressed concern over the scarcity of room for self-expression by colonised/destination peoples and its far-reaching effects on their self-definition, such an approach has recently been subjected to critical scrutiny by those who call attention to the various means by which subordinated peoples challenge or appropriate dominant imagery. Uncritical assumption of the effectiveness of dominant discourse has been held not only analytically limited but also ethically problematic, for it regards 'the "oppressed" as the unfortunately deluded' (Lather 1991: 137) and refuses to take full account of their viewpoints. Similarly divergent views are found in the literature on Pacific islands tourism. Although few empirical studies have specifically focused on this question, mentions have been made in the past of the significant impact of touristic imagery, especially of the eastern Pacific (see, for example, Cohen 1982; Macnaught 1982; Stanton 1989; Bruner 1991; Douglas and Douglas 1996). On the other hand, increasing attention to indigenous agency, as well as to the plurality of representation, can be seen in the work of more recent researchers (Edmond 1999; Bell 1999; Douglas 1999; Jahnke 1999; Thomas 1999a,b; Nicole 2001). Although most of these researchers are concerned with colonial rather than touristic imagery, Nicole (2001), for instance, has extensively discussed touristic appropriation of colonial representation of Tahiti, to which, he points out, the indigenous Polynesian Maohi have responded with various strategies of resistance as well as collaboration.

This chapter offers some tentative suggestions on the question of local response and agency, based on semi-structured interviews conducted with 41 Fijian tourism workers between March and August 2001 (Kanemasu 2005). The workers, mostly in direct service occupations and selected by snow-balling, were requested to discuss a set of predetermined guideline topics designed to explore their views and experiences in relation to contemporary touristic imagery of Fijians and their work role. Promotional photographs of Fijians were used to facilitate this. The results of the interviews were analysed in an interpretive manner, with an emphasis on in-depth investigation of the range and details of the responses rather than measurement of their statistical distributions. The results can be described as a complex mixture of diverse responses to the touristic imagery and role. They can be broadly classified as affirmative and critical responses, which coexist in many workers' accounts, although the affirmative responses are consistently salient and articulate in most accounts, whereas the critical responses receive varying degrees of emphasis and focus.[1] Together they allow a glimpse of the workers' ideas and experiences, closely linked to the touristic imagery and the dynamics of hegemony upon which Fiji's mass tourism rests.

Voices of the subordinate: workers' response

Affirmative response: foundation of hegemonic rule

The Fijian workers' most salient response was affirmation of amiability, the central component of the imagery, on the basis of its association with indigenous virtues. Amiability, in the workers' view, is rooted in the most basic aspects of their way of life, 'something you're brought up with'.[2] Many commented that this gave a unique and valuable quality to their work – that friendliness to strangers was not an alien corporate practice but part of their cherished tradition – for which tourists continued to return. The famous Fijian smile is not 'plastic', a calculated façade, but 'natural', 'genuine', 'the way we are': in one worker's expression, 'We're not taught to smile. We smile. We're not taught to sing. We sing.' This, they explained, renders Fijians ideally suited to the work of serving tourists: Fijians have 'the natural flair' for hospitality work. They affirmed the allocation of direct service roles to Fijians as a necessary consequence of their cultural peculiarity. Indeed, for one worker, tourism work was not merely a job for which her cultural background came in useful but was the proud display of that culture itself: 'It [tourism work] goes back to showing Fijians . . . what real Fijians are, you know.' Hence almost all noted with emphatic or taken-for-granted approval the hegemonic status of the imagery of Fijian amiability: 'Friendly – that's what we are known for, isn't it?' The workers proudly endorsed it as part of their self-definition.

The workers' accounts suggest that the imagery, if it originally emerged as an externally derived definition of Fijians, has effectively incorporated elements of what they regard to be their unique values and way of life, and consequently sustained and enhanced its legitimacy and hegemonic status. The 'cultural' interpretation of amiability has served to consolidate the hegemonic status of the imagery and further naturalise Fijians' role in embodying it. This in turn gives normative sanction to their designated position in the industry, which allows them marginal economic benefit and little control over the overall enterprise: Fijians are supposedly in the position that best suits them – to serve with friendliness. Just as Third World women have been supposed to possess 'naturally' nimble fingers and docile dispositions ideally suited to factory work (Elson and Pearson 1981), the 'natural' amiability of Fijians legitimises their direct service role and the status it assigns to them. In the case of Fijian tourism workers, the positive normative value attached to the notion of amiability indeed seems to encourage their *active* consent, not just acquiescence, to their prescribed position in the industry. Their touristic role is not merely tolerated but actively approved of in the name of indigenous virtues.

When touristic ideologies fail to secure such hegemonic status, active consent of the local population to their designated role is less likely to be obtained, and their discontent may be translated into overt oppositional actions, rendering coercive measures necessary, as demonstrated by the case of Caribbean tourism in the past (Young 1973; Turner and Ash 1975). Although scholars and officials alike feared that Fiji might possibly follow the same path (see, for example, Belt, Collins and Associates 1973; Fong 1973; Thompson 1973; Central Planning Office 1975; Samy 1980; Britton 1983), the touristic imagery of Fijians has successfully become hegemonic and contributed to creating an ideological terrain conducive to winning more or less 'spontaneous' consent of the workers to their position. Fiji's mass tourism can thus be said to rest upon a hegemonic order – rule by consent of its subordinates – secured by hegemonic ideologies as well as by other means.[3]

At the same time, however, one may question whether this hegemony is entirely as secure as it seems. It is debatable, for instance, whether the workers' affirmation can be equated with unproblematic acceptance of touristic amiability. Many workers' accounts indicate, upon scrutiny, that the meanings they attach to amiability are subtly different from its touristic definition. Whereas the touristic imagery defines amiability as part and parcel of the act of serving, a passive gesture of the subordinate, many workers affirmed it in a rather different context – in the context of reciprocity. They described reciprocal interaction with visitors as one of the genuine satisfactions they derived from their work. They found it deeply gratifying when their amiability was not a one-way transaction but returned by visitors at a personal level, particularly when it developed into a lasting friendship. They discussed numerous examples of visitors who spent time with them outside the hotel/resort, shed tears upon departure, continued to write to them or continued to return to Fiji especially to see them. Some of these tourists, who have now become the workers' 'family friends', are invited to their homes on every visit to Fiji. Some interviewees spoke of the joy of watching these tourists' children grow over the years. Amiability in such a context is no longer part of the passive and anonymous role of serving but part of an active and reciprocal relationship, which defines them as an interacting subject, not objects of the 'tourist gaze' (Urry 1990).

Furthermore, in contrast with the colonial/touristic definition of Fijian amiability as a manifestation of voluntary subjection, amiability for many workers is the basis of a positive self-definition – self-*affirmation* rather than self-subjection. The workers described their amiability as empowering: not only is it an indispensable, valuable resource on which the whole industry depends, it is at times capable of producing significant effects on the strangers they encounter. An illustrative account was given by Jone, a young tourist coach driver, who spoke of the deep sense of reward he felt when he knew he made people truly happy. He recounted his encounter with a couple on a sightseeing tour of which he had been in charge. He noticed at the beginning of the tour that they had had a fight, since they refused to talk to each other, the husband sitting at the back of the coach and the wife at the front. He tried hard to make them laugh by telling his favourite jokes and managed to make friends with both. Towards the end of the day, the husband secretly told Jone that it was his wife's birthday and asked him to sing 'Happy Birthday' for her. When Jone and a tour attendant sang to surprise her, they found the couple crying and making up: 'It was the first time tourists cried in front of me.' The experience made him feel rewarded and empowered, that 'I can change bad problems to good problems.' Marika, a veteran porter, offered a similar perspective. He noted that although some tourists were 'racists' and ignored his greeting, when he refused to withdraw and continued to try to engage them in friendly interaction, many demonstrated 'a new attitude', perhaps with the realisation that Fijians were 'human beings just like them'. Hence, amiability for Marika was not an 'offering of an inferior' (Goffman 2001: 48) but, on the contrary, a resource that allowed him to challenge the power relationship imposed upon him. It is not a manifestation of passivity or self-subordination but an ability to interact actively and meaningfully with others and sometimes even to effect profound changes in them. The meanings of amiability for these workers are grounded in positive self-evaluation that runs counter to the meek obedience of 'the good savage'. Indeed, from these workers' perspective, amiability may reflect not self-subjection to racism but a strength that could even overcome it.

Critical response: discontent and resistance

Although the notion of amiability was almost unanimously endorsed, the workers' accounts suggest that they may be privately critical of other dimensions of the imagery. As noted earlier, this imagery is centred around the notion of savagery – overwhelmingly soft primitivistic but also accentuated by elements of hard savagery. The workers identified and often critically discussed representations of such (especially hard) savagery. A promotional photograph of a Fijian man dressed in a grass skirt (Figure 6.2), for instance, was critically appraised

Figure 6.2 Promotional image of Fijian man in grass skirt.

by a number of workers, who objected to what they considered to be inaccuracy of the representation and its underlying meanings; in particular, its definition of the Fijian as a primitive and its association with cannibalism and violence. For these workers, over-emphasis on premodern lifestyle is not merely inaccurate but entails a latent, antagonistic typification of Fijians. One interviewee observed that the image 'symbolises cannibalism' and that the man, 'opening his mouth like this, jumping', is presented as a wild and warlike savage. Another commented: 'This gives a negative idea of Fiji . . . cannibalism and other things. If you don't know our history and the first thing you see is this, you'd have a negative idea.' A few responses took the form of jokes and expressions of amusement. A young man exclaimed with laughter: 'Cannibal! Cannibal! – You don't see every Fijian dressed like this.' A young woman joked in a similar manner: 'I hope he won't eat me with his big open mouth!' These youths found the hard savagery they detected in the image far distanced from their self-definition to the point of being laughable and fit for a joke. Moreover, in light of the Pacific island tradition of mimicry and clowning as a means of critical expression (Mitchell 1992; Hereniko 1994; Smith 1998), such responses may be seen as not just indications of light-hearted amusement but subtle and creative expressions of opposition and criticism.

Self-subordination

The workers may also be critical of another crucial dimension of the imagery: self-subordination. This was most evident in their responses to a promotional photograph of a Fijian waiter serving tourists at the sea, fully dressed and knee-deep in the water (Figure 6.3). A number of workers found this image questionable because of its manifest inaccuracy and/or the latent theme they derived from it. Many described the image as 'only for tourism' or 'purely for advertisement' – removed from 'reality', their own definition of Fijians. At a manifest level, they found the image contrived and exaggerated: 'This is in the middle of the sea. We don't go out to the sea and serve tourists like this.' At a latent level, more significantly, it is not only inaccuracy but its underlying theme that the workers found problematic. The exaggeration of the waiter's service stretches his amiability to servility, projecting an image of the Fijian who even steps into the sea and wets his clothes to attend on tourists who command his service while lying down at leisure: in a veteran porter's words, 'Some people would think it is stupid, the tourists are lying down and he is serving like this.' Some young workers responded by laughing and joking, which, as noted above, may be seen as an indirect means of challenging and critically commenting on what is objectionable to them. The youths playfully questioned and mocked the definition of the Fijian as an eagerly serving subordinate:

Figure 6.3 Promotional poster featuring Fijian waiter and tourists.

(Laughing, to her friend, who also participated in the interview:) 'Do you think a Fijian man would wear a flower, go out to the sea and serve tourists like this?' (Her friend:) 'No!'

(Bursting into laughter:) 'No, no, no, that's too much. We usually wait in the *bure* [house] and tourists come to us. He wouldn't go into the water like that, especially if he has to wet his *sulu!*'

(Bursting into laughter:) 'Baking powder![4] . . . This man must have thought, "They're taking a picture, so I'll get my *sulu* wet."'

These responses assume much significance since the photograph figured prominently in a recent Fiji Visitors Bureau image-building campaign (Kanemasu 2005). Pictorial images of eagerly serving Fijian workers, along with those of smiling Fijians, have become the central advertising images of Fiji's tourism. These workers, however, considered such fusion of amiability and self-subordination inaccurate, far removed from their reality, 'stupid', 'too much', or 'baking powder' as a young woman summed it up in a local slang expression.

Institutionalisation of amiability

One of the peculiarities of the workers' daily duties is the role of acting out the touristic imagery in face-to-face interaction with visitors. The presentation of the workers, including their personal appearance, bodily gestures, facial expressions, speech patterns and other minute details of their expressive equipment, is scrutinised and defined by the industry. Above all, the central emphasis of the corporate effort is placed on smiling. According to the workers, smiling is enforced with slogans, practised and perfected through training programmes and continually emphasised by supervisors and managers. The famous Fijian greeting of *Bula!* (hello) with a smile is a primary guest-relations routine, the observance of which by all workers is strictly supervised at hotels/resorts. Amiability, which most workers affirmed as 'natural' and part of their way of life, tends to lose its very naturalness when it is thus institutionalised, to the extent that it becomes a rather singular touristic performance. As an experienced hotel worker put it:

> The hotel tells us to do that [smiling and saying *Bula!*]. But we don't do it among ourselves. If you say *Bula!* to another local [in the same manner as it is done at hotels/resorts], he'll think, 'Is this guy crazy or what?'

Touristic amiability is hence a carefully designed and controlled emotional display the workers are required to maintain throughout their working hours, divorced from its contexts such as interactional situations (such as meeting the same tourist a number of times and giving the same greeting each time), sociocultural relevance (for example, it may have become a somewhat 'unnatural' display of amiability) and the workers' inner emotional state. Much of this involves what Hochschild (1983: 7) calls 'emotional labour', or 'the management of feeling to create a publicly observable facial and bodily display' for a wage. Although maintenance of an institutionalised personal front is a common requirement in service occupations, and indeed in social life in general according to Goffman (1959), it assumes particular significance here, since Fijian amiability is defined as the primary competitive advantage of Fiji's tourism. Maintenance of its display is considered vital and has become the workers' principal daily duty.

The workers respond in a variety of ways to such enforcement of institution-alised amiability. Some accommodate it actively on the basis of its perceived connection with their cultural tradition – even if its spontaneity may be somewhat lost – and the expressive satisfaction they derive from it, or take a pragmatic and instrumental view that it is part of their professional duties for which they are paid. Active accommodation may be described as similar to what Hochschild (1983: 38–42) calls 'deep acting' in emotional labour, 'a natural result of working on feeling'. For these workers, even if they may have initially found it difficult, it has become more or less unproblematic that 'We say *Bula!* and give our expansive smile [laugh]'. Others seem less active or voluntary in their compliance. In fact, these workers expressed varying degrees of discontent with the loss of autonomy of their emotional display and the coercion they detected in it ('You just have to say it [*Bula*] . . . It's like the boss is standing over you with a stick'). Although institutionalised amiability does not preclude possibilities of reciprocity or mean-ingfulness, it is often a one-way offering of the workers, which seems to render their amiability a rather passive act of serving, contrary to the self-affirmative meanings they endorsed. The workers are denied control over their own facial and bodily display, as though their interaction with visitors' amiability tends to lose its genuine meaningfulness for them. What could be a source of empowerment here assumes an aspect of disempowerment.

This discontent is expressed in at least two ways through the workers' outward response. Some opt to accommodate the enforcement passively or mechanically. This was vividly described by a young woman's comment that smiling and saying *Bula!* was for her a 'ritual', the execution of which was 'automatic whenever you see a tourist coming'. For those who are discontented with the enforcement yet also pragmatically aware of the need to comply with it, 'switching' touristic amiability 'on and off', or what Hochschild (1983: 37–8) calls 'surface acting', becomes a way of dealing with the existing conditions. Others, although stopping short of open opposition, nevertheless covertly resist it by seeking to maintain certain autonomy over their facial and bodily display. A young hotel worker explained, for instance, that the greeting was done 'to show we are happy. So we don't say *Bula* when we are not happy.' Whereas Fijian tourism workers' unfailing amiability has been widely commended by the industry, government officials and visitors alike, what these workers' accounts suggest is that underneath this dexter-ity may lie inner discontent and a sense of alienation.

Inequity of power

The workers' role brings to the fore another possible contradiction between their self-definition and the touristic definition of Fijians. Although the workers critically responded to the definition of the Fijian as a self-subordinated server, their everyday work nevertheless places them in an emotionally (and economically) subordinate position: they may be daily subjected to unequal 'emotional exchange' (Hochschild 1983: 84–5). Moreover, as noted by many in the past (e.g. Young 1973; de Kadt 1979; Farrell 1982; Baum 1997), worker–tourist interaction in societies like Fiji is embedded in inequity that is intersected by class, race, global location and often sex, which situates workers as the subordinate party not merely because of their occupational role but also because of their status as a formerly colonised, (largely) working-class people (and often women) of colour: inequitable exchange between workers and visitors reproduces and is reproduced by wider relations of domination. The significance of tourism work lies in the fact that it aggravates the historically entrenched inequity of emotional exchange by formally enforcing its acceptance with displays of spontaneity on the part of workers. Self-subordination is thus daily lived by workers in the context of multiple relations of domination. Many of the interviewees were keenly aware of this and illustrated it with various examples of their experience of inequity. They commonly discussed them as an inevitable part of their work, referring to the motto 'The customer is always right', yet, notably, many felt that such exchange was at the same time framed by racism and colonialism ('No "thank you", no "please", . . . just because of the colour of my skin'; 'You know, Europeans. They look down on us and think we should be at their service'). Some young women added sexism, especially sexual harassment, as another of such interlocking relations of domination. These workers were, then, critically aware of multiple dimensions of their subordination.

The workers had a wide range of responses to this inequity. Some accommodated it more or less actively, owing to pragmatic and instrumental identification of it as simply part of their job; after all, the customer is always right. Others accommodated it, but only passively ('What can you do?'; 'You just have to keep it [discontent] to yourself'). Yet others opted to employ means of resistance. Although few spoke of overtly resisting emotional subjugation, inner discontent at times took the form of covert resistance, such as facial and bodily display of displeasure ('making faces at tourists'). These workers may also secretly joke about dominating visitors: when they feel that they are emotionally subjugated or otherwise unjustly treated by visitors, they may criticise or 'tease' these visitors in their presence by speaking in the Fijian language. It is 'light-hearted mischief' which helps them deal with stress and frustration, but it is also often a means of covert resistance. As an experienced hotel worker explained:

> When I talk to the guests and they don't want to talk back, you know, I always say to the staff, 'This person must be dumb' [in Fijian]. It just comes out of my mouth. They just look at you and ignore you, when you want to help them. It's something that really makes me angry.

Thus the workers may express their resistance in a covert, contained form, not directly detrimental to their immediate interests of job security. This may not appear to be an expression of resistance in the visitors' eyes, yet it is from the workers' point of view: by playing a joke on, or retorting to the dominant other who compels them to emotional subjection, they privately 'get even' and thereby refuse to resign themselves to their subordination. Once again, these responses may be seen in the context of Pacific island communities where mimicry has played a critical role of counter-hegemonic commentary: the inequity of the relationship is here jokingly but firmly disapproved by indirect means. Indeed, according to Goffman (1959: 188), such acts may allow the actor to 'show not only that [s/]he is not bound by the official interaction but also that [s/]he has this interaction so much under control that [s/]he can toy with it at will'. Lastly, a few workers suggested dialogue as another response. These workers, though privately critical of the inequity, did not simply accommodate or resist it but attempted to engage the other in dialogic interaction, seeking to reach an understanding with him/her. Marika, for instance, as mentioned earlier, refused to withdraw from offensive tourists and continued to try to engage them in friendly interaction, through which he felt he often succeeded in redefining his relationship with them. In such a context, maintaining amiability in the face of open aggression or disregard by the other becomes not an act of submission but an attempt at dialogue.

Popular response: dynamics of hegemonic rule

These accounts suggest that, despite their emphatic endorsement of amiability, the workers may privately question other dimensions of the imagery, such as (hard) savagery and self-subordination. The criticisms remain largely overshadowed and suppressed by the conspicuous affirmation of amiability, yet they indicate the subtlety of the workers' viewpoints that may not commonly find expression. The workers' consent to their touristic role similarly reveals its complexity upon scrutiny. Although they approve of the allocation of direct service positions to Fijians as a logical outcome of their cultural peculiarity and find it genuinely satisfying when their amiability is reciprocated by visitors, they may be discontented with other dimensions of their role, particularly the institutionalisation of amiability and the inequity of power that defines their relationship with visitors. Some accommodate these actively, yet others do so only passively and mechanically. Although open resistance is unusual, some may employ covert forms of resistance. Indeed, the workers may employ alternatives to total conformity, even as they apparently embrace the imagery and their designated role. The notion of amiability itself may be given alternative meanings, which allow them to maintain counter-hegemonic self-definitions: amiability defined in the context of reciprocity and self-affirmation is a positive and empowering asset. Similarly, in response to the unequal emotional exchange with visitors, they may reverse the logic of touristic amiability and use it to achieve dialogic relationship with those who seek to subordinate them, the exact opposite of what it is designed to produce. They may thus embrace the touristic imagery and role not to resign themselves to their designated position but to attempt to redefine it within the limits of the existing conditions.

These responses, to be sure, may not in any significant way undermine the existing hegemonic arrangements, since they generally do not take the form of articulate oppositional ideas and actions. However, what they reveal most importantly is that the hegemony is continuously challenged, if privately or covertly: resistance may be latent, but not absent. Hegemony, as Williams (1977: 113–14) stresses, is:

> never either total or exclusive. At any time, forms of alternative or directly oppositional politics and culture exist as significant elements in the society . . . It would be wrong to overlook the importance of works and ideas which, while clearly affected by hegemonic limits and pressures, are at least in part significant breaks beyond them, which may again in part be neutralised, reduced, or incorporated, but which in their most active elements nevertheless come through as independent and original.

Even if the workers willingly embrace the dominant imagery, seem content in their designated position, and readily affirm that they are 'naturally' suited to such a position, they may also be privately critical and devise ways of coping with their discontent or expressing their dissent in creative, covert forms. The argument presented by Fanon (1967) decades ago, that the outward passivity of colonised peoples disguises their inner resistance, continues to bear relevance to contemporary discussion of hegemony.

Conclusion

The workers' accounts illuminate the complexity of local response to touristic imagery. The ideological effects of such imagery may be considerable, yet are by no means unproblematic. As noted by many observers in the past, touristic representation of destination/indigenous peoples, if seemingly innocent, is often implicated in relations of domination. In the case of Fiji, the taken-for-granted touristic definition of the indigenous population is deeply embedded not only in its marketing value but also in the country's colonial/post-colonial hegemonic order, and more specifically, the structure and politics of contemporary tourism. The lucrative industry rests on its existing structure that places the vast majority of indigenous workers in a subordinate position. Such arrangements are secured by persuasive rather than coercive means, with the apparently privileging imagery playing a notable role. This has significant implications for the indigenous people's identity construction as well as participation in the industry, since their options for subverting touristic discourse tend to be limited. There is, then, a significant linkage between touristic representation, ideology and identity, which calls for critical scrutiny. At the same time, subordinated peoples, far from being 'ideological dupes', respond to hegemonic pressures in a variety of ways that not only secure the durability of such hegemony but also point toward its limitations and essential fluidity. The gratifying overtones of the imagery of amiable Fijians and its increasing association with indigenous tradition seem to secure the workers' endorsement of the imagery and the touristic role it assigns to them, making the hegemonic order appear almost uncontested. Yet closer examination reveals that this hegemony stands in the face of creative and potentially counter-hegemonic responses of its subjects, which, if ultimately contained or neutralised, nevertheless crystallise their agency. The touristic imagery and role are, then, not simply imposed upon the workers but mediated by an ongoing struggle between conflicting ideological elements and interests.

The workers' accounts bring to light what lies beneath the renowned and conspicuous dexterity with which they daily display touristic amiability, that is, what it means to them to 'consent' to the touristic imagery and role. The subtlety of their accounts warns against totalising and deterministic analysis, which is analytically incomplete and ethically problematic. The ideological effects of touristic imagery need to be studied as a lived process involving human subjects whose ideas and actions constitute its crucial dynamics and complexity. Fijians may continue to smile in the posters and at the resorts just as they are expected to, but it is their inner emotions, private viewpoints and everyday actions that protect as well as potentially challenge the hegemonic order in Fiji's mass tourism.

Notes

1 The results do not suggest consistent gender or age differences except in relation to the workers' experiences of and views on their work role, where their gender, age, as well as occupational positions seem to be of some significance (Kanemasu 2005). These are, however, beyond the scope of this chapter and are not included in the following discussion.
2 The interviewees' words and statements are indicated by quotation marks. They are reproduced in the exact forms as recorded in the interview accounts except for minor alterations of editing, and the interviewees have been given fictitious names.
3 It is not the contention of this chapter that the existing structure of the industry is sustained solely by ideological means. The workers may consent to their positions, for instance, on account of the scarcity of alternative employment with comparable or preferable economic rewards. The consent can thus be pragmatically motivated by material interests.
4 'Baking powder' is a local slang expression that denotes exaggeration. It derives from an analogy with the effect of baking powder in causing batter to rise, and is often used as a method of playful mockery when exaggeration of actions and achievements is detected. Here, the term is used to mock the depiction of amiability that is magnified into self-subordination.

References

Baum, T. (1997) 'Making or Breaking the Tourist Experience: The Role of Human Resource Management', in C. Ryan (ed.), *The Tourist Experience: A New Introduction*, London: Cassell.

Bell, L. (1999) 'Looking at Goldie: Face to Face with "All 'e Same T'e Pakeha"', in N. Thomas and D. Losche (eds), *Double Vision: Art Histories and Colonial Histories in the Pacific*, Cambridge: Cambridge University Press.

Belt, Collins and Associates (1973) *Tourism Development Programme for Fiji*, Washington, DC: Tourism Project Department, International Bank for Reconstruction and Development.

Britton, S.G. (1983) *Tourism and Underdevelopment in Fiji*, Development Studies Centre, Monograph 31, Canberra: Australian National University.

Bruner, E.M. (1991) 'Transformation of Self in Tourism', *Annals of Tourism Research*, 18: 238–50.

Central Planning Office, Government of Fiji (1975) *Fiji's Seventh Development Plan 1976–1980*, Suva.

Cohen, E. (1982) *The Pacific Islands: From Utopian Myth to Consumer Product: the Disenchantment of Paradise*, Aix-en-Provence: Centre des Hautes Etudes Touristiques.

Douglas, B. (1999) 'Art as Ethno-historical Text: Science, Representation and Indigenous Presence in Eighteenth and Nineteenth Century Oceanic Voyage Literature', in N. Thomas and D. Losche (eds), *Double Vision: Art Histories and Colonial Histories in the Pacific*, Cambridge: Cambridge University Press.

Douglas, N. and Douglas, N. (1996) 'Tourism in the Pacific: Historical Factors', in C.M. Hall and S.J. Page (eds), *Tourism in the Pacific: Issues and Cases*, London: International Thompson Business Press.

Durutalo, A. (1997) *Provincialism and the Crisis of Indigenous Fijian Political Unity*, unpublished PhD thesis, University of the South Pacific.

Durutalo, S. (1985) *Internal Colonialism and Unequal Regional Development: the Case of Western Viti Levu, Fiji*, unpublished thesis, University of the South Pacific.

—— (1986) 'Na Lotu, Na Vanua, Na Matanitu: The Paramountcy of Fijian Interest and the Politicization of Ethnicity', paper delivered at United Nations University Symposium held at University of the South Pacific, Suva, 11–15 August.

Edmond, R. (1999) 'Missionaries on Tahiti, 1797–1840', in A. Calder, J. Lamb and B. Orr (eds), *Voyages and Beaches: Pacific Encounters 1769–1840*, Honolulu, HI: University of Hawai'i Press.

Elson, D. and Pearson, R. (1981) 'The Subordination of Women and the Internationalisation of Factory Production', in K. Young, C. Wolkowitz and R. McCullagh (eds), *Of Marriage and the Market*, London: CSE Books.

Fanon, F. (1967) *The Wretched of the Earth*, Harmondsworth: Penguin Books.

Farrell, B.H. (1982) *Hawai'i, the Legend that Sells*, Honolulu, HI: University of Hawai'i Press.

Fiji Visitors Bureau (2003) *Fiji Islands Travel and Accommodation Guide*. Online. Available <www.bulafiji.com> (accessed 28 September 2003).

Fong, A. (1973) 'Tourism: A Case Study', in A. Rokotuivuna, J. Dakuvula, W. Narsey, I. Howie, P. Annear, D. Mahoney, A. Fong, C. Slatter and B. Noone (eds), *Fiji: A Developing Australian Colony*, Melbourne: International Development Action.

Gramsci, A. (1971) *Selections from the Prison Notebooks of Antonio Gramsci*, ed. Q. Hoare and G.N. Smith, New York: International Publishers.

Goffman, E. (1959) *The Presentation of Self in Everyday Life*, Harmondsworth: Penguin Books.

—— (2001) *Gender Advertisements*, Cambridge, MA: Harvard University Press.

Halapua, W. (2003) *Tradition, Lotu and Militarism in Fiji*, Lautoka: Fiji Institute of Applied Studies.

Hereniko, V. (1994) 'Clowning as Political Commentary: Polynesia, Then and Now', *The Contemporary Pacific*, 6: 1–28.

Hochschild, A.R. (1983) *The Managed Heart: Commercialisation of Human Feeling*, Berkeley, CA: University of California Press.

Jahnke, R. (1999) 'Voices beyond the *Pae*', in N. Thomas and D. Losche (eds), *Double Vision: Art Histories and Colonial Histories in the Pacific*, Cambridge: Cambridge University Press.

de Kadt, E. (1979) 'The Encounter: Changing Values and Attitudes', in E. de Kadt (ed.), *Tourism: Passport to Development?*, New York: Oxford University Press.

Kanemasu, Y. (2005) *From the Cannibal Isles to the Way the World Should Be: A Study of Ideology, Hegemony and Resistance*, unpublished PhD thesis, University of New South Wales.

Korth, H. (2000) 'Ecotourism and the Politics of Representation in Fiji', in A.H. Akram-Lodhi (ed.), *Confronting Fiji Futures*, Canberra: Asia Pacific Press.

Larrain, J. (1996) 'Stuart Hall and the Marxist Concept of Ideology', in D. Morley and K. Chen (eds), *Stuart Hall: Critical Dialogues in Cultural Studies*, London: Routledge.

Lather, P.A. (1991) *Getting Smart: Feminist Research and Pedagogy with/in the Postmodern*, New York: Routledge.

Lawson, S. (1990) 'The Myth of Cultural Homogeneity and Its Implications for Chiefly Power and Politics in Fiji', *Society for Comparative Study of Society and History*, 32: 795–821.

—— (1991) *The Failure of Democratic Politics in Fiji*, Oxford: Clarendon Press.

McLellan, D. (1995) *Ideology*, Minneapolis, MN: University of Minnesota Press.

Macnaught, T.J. (1982) 'Mass Tourism and the Dilemmas of Modernization in Pacific Island Communities', *Annals of Tourism Research*, 9: 359–81.

Mitchell, W.E. (ed.) (1992) *Clowning as Critical Practice: Performance Humor in the South Pacific*, Pittsburgh, PA: University of Pittsburgh Press.

Nicole, R. (2001) *The Word, the Pen, and the Pistol: Literature and Power in Tahiti*, Albany, NY: State University of New York Press.

Norton, R. (1990) *Race and Politics in Fiji*, second edition, St Lucia: University of Queensland Press.

Plange, N. (1996) 'Fiji', in C.M. Hall and S.J. Page (eds), *Tourism in the Pacific: Issues and Cases*, London: International Thompson Business Press.

Said, E.W. (1979) *Orientalism*, New York: Vintage Books.

Samy, J. (1980) 'Crumbs from the Table: The Workers Share in Tourism', in F. Rajotte (ed.), *Pacific Tourism: As Islanders See It*, Suva: Institute of Pacific Studies, University of the South Pacific.

Smith, V. (1998) *Literary Culture and the Pacific: Nineteenth-century Textual Encounters*, Cambridge: Cambridge University Press.

Stanton, M.E. (1989) 'The Polynesian Cultural Centre: A Multi-Ethnic Model of Seven Pacific Cultures', in V.L. Smith (ed.), *Hosts and Guests: The Anthropology of Tourism*, second edition, Philadelphia, PA: University of Pennsylvania Press.

Thomas, N. (1999a) 'Introduction', in N. Thomas and D. Losche (eds), *Double Vision: Art Histories and Colonial Histories in the Pacific*, Cambridge: Cambridge University Press.

—— (1999b) *Possessions: Indigenous Art/Colonial Culture*, London: Thames and Hudson.

Thompson, J.B. (1984) *Studies in the Theory of Ideology*, Cambridge: Polity Press.

Thompson, P. (1973) 'Some Real and Unreal Social Effects of Tourism in Fiji', in *Tourism in Fiji: The Ray Parkinson Memorial Lectures*, Suva: University of the South Pacific.

Turner, L. and Ash, J. (1975) *The Golden Hordes: International Tourism and Pleasure Periphery*, London: Constable.

Urry, J. (1990) *The Tourist Gaze: Leisure and Travel in Contemporary Societies*, London: Sage.

Varley, R.C.G. (1978) *Tourism in Fiji: Some Economic and Social Problems*, Bangor Occasional Papers in Economics 12, Bangor: University of Wales Press.

Williams, R. (1977) *Marxism and Literature*, New York: Oxford University Press.

Young, G. (1973) *Tourism: Blessing or Blight?* Harmondsworth: Penguin Books.

7 On the beach

Small-scale tourism in Samoa

Regina Scheyvens

A tourist strolling along a pristine beach in Samoa can enjoy the same experience as tourists to scores of similar destinations in southeast Asia and the Pacific: the feeling of warm sand between their toes, palm trees to shade them from the tropical sun, and azure waters incessantly lapping at the shore just metres away. What may make the tourist's experience different in Samoa, however, is that here they would be far more likely to stumble upon a beach *fale* (a traditional, open-sided hut) as they strolled rather than a hotel or resort.

A unique style of small-scale tourism development centred on basic beach huts has evolved in rural areas of Samoa. Beach *fale* provide basic, budget accommodation for guests and an alternative means of income generation for Samoans who have chosen to remain in the villages rather than moving to Apia, the capital, or overseas in search of better economic options. Owned mainly by families, beach *fale* have grown from a mere handful of enterprises in the early 1990s to become the dominant feature of the coastal landscape in Samoa. Samoa also has a small number of high-class hotels and resorts, as well as many middle-of-the-range motels and hotels, but it is the growth of beach *fale* that has changed the face of Samoan tourism most significantly in recent years.

This chapter will provide an overview of debates concerning the value of small-scale tourism before documenting the growth of beach *fale* in Samoa and analysing their significance in terms of local development. A key argument advanced is that, although small-scale tourism cannot be seen as a panacea to development challenges in small island states such as Samoa, it can contribute very effectively to improving social and economic wellbeing in certain contexts and, as such, it should be valued and supported by government.

Is small beautiful in the tourism sector?

Tourism academics have been debating the value of small-scale tourism for many years. Some suggest that small-scale tourism may enhance local ownership and control over tourism, increasing the likelihood that benefits stay within the local area and reducing leakages (Brohman 1996; Cater 1993; Guthunz and von Krosigk 1996: 28; Singh 2003; Woodley 1993). For example Wilson (1997), discussing tourism in Goa prior to the growth of the charter-package trade, found that small-scale entrepreneurs were able to cater effectively for the needs of domestic tourists and international backpackers, thus the industry was characterised by 'wide local ownership of resources and the broad distribution of benefits throughout the local community' (1997: 63). Although mass tourism can also contribute to community development (see, for example, Thomlinson and Getz 1996, and Dombroski, this volume), it is certainly much easier for local people to take a controlling role in small-scale ventures than in large, capital-intensive enterprises (Hampton 1998). Thus, in the Cook Islands, where successive governments have chosen to follow a model of small to medium sized tourism enterprises, the local Cook Islands Maori have high levels of ownership of tours, restaurants and small-scale accommodation enterprises (Milne 1997). This ownership and sense of control over tourism is a developmental benefit in its own right. As Berno (2003) notes in a fascinating psycho-social study of local perceptions of tourism in the Cook Islands, when resident peoples feel they exert control over tourism they are less likely to experience what others may see as negative social or cultural impacts. Another benefit of small-scale tourism is that less powerful groups in society have more chance of participating in smaller enterprises. Thus, for example, 'the ability of women to influence the sustainability of tourism development increases as one moves *down* the global–local nexus toward the community and household/firm levels' (Milne 1998: 41).

Case studies of particular countries where small-scale tourism provides an important part of the tourism product have often been very positive. For example, Duval (1998: 44) explains how on St Vincent small alternative tourism ventures provide a number of advantages ranging from economic self-reliance and dynamism in response to market trends through to 'a new awareness of resource management issues' and 'positive socio-cultural interactions between hosts and guests'. Thus, rather than expanding what Ioannides and Holcomb refer to as 'the "status" gap between hosts and guests' (2003: 45), a common phenomenon in up-market tourism, small-scale tourism can help to break down these barriers.

Despite these potential benefits, few developing country governments have chosen to pursue a strategy that prioritises small-scale development. Rather, those that have moved beyond wholesale encouragement of mass tourism tend to be lured instead by notions of 'high-value, low-volume' up-market tourism. However this may cause more environmental harm than small-scale tourism because of its heavy demands on land, water and energy (Ioannides and Holcomb 2003). In addition, the economic benefits are not as high as they may at first seem as there can be extensive leakages from such tourism developments due to the heavy reliance on imported products ranging from computer systems and furnishings to food and drink, and dependence on expatriate management staff. Repatriation of profits by foreign-owned companies compounds this problem. Thus, notes Harrison, 'In the Pacific, the more developed the tourism industry, and the more it caters for high-spending tourists or for tourists in great numbers, the more likely it will be owned and operated by overseas interests' (2003: 7).

Nevertheless a number of people contest the notion that small-scale tourism can deliver the wide range of benefits documented above. A key problem is that small-scale initiatives are often not viable, being established without adequate attention to publicity or marketing, lacking connections to mainstream tourism enterprises, and run by people who have little understanding of tourism enterprises and lack financial management skills (Butler 1990). Thus small enterprises can be somewhat weak: 'small is beautiful in the context of ecotourism . . . [but] small is also vulnerable' (Thomlinson and Getz 1996: 197). Harrison (2003) and Sofield (2003) further note that the relationship between tourism and community development is both complex and problematic, with the benefits often being secured by local elites and a small number of people with business experience. Meanwhile the broader local community may bear the brunt of this more invasive form of tourism, especially where 'small scale' equates with homestay-style accommodation and visitors expect to mingle as much as possible with 'the locals'. Small-scale tourism ventures run from home are also more likely to increase the workloads of women and children, who are typically responsible for cooking for guests and doing the cleaning up. Such factors led Wall and Long to conclude that 'small-scale, indigenous, tourism development, which is advocated by many proponents of "appropriate" tourism, is not without associated problems and will require compromises and trade-offs to be made' (1996: 45).

Although small-scale tourism can sometimes work well, 'small can also be insignificant', constituting just a 'drop in the ocean' when compared with mainstream tourism at different scales (Wheat 1994: 2). Essentially, although small-scale ventures can be drawn on as examples of good practice, they are actually quite tokenistic in many cases, leading some commentators to suggest the more important task is to work on changing the mainstream tourism industry (Harrison 2003).

Small-scale tourism initiatives will necessarily evolve over time, often into something that replicates conventional, mass tourism (Butler 1980; Doxey 1975; Williams 1982). From the 1970s onwards various models of tourism development suggest that, whereas in its early stages tourism may be somewhat organic, with local entrepreneurs responding to tourist demand by setting up small-scale enterprises and the local population being largely 'euphoric' in terms of their attitude to tourism, over time the situation will change quite dramatically. Tourist numbers will grow, facilities provided for them will become more mainstream (for example, hotels and resorts rather than homestays and bungalows) and require more outside investment and control, impacts on local society and the environment will increase, and the 'hosts' will eventually come to resent the heavy presence of tourists in their communities.

Certainly, with small-scale enterprises owned by local people, there is always a danger that what begins as an effective, locally based response to increasing demand for certain tourism experiences from intrepid tourists venturing off the beaten track will, as it becomes more successful, be taken over by outside interests. This has occurred to some extent, for example, in the Gili islands off the coast of Lombok in Indonesia. Tourism development on these small islands was controlled almost exclusively by local people in the 1980s; however by 1999 only around half the local businesses were under local ownership and fewer than one-quarter of direct tourism jobs were held by local people. Those jobs that were available mainly involved lower-skilled manual labour. Lack of education (40 per cent of local people had never been to school) and lack of business and tourism experience on the part of the local population made it easier for non-islanders to adapt to meet the changing demands of tourists. Outsiders also had greater access to capital to invest in their businesses (Wong 2001).

When small-scale tourism enterprises are situated within a small island state, this can compound the constraints they face (Pearce 1987). Tourism development in small islands is often impeded by inadequate transportation links, lack of accessibility to remote locations, lack of appropriate skills among the local population, and inadequate amounts of local capital (Harrison 2003: 7). This has, in the past, led commentators to suggest that tourism development in small island states is controlled by outside interests (Britton 1983). However, local governments and people can play a significant role in managing the development of their tourism industries (Milne 1997: 297). Milne (1997) thus suggests that where governments carefully manage tourism, where there are high levels of local ownership and strong economic linkages between tourism and other local industries, the benefits can be great. These conditions do in fact exist in Samoa, so the chapter now turns to examine tourism in Samoa with relation to the beach *fale* sector.

The nature of tourism development in Samoa

Samoa is a small island state in the Pacific with 2934 square kilometres of land divided between two main islands (Upolu and Savaii) and home to about 177,000. Many more thousands of Samoans live abroad, however, mainly in New Zealand, Australia and the USA, and make a major contribution to their country's economy through remittances. Samoa has thus been categorised as a MIRAB state, that is, one that is dependent upon Migration, Remittances and Aid, and with a consequent significant Bureaucracy (Bertram 2006); however, it can be argued that tourism is of growing significance to the economy and that the MIRAB label is no longer appropriate.

After periods of colonialism by Germany and New Zealand, Samoa emerged in 1962 as the first Pacific island state to gain independence. Samoans have a long history of resistance to outside interference and until relatively recently they were reluctant to trade on their country's natural beauty and cultural features by encouraging tourism development. Thus, Samoans have often rejected offers by foreign investors seeking to build hotels and resorts in prime beachside locations. Over 80 per cent of land is under communal ownership in Samoa, which means it cannot be sold or transferred to foreigners, hence developers can gain access to it only through a 30-year lease or joint ventures. In practice, 'the communal nature of land holding and consensus decision-making . . . hinder the smooth development of tourist initiatives' (Fairbairn-Dunlop 1994: 132). The importance of land to Samoan identity, along with the Samoan people's commitment to protecting *fa'a Samoa* (the Samoan way of life), has led landowners to eschew many advances from international investors wanting to establish large, up-market resorts and hotels, and instead to develop their own small-scale forms of tourist accommodation (Fairbairn-Dunlop 1994). This has created a context in which the tourism industry in Samoa is dominated by small-scale, locally owned and operated initiatives. Only one of the five hotels with over 50 rooms is foreign owned, and 'Tourism in Samoa is almost exclusively a family business' (Twining-Ward and Twining-Ward 1998: 266).

From the 1960s, when the first tourists started trickling in to Samoa, through to the present day, the Samoan people and successive governments have had an ambivalent attitude to tourism development. There is 'concern that foreign-initiated tourism development may have adverse consequences upon the dignity, self-reliance, traditional customs, authority structure and morals of rural people' (Meleisea and Meleisea 1980: 42). Thus, active promotion of tourism by the government did not begin until the 1990s, when they were spurred on to find development alternatives after the devastation caused by two cyclones (in 1990 and 1991) and taro leaf blight (in 1993) which destroyed almost the entire crop of this staple – and main foreign exchange earner – on both main islands (Scheyvens 2005). Tourism has since rapidly grown to become Samoa's main industry, contributing four times more to the economy than agriculture (Twining-Ward and Twining-Ward 1998). Arrivals grew from less than 48,000 in 1990 to over 92,000 by 2003, but this includes returning Samoans, who are not primarily tourists.

As in many other countries the government in Samoa chose to embrace the concept of sustainable tourism, but they have done this in ways which directly reflect respect for the wellbeing of Samoan people:

> What is distinctive about the Samoan case is the way in which the country's strong social and cultural traditions – the fa'a Samoa – have been incorporated in government tourism policies and the ways in which these policies are being implemented through the NTO [National Tourism Office] in their attempts to foster local participation in the development process.
>
> (Pearce 1999: 154)

Support for sustainable tourism is the undercurrent running through the 1992–2001 Tourism Development Plan, which stressed that 'tourism in Samoa needs to be developed in an environmentally responsible and culturally sensitive manner, follow a policy of "low volume, high yield", and attract discerning and environmentally aware visitors' (Government of Western Samoa and Tourism Council of the South Pacific 1992, cited in Twining-Ward and Twining-Ward 1998: 263). Although it is clear that the Samoan government has never endeavoured to attract large numbers of mass tourists, it also never anticipated the growth of tourism at the budget end of the spectrum: small-scale, family-operated beach *fale*. As can be seen from Map 7.1, beach *fale* enterprises have come to dominate the tourist accommodation options in a number of rural, beachside locations.

Although a handful of beach *fale* enterprises were scattered around the coastline in the 1980s, it was not until the 1990s that they began to pop up in large

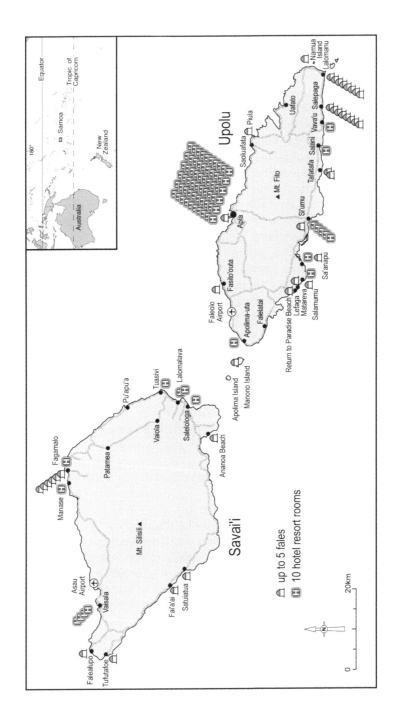

Map 7.1 Fale in Samoa.

numbers, families often spurred on by the success of a neighbouring enterprise to set up their own beach *fale* business. There were 44 registered beach *fale* operations by December 1999, mostly on the island of Upolu, where the country's capital, Apia, is located. The exact number of functioning commercial beach *fale* enterprises is unclear, however, as only 31 beach *fale* establishments were promoted in the Samoan Tourism Authority's 2003 Visitor Guide, yet as the 2002–6 Tourism Development Plan points out, there are over 200 actual beach *fale* in Samoa (Tourism Resource Consultants 2002: 58). Individual enterprises comprise between three and thirty *fale*, with the vast majority having fewer than ten *fale*. Most follow a similar formula, although a few have separated themselves out in recent years by offering slightly more sophisticated *fale* (for example, with verandas and solid walls). In general for around US$20 per night a guest will receive accommodation in a basic, oval shaped, open-sided hut with bedding, lighting and a mosquito net, with two meals and shared use of bathroom and dining facilities.

Beach *fale* attract a wide range of clientele, with both Samoan and foreign tourists being very important markets. Samoan guests include couples, families and youth groups, who typically take weekend day trips from Apia to nearby beaches where they pay a small fee to cover use of the beach and a beach *fale* (which provides shade and a comfortable resting place). Longer-stay visitors to beach *fale* include government and non-governmental agency staff who are sent on retreats to work on their mission statement or review goals and objectives, and family or school groups celebrating a reunion. Foreign visitors are also a diverse group which includes a lot of young and middle-aged couples from Australia and New Zealand, surfers, international backpackers of all ages, and clientele from up-market hotels like Aggie Grey's who desire a unique cultural experience for a day or two.

A key factor constraining the success of small-scale community tourism initiatives is a lack of publicity (Moscardo and Pearce 2003). This is the case in Samoa also, especially for the smallest ventures and those catering primarily for day visitors rather than overnight visitors. In practice, most beach *fale* enterprises catering for day visitors do no marketing of their own apart from a hand-painted billboard along the road. Those hosting overnight visitors are more likely to print brochures or posters, sometimes available in the Tourist Information *Fale* in Apia, or to have an email address or, for a few, their own website. Information on beach *fale* is also provided on the Samoan Tourism Authority's website and the websites of a few organisations providing information to backpackers (see for example www. pacific-travel-guides.com/samoa-islands/accommodation/beach-fales.html). As noted by Steve Brown, who runs Ecotour Samoa, many guests seeking 'alternative' travel experiences do all their bookings on the web, so the importance of advertising beach *fale* well using this medium is obvious (Steve Brown, Ecotour Samoa, personal communication June 2003). For many enterprises, however, their custom comes from either word of mouth or guide books, specifically Lonely Planet's *Samoan Islands* (Bennett *et al*. 2003) and *South Pacific Handbook* published by Moon Travel (Stanley 2004).

The Samoan Tourist Authority (STA) has made an effort to actively market beach *fale*, listing links to some beach *fale* on their website and stocking brochures for beach *fale* in their Information *Fale* in Apia. However, these brochures often ran out and were available for only a few establishments. Conveniently, rather than promoting any individual tourism business, the STA encouraged visitors to travel around Upolu and Savaii and see for themselves where they would like to stay. While based in Apia visitors can easily do a circuit of Upolu (by tour bus or hired car) in a day. STA does not have funding to set up websites for beach *fale* enterprises but, when consulted, it does advise on ways in which beach *fale* operators planning a brochure or website can best market their products. Overall, therefore, it is the more successful, established beach *fale* enterprises that are both marketing themselves effectively and being marketed by others.

Research on beach *fale* tourism in Samoa was conducted with the assistance of a local researcher, Bronwyn Tavita Sesega, for two weeks in June/July 2003. The methodology was qualitative, with the main technique of data collection being semi-structured interviews with a range of stakeholders including beach *fale* owners and operators (from 13 enterprises), clients of beach *fale* (over 20 foreign tourists and 10 Samoan tourists), officials from the Samoan Tourism Authority, and tourism industry representatives. In many cases these were individual interviews, but in some cases family members of owners wished to participate in the interviews at the same time, and often beach *fale* clients were interviewed in small groups. Bronwyn and I were also participant observers of beach *fale*, travelling around both Upolu and Savaii and staying in beach *fale* every night. We interviewed beach *fale* owners in areas where tourism has become a central economic activity, as well as some more remote locations where constraints to business success were more apparent.

The majority of tourists interviewed in this research were extremely positive about beach *fale* tourism. Typical comments from a beach *fale* visitor's book were as follows: 'Great place'; 'fantastic food'; 'wonderful beach'; 'friendly family'. Many guests commented that they would like to stay longer and/or to return one day. The location of beach *fale* on what would be prime coastal property with premium values in most parts of the world certainly adds to the value of the *fale*: 'beach *fale* offer the best views, every morning and every evening. You certainly sleep really well in the fresh open air and wake up early ready to hop in the water [for a surf]' (British man, 22 July 2003), and 'Thailand, Laos and Cambodia had huts made of similar materials [to beach *fale*] – but these are right on the beach. I love this – you can fold up the walls and get a lot of sunshine. This is more unique' (Swedish woman, 24 June 2003).

One respondent referred to beach *fale* as '5 star hotels the Samoan way', and many tourists remarked that staying in an open *fale* gave them a sense of the affinity between culture and nature in Samoan society. For example: 'Sinalei [an exclusive resort] is luxurious and it's great for people who just want to party and get laid, but it's so separate from the culture. Staying in beach *fale*s you are *in* the culture; you can't help but be a part of it' (Netherlands man, 27 June 2003).

Contribution of beach *fale* tourism to local development

Beach *fale* have brought a number of positive developments to rural areas of Samoa including the rejuvenation of villages, enhanced local pride and economic improvements. Multiplier effects are experienced in the local economy because tourists staying in beach *fale* do not demand luxurious, imported goods, meaning owners do not require large amounts of capital to start their businesses and they can maximise the use of local products and services. Beach *fale* are constructed using mainly local materials and expertise, and their owners often purchase items such as fruit, vegetables, seafood and mats from other villagers and hire village labour during busy periods. Many owners purposefully ensure that benefits flow beyond their immediate family: 'Our idea is to help the community. We buy food products from villagers – if you need fish, arrange it with a fisherman. The same goes for building materials – buy locally' (Beach *fale* operator, Upolu, June 2003).

Beach *fale* owners often make donations to church and community causes, and support community groups. For example, they may employ the church youth group or school groups to entertain visitors at *fiafia* nights, at which traditional dance is performed (David, beach *fale* operator, June 2003). Guests also contribute to the wider village through various means such as purchases made from village shops, use of local transport, and contributions to the collection plate when they attend a church service.

Foreign tourists interviewed generally spent between US$35 and $50 per day, per person, and were staying in Samoa for periods of between one and four weeks. This includes their beach *fale* accommodation but it also allows for some 'luxuries', such as Vailima (the local beer) and going diving or hiring a motorbike for one or two days. Some *fale* owners prefer international guests since they stay longer and spend more money than domestic visitors (Scheyvens 2005: 195). For a two-week visit, a foreign guest would therefore spend around US$600, much of it staying within rural communities.

Many families establish beach *fale* enterprises to diversify their livelihood options. Rural families typically rely on a variety of sources of income, from sales of produce from their plantation or from fishing, to wage labour and remittances from family members living in Apia or abroad. Thus if the price for bananas goes down or a relative who had been remitting cash loses her job in Auckland, economic prospects for the family are still promising if they own a beach *fale* enterprise. As Table 7.1 shows, visitor numbers at a typical beach *fale* can vary significantly across the year. Whereas during busy periods beach *fale* may provide most of a family's income, even successful beach *fale* ventures may need to call on other livelihood strategies at less popular times. One beach *fale* owner, for example, had a ST$24,000 (US$9000) loan, which she had used to construct a large communal eating house. When I asked if she ever had a problem with repayments, she mentioned that her brother had a big plantation: 'If there's no tourists, taro talks!' (Luisa, beach *fale* operator, June 2003). She continued: 'This is the Samoan way – we work together and help each other.'

In addition, beach *fale* tourism has resulted in capacity-building for beach *fale*

Table 7.1 Visitor numbers for Heavenly Beach *Fale*, 2002

Month	Visitors
January	91
February	54
March	40
April	66
May	9
June	66
July	107
August	110
September	85
October	101
November	51
December	39
Total	819

Source: Tourism Resource Consultants (2002).

owners. The New Zealand Agency for International Development (NZAID) has funded training workshops, organised by the Samoan Tourism Authority specifically for beach *fale* owners, and publication of a Beach *Fale* Owner's Manual, written in both Samoan and English. More women than men have attended training sessions for beach *fale* owners, a reflection of who is most involved in the everyday operations of beach *fale* ventures. The establishment of beach *fale* enterprises has also led to expansion of gender roles for some men as the pressure for families to have 'all hands on deck' to meet the needs of tourists during peak periods has led to an expansion of men's capacity in new work domains. Thus, for example, men have taken on non-traditional roles such as cleaning toilets and bathrooms, preparing bedding and washing dishes.

A significant benefit of beach *fale* tourism that has not been officially acknowledged is that it has restored the pride of many villagers in their home environment. Most Samoans feel genuinely honoured when people from all over the world come to visit their village and some of them show interest in Samoan culture; consequently community members contribute enthusiastically to village beautification efforts. Furthermore, the economic rejuvenation of some villages through beach *fale* tourism has reduced rural–urban migration, as young people feel they now can stay in their home village and have a viable future, and particularly so in the villages most distant from Apia, where *fale*s are clustered. Several people interviewed commented that communities with successful beach *fale* ventures were now more vibrant and attractive places to live in, with more jobs on offer, and that young people were moving back to the villages from Apia: 'Beach *fale* tourism has helped to boost the morale of communities and helped people to cater for their day to day needs . . . . It's also helped them to improve their surroundings, their gardens etc.' (Sione, pastor, June 2003). Thus, as one tourist observed, 'traditional culture' should not be seen as static, and there need not be a negative relationship between tourism and culture: '[Beach *fale*] could be a disruption to the traditional way of life . . . but what is this? And they could help the traditional way of life to survive because beach *fale* keep people in the villages' (NZ man, 44 years of age, June 2003).

In addition to the economic and social benefits of beach *fale* tourism, beach *fale* also contribute to development in a political sense in that they are owned by local families and community groups, which means that prime beach sites remain under the management and ownership of local people: 'The beach *fale* industry is about Samoan people owning and manipulating tourism to fit in with their practices and traditions, not the other way around' (Park 2003: 44). This challenges foreign domination of the tourism sector, which is characteristic of tourism in many small island states. Such local control is possible because of the communal land tenure system described earlier, the strength of *fa'a Samoa* and people's support for ensuring protection of their culture and land.

Traditional institutions also exert considerable control over this small-scale tourism development in the villages. *Matai* (chiefs) play a key role in terms of leadership and decision-making in rural communities. Beach *fale* tourism is overseen by the *fono* (council of *matai*) of each village, which sets rules to ensure that village life is not adversely affected by beach *fale* operations and that visitors feel welcome and safe. Thus, for example, beach *fale* operators are expected to instruct their guests in cultural protocol with respect to how they must dress when entering a village (village housing is usually located a short distance away from beach *fale* enterprises), that they must wait quietly during evening prayer time rather than wandering around the village, and at what time noise from the beach *fale* should stop in the evenings. If the *fono* decides that a *fiafia* must finish by 10 p.m., or that there should be no noise from beach *fale* after 11 p.m., operators have to ensure compliance. One beach *fale* manager noted that, if a business next door had guests who were making noise late at night, she could be fined for not reporting them (Lucy, beach *fale* manager, June 2003). Larger beach *fale* establishments thus employ security guards both to protect guests and their possessions and to preserve the peace by ensuring their behaviour does not get unruly. However, the *fono* may also place restrictions on access of local people to the beach in order to protect the privacy and security of guests, something that could potentially cause resentment in the future among local people not benefiting from tourism.

Although beach *fale* constitute a dynamic, innovative response to the need to find an alternative source of income in rural areas, some problems have arisen because they have emerged in a somewhat spontaneous fashion, rather than being carefully planned. Whereas in some villages there may be only one beach *fale* enterprise offering a handful of *fale* as guest accommodation, in other locations (see Map 7.1) it appears that every second family has a beach *fale* enterprise and beach *fale* are thus scattered all along the coastline (see Figure 7.1). Beach *fale* are, by nature, located very close to fragile marine ecosystems. A major issue for most beach *fale* operators then is effective disposal of sewage, which can otherwise leach into marine areas. Most use septic tanks; however, some respondents felt that septic tanks were an environmental disaster waiting to happen because of

Figure 7.1 The coastal landscape dotted with beach *fale.*

the risk of leakage. Access to safe drinking water has been identified as another environmental issue in some areas (Twining-Ward and Butler 2002), along with provision of adequate fresh water for villagers and visitors. Another concern is disruption to ecosystems because clearance of natural vegetation on the beaches, in order to create the 'pristine' expanses of white sand desired by tourists, can destabilise coastal areas leading to heightened erosion. The 2002–6 Tourism Development Plan highlighted some additional environmental issues:

- poorly maintained toilets in close proximity to the sea
- only 2 per cent of beach *fale* use secondary or tertiary wastewater treatment
- deterioration of sea water quality (algae bloom).

(Tourism Resource Consultants 2002: 92, 116)

Although such legitimate concerns about the environmental impacts of beach *fale* tourism have been expressed, most local people felt these effects were benign compared with those caused by other forms of tourism, since beach *fale* in general place small demands on the natural resource base. For example, guests have cold showers instead of warm, so no energy is required for heating and this means that showers are shorter, so less fresh water is wasted. Similarly, beach *fale* tourists are happy to swim in the sea, rather than expecting a freshwater swimming pool. There are neither fans nor air conditioning in the *fale*, as tourists rely on the cool sea breeze, which filters through their open *fale*. Beach *fale* can directly benefit the local environment too, in part because tourists' interest in an attractive natural environment and well-preserved corals has led local people to place higher value on conservation of natural resources (Park 2003: 66). Tourism has certainly provided an impetus for the creation of marine protected areas and mangrove conservation areas: 'Local people now appreciate the environmental significance of different areas, because these areas attract tourists' (Mary, beach *fale* operator, June 2003).

The undervaluation of beach *fale* tourism

Although concerns about environmental sustainability of beach *fale* and restrictions on the movement of villagers in beach *fale* villages exist, on the whole beach *fale* tourism has contributed greatly to the wellbeing of rural people in Samoa. Yet this has been largely overlooked or undervalued in official reports on Samoa. Beach *fale* are seemingly invisible to those estimating the number of beds available for tourists in Samoa, for example, as they do not include beach *fale* accommodation in their calculations (Pearce 1999: 145). Similarly, an Asian Development Bank report, *Samoa 2000* (ADB 2000), which purported to 'provide a comprehensive analysis of current economic and key sector developments in Samoa', does not even mention that beach *fale* exist, despite an entire chapter devoted to tourism. Consequently claims in the report that domestic tourism in Samoa is very small are not accurate because the beach *fale* sector is absent. Outside consultants and investors, meanwhile, have voiced frustration with the communal land tenure system and its requirement for consensus in decisions concerning land use, as this has impeded the development of large resorts in prime beach-side locations; instead, small clusters of beach *fale* occupy some of the best coastal sites in Samoa. Other commentators have overlooked the value of beach *fale* because they are more interested in development of high-class tourist facilities, which they feel will earn the country more foreign exchange and enhance its reputation as a provider of higher quality tourism (see ADB 2000; Pearce 2000). For example, one Samoan private sector stakeholder felt that there were too many beach *fale* around the two main islands now, many of which were 'eyesores': 'Since beach *fale* came up in Samoa we've had a sort of unchecked rash of huts [develop] all over the place. A perfect example is Aleipata [Lalomanu]: that whole strip used to be a beautiful place but now it's just *littered* with beach *fale*' (Tour company owner, June 2003).

Even though the government has provided support to the beach *fale* sector in terms of training for beach *fale* owners and administering the tourism development fund, which a number of beach *fale* have accessed, the significance of this small-scale form of tourism has been underrated. For example, the Tourism Development Plan for 2002–6 suggests that there is already an oversupply of beach *fale* leading to low occupancy rates, and states that it is at the high-end level of 'quality accommodation' that more investment needs to be made (Tourism Resource Consultants 2002: 73). It predicts that between 260 and 600 additional staff will be required in hotels by 2006 (compared with 1998), whereas only five to ten new staff will be required by beach *fale* establishments: 'there are sufficient *fale* to well and truly cater for current and projected demand, but many of these could be upgraded, managed and marketed better' (Tourism Resource Consultants 2002: 61). The government also passed legislation in 2003 to provide generous tax breaks to companies developing large-scale hotels and resorts, whereas no similar incentives were offered to small- or medium-scale tourism entrepreneurs (Scheyvens 2005). This stance of developing accommodation for higher spending tourists echoes that of the earlier tourism plan, a plan that did not produce the anticipated results.

As mentioned earlier, beach *fale* are but one of several livelihood activities in which a family may engage and, as such, they do not need hotel-style occupancy rates of 70 per cent to be able to declare their business a success, since family labour is unpaid. For many owners, success is measured rather in terms of the small amount of income they can generate from their beach *fale* establishment, the status of owning their own small business and to a lesser extent the relationships with tourists that develop through their work.

Conclusion

It is inspiring to see what can result when a country like Samoa takes a development path that leads to small- and medium-scale tourism development. The development of small-scale beach *fale* establishments has provided many families with an alternative livelihood option and led to economic and social revitalisation of some areas. Small-scale tourism is a form of tourism that is well integrated with the diverse livelihood systems of many communities, offering an economic activity that can absorb labour and resources from other sectors (such as fishing and agriculture) during peak periods but also release labour and resources to support alternative livelihood activities at other times. Small-scale tourism can thus offer a very good option where communities have a product that is attractive to tourists and where tourism complements existing livelihood activities.

The tourism industry has evolved in a unique way in Samoa with respect for *fa'a Samoa* and the land tenure system effectively limiting foreign involvement and large-scale growth of the industry. Although land tenure and traditional customs may be constraints to large-scale tourism development, that they have minimised foreign involvement in Samoan tourism may have been to the overall benefit of the country and its people: 'these constraints . . . have also resulted in a more socially equitable and ecologically sustainable tourism industry than is found in other Pacific island countries' (Twining-Ward and Twining-Ward 1998: 270). In addition, visitor satisfaction levels are high (ibid.). Thus, rather than assuming that growth of the industry and attracting higher spending tourists should be key goals, this study suggests that it could be in Samoa's interests to promote small-scale tourism development and cater for a diverse range of tourists, including domestic tourists and those travelling on a budget. Many international tourists who come to Samoa are attracted at least partly because of what a small-scale, locally controlled tourism industry can offer, namely, low to moderate prices, friendly service, basic accommodation in stunning locations and a cultural experience.

Although the benefits of small-scale tourism to Samoan development are clear, tourism policy should not focus exclusively on support for small-scale tourism. Small, alternative-style tourism enterprises such as the Samoan beach *fale* often work most effectively in complementing coexisting mainstream, sometimes mass, tourism enterprises (Harrison 1996; Weaver 2002), but are unlikely to replace other forms and cannot be expected to deliver direct benefits to large numbers of people (Harrison 2003):

Because many islands do not have the option of completely turning their back on mass tourism, a major policy issue is how one can 'add value' to more traditional forms of travel in order to maximize economic benefits while reducing negative impacts. In this respect the continued encouragement of mixed accommodation types and the development of local activities that can be integrated into mass tourism packages is essential.

(Milne 1997: 298)

Most countries will benefit from having a diversified tourism sector that offers a range of tourism products at different scales, thus reducing risk if one sector of the market is hit hard by external forces that cannot be controlled.

Thus far the Samoan government has supported a strategy of cautious growth and sustainable development of tourism, which provides the context in which a high degree of local ownership and control over the tourism industry has occurred. However, it has recently established initiatives to increase foreign investment in tourism (Scheyvens 2005). This can be seen in the latest tourism development plan, which, although acknowledging the existence of the beach *fale* sector, does not see this sector as being of high value and thus encourages development of more upper-end accommodation options. Although there are certainly key officials in the Samoan Tourism Authority who support small-scale tourism, it appears that pressure from outside institutions has instigated a move towards more large-scale tourism ventures involving foreign investment.

If governments of small island states want to promote a form of development that balances economic goals with environmental and sociocultural goals, they should not dismiss the role that small-scale tourism can play and neither should they assume that small-scale entrepreneurs should be left to their own devices. Small-scale tourism often emerges in a dynamic, bottom-up fashion in response to demand from tourists looking for unique experiences. For it to be successful, however, it is often important that states intervene to provide both support and regulations (Sofield 2003: 346). They may need to legislate to protect traditional land rights, to protect the environment, or to encourage joint venture arrangements. It can be detrimental to overall development when local people lose control over tourism development in their own area because of a lack of business experience, lack of knowledge of the tourism industry, or lack of capital. Government can also provide vital support in the way of raising awareness about tourism for the general public, and training to up-skill small-scale tourism providers. If a government envisages that tourism is part of an overall development strategy, it cannot afford to overlook small-scale tourism ventures, or leave them to be supported by under-resourced local government offices (Dahles and Bras 1999). What makes beach *fale* tourism stand out from dominant forms of tourism in Asia and the Pacific is that this is a form of development that entrenches indigenous ownership and control, while building upon local skills, knowledge and resources. Small-scale tourism should not be undervalued, as it has the potential to contribute to a broad range of positive development outcomes.

References

Asian Development Bank (ADB) (2000) *Samoa 2000*, Manila: ADB.

Bennett, M., Talbot, D. and Swaney, D. (2003) *Samoan Islands*, fourth edition, Melbourne: Lonely Planet.

Berno, T. (2003) 'Local Control and the Sustainability of Tourism in the Cook Islands', in D. Harrison (ed.), *Pacific Island Tourism*, New York: Cognizant Communication.

Bertram, G. (2006) 'The MIRAB Model in the Twenty-First Century', *Asia-Pacific Viewpoint*, 47: 1–13.

Britton, S.G. (1983) *Tourism and Underdevelopment in Fiji*, Canberra: Development Studies Centre, Australian National University.

Brohman, J. (1996) 'New Directions for Tourism in the Third World', *Annals of Tourism Research*, 23: 48–70.

Butler, R. (1980) 'The Concept of a Tourism Area Cycle of Evolution: Implications for the Management of Resources', *Canadian Geographer*, 24: 5–12.

—— (1990) 'Alternative Tourism: Pious Hope or Trojan Horse?', *Journal of Travel Research*, 28: 40–5.

Cater, E. (1993) 'Ecotourism in the Third World: Problems for Sustainable Tourism Development', *Tourism Management*, 14: 85–90.

Dahles, H. and Bras, K. (eds) (1999) *Tourism and Small Entrepreneurs: Development, National Policy and Entrepreneurial Culture – Indonesian Cases*, New York: Cognizant Communication.

Doxey, G.V. (1975) 'A Causation Theory of Visitor–Resident Irritants: Methodology and Research Inferences', in *The Impact of Tourism*, Proceedings of the Sixth Annual Conference of the Travel and Tourism Research Association, San Diego.

Duval, D.T. (1998) 'Alternative Tourism on St Vincent', *Caribbean Geography*, 9: 44–57.

Fairbairn-Dunlop, P. (1994) 'Gender, Culture and Tourism Development in Western Samoa', in V. Kinnaird and D. Hall (eds), *Tourism: A Gender Analysis*, Chichester: Wiley.

Guthunz, U. and von Krosigk, F. (1996) 'Tourism Development in Small Island States: From Mirab to Tourab?', in L. Briguglio, B. Archer, J. Jafari, and G. Wall (eds), *Sustainable Tourism in Islands and Small States: Issues and Policies*, London: Pinter.

Hampton, M.P. (1998) 'Backpacker Tourism and Economic Development', *Annals of Tourism Research*, 25: 639–60.

Harrison, D. (1996) 'Sustainability and Tourism: Reflections from a Muddy Pool', in L. Briguglio, B. Archer, J. Jafari, and G. Wall (eds), *Sustainable Tourism in Islands and Small States: Issues and Policies*, London: Pinter.

—— (2003) 'Themes in Pacific Island Tourism', in D Harrison (ed.), *Pacific Island Tourism*, New York: Cognizant Communication.

Ioannides, D. and Holcomb, B. (2003) 'Misguided Policy Initiatives in Small-Island Destinations: Why Do Up-Market Tourism Policies Fail?', *Tourism Geographies*, 5: 39–48.

Meleisea, M. and Meleisea, P.S. (1980) 'The Best Kept Secret: Tourism in Western Samoa', in *Pacific Tourism as Islanders See It*, Suva: Institute of Pacific Studies, University of the South Pacific.

Milne, S. (1997) 'Tourism, Dependency and South Pacific Microstates: Beyond the Vicious Cycle?', in D.G. Lockhart and D. Drakakis-Smith (eds), *Island Tourism: Trends and Prospects,* London: Pinter.

—— (1998) 'Tourism and Sustainable Development: Exploring the Global–Local Nexus', in C. Hall and A. Lew (eds), *Sustainable Tourism: A Geographical Perspective*, New York: Longman.

Moscardo, G. and Pearce, P. (2003) 'Presenting Destinations: Marketing Host Communities', in S. Singh, D.J. Timothy and R.K. Dowling (eds), *Tourism in Destination Communities*, Wallingford: CABI Publishing.

Park, K.M. (2003) *Beach Fale: Indigenous Initiatives in Tourism and Development in Samoa*, unpublished MA thesis, Auckland University.

Pearce, D. (1987) *Tourism Today: A Geographical Analysis,* New York: Longman.

—— (1999) 'Tourism Development and National Tourist Organizations in Small Developing Countries: The Case of Samoa', in D.G. Pearce and R.W. Butler (eds), *Contemporary Issues in Tourism Development*, London: Routledge.

—— (2000) 'Tourism Plan Reviews: Methodological Considerations and Issues from Samoa, *Tourism Management*, 21: 191–203.

Scheyvens, R. (2005) 'Growth of Beach *Fale* Tourism in Samoa: The High Value of Low-cost Tourism', in C.M. Hall and S. Boyd (eds), *Nature-Based Tourism in Peripheral Areas: Development or Disaster?*, Clevedon: Channel View.

Singh, T.V. (2003) 'Tourism and Development: Not an Easy Alliance', in R.N. Ghosh, M.A.B. Siddique and R. Gabbay (eds), *Tourism and Economic Development: Case Studies from the Indian Ocean Region*, Aldershot: Ashgate.

Sofield, T. (2003) *Empowerment for Sustainable Tourism Development*, New York: Pergamon.

Stanley, D. (2004) *South Pacific Handbook*, Emeryville, CA: Moon Travel Handbooks.

Thomlinson, E. and Getz, D. (1996) 'The Question of Scale in Ecotourism: Case Study of Two Small Ecotour Operators in the Mundo Maya Region of Central America', *Journal of Sustainable Tourism*, 4: 183–200.

Tourism Resource Consultants (2002) *Samoa Tourism Development Plan 2002–2006: A Focused Future for Tourism in Samoa*, Wellington: Tourism Resource Consultants.

Twining-Ward, L. and Butler, R. (2002) 'Implementing STD on a Small Island: Development and Use of Sustainable Tourism Development Indicators in Samoa', *Journal of Sustainable Tourism*, 10: 363–87.

Twining-Ward, L. and Twining-Ward, T. (1998) 'Tourism Development in Samoa: Context and Constraints', *Pacific Tourism Review*, 2: 261–71.

Wall, G. and Long, V. (1996) 'Balinese Homestays: An Indigenous Response to Tourism Opportunities', in R.W. Butler and T. Hinch (eds), *Tourism and Indigenous Peoples*, London: International Thomson Business Press.

Weaver, D. (2002) 'Mass Tourism and Alternative Tourism in the Caribbean', in D. Harrison (ed.), *Tourism and the Less Developed World: Issues and Case Studies*, Wallingford: CABI Publishing.

Wheat, S. (1994) 'Is There Really an Alternative Tourism?' *In Focus*, 13: 2–3.

Williams, T.A. (1982) 'Impact of Domestic Tourism on Host Populations: The Evolution of a Model', in T.V Singh, J. Kaur and D.P Singh (eds), *Studies in Tourism Wildlife Parks Conservation*, New Delhi: Metropolitan.

Wilson, D. (1997) 'Paradoxes of Tourism in Goa', *Annals of Tourism Research*, 24: 52–75.

Wong, P.P. (2001) 'Small-Scale Tourism and Local Community Development: The Case of Gili Islands, Lombok, Indonesia', in *Island Tourism in Asia and the Pacific*, Madrid: World Tourism Organization.

Woodley, A. (1993) 'Tourism and Sustainable Development: The Community Perspective', in J. Nelson, R. Butler and G. Wall (eds), *Tourism and Sustainable Development: Monitoring, Planning, Managing*, Waterloo, Ont.: Heritage Resources Center, University of Waterloo.

8   After the bomb in a Balinese village

Susan Tarplee

The bombings that occurred on 12 October 2002 in the popular Balinese resort of Kuta have been scrutinised within the Asia-Pacific region, reflecting concerns about regional security, and have been seen as evidence of the growing influence and risks of militant Islam in southeast Asia. However, apart from the analysis of the bombings from a security perspective, since the bombings occurred little assessment has been made of the ongoing impact of the bombings on the communities that were and are dependent upon tourism for their livelihoods. This is a reflection of the broader lack of understanding about the effects of downturns in tourism on local populations who rely on the industry.

As a result of two bombs, detonated almost simultaneously outside two nightclubs along the busy strip of Legian Street in the centre of Kuta, 202 people were killed. Immediately following the bombings, there was a mass exodus of tourists from Bali and a widespread cancellation of bookings, as well as a suspension of services by many operators organising trips to the island (Allen and Strutt 2003; Hitchcock and Darma Putra 2005). The evidence of terrorist activity on the primarily Hindu island was a catastrophe for Bali, which had long been viewed as a safe haven for many tourists wishing to visit Indonesia. Indeed, when the Australian government had issued travel warnings against Indonesia in the years prior to the 2002 bombings, Bali had been explicitly excluded.

This chapter focuses on the rural village of Tegaljadi in Bali's Tabanan district to demonstrate the effect of localised tourism slumps upon households, suggest effective mechanisms for reducing vulnerability and provide an insight into the differential impacts of downturns upon household members.

Vulnerability and tourism

Terrorist acts and similar shocks can have devastating consequences for tourism destinations, affecting both the profitability of the tourism industry and the wider economies of the target areas. Existing literature about these effects focuses rather narrowly on the macro-economic scale and especially on responses by industry leaders and governmental bodies; this comes at the expense of an understanding of the impacts of disasters for those who rely on tourism for employment (see

Pizam and Mansfeld 1996). Although it is acknowledged by some existing studies that simply focusing on the macro level is insufficient to gain a true picture of the impacts of tourism downturns, there have been few attempts to expand studies of tourism downturns to encompass responses at a household or community level (see Beirman 2003: 75; Wahab 1996: 178).

Vulnerability, according to Moser and McIlwaine, is 'a dynamic concept referring to insecurity in the well-being of individuals, households, or communities in the face of a changing environment' (1997: 17). Although vulnerability can be examined from an individual or household level, it is understood to be a condition that is created by the wider social situation of (especially developing world) societies (Adger and Kelly 2001: 21). Vulnerability is not a static entity, but rather a process that changes in response to external factors such as the socioeconomic and political circumstances of a society (Maskrey 1989: 2). It is generally created at the macro-economic level through global economic relations that perpetuate economic differences between the developed and developing world. Thus mitigation efforts need to tackle the underlying causes of vulnerability in a society, rather than just the risks or symptoms of a particular disaster. These causes include access to and ownership of resources, settlement patterns and the 'real causes' of poverty and underdevelopment (Maskrey 1989: 39).

Tourism has become an important source of revenue for many regions that would otherwise be on the periphery of the world economy, as tourists tend to seek and favour 'distinctive' areas that promise 'unique' experiences (Christaller 1964, cited in Robinson 1976: xxiii). These are often the very regions that suffer many of the world's disasters, as Maskrey (1989: 3) states:

> Large numbers of people live on the social and territorial periphery of the global economic relationships which do not allow them access to the basic resources, such as land, food and shelter, necessary to stay alive. The empirical evidence, from a large number of case studies, points to the fact that it is these groups who most often suffer disaster. Vulnerable conditions are far more prevalent in the Third World than in the First World.

This chapter attempts to identify causes of vulnerability amongst those in a single rural village affected by Bali's tourism crisis, the attempts being made to reduce this vulnerability, attitudes towards tourism and how these views are being shaped and changed by crisis and vulnerability within the tourism industry. In some respects it complements examination of the crisis at a larger scale (Hitchcock and Darma Putra 2005).

Bali and tourism

The substantial economic impact of tourism in Bali is widely acknowledged. Officially, tourism is Bali's second economic priority, behind agriculture, and has been since the early 1970s with the Indonesian government's adoption of the Bali Tourism Development Master Plan (Picard 1996: 49). Since that time, the importance of agriculture to the economy has progressively decreased whilst tourism and related sectors have rapidly grown in size and economic influence (Picard 1996: 57). Because of the 'multi-sectoral' impact of tourism, it is difficult to quantify its exact economic impact in Bali (ibid.); however, according to the UNDP and World Bank (UNDP *et al.* 2002: 1), in Bali 'tourism now impacts the livelihoods of the whole island, including the poorer North and East, through linkages such as migration, remittances, and general spending across all sectors', including manufacturing and agriculture. Thus, any study of tourism in Bali cannot focus simply upon those directly employed by hotels or other services directly linked to tourism but must also encompass employment in related sectors including primary and manufacturing industries supplying products to tourism providers.

This study, based in a rural area of Bali, was able to gain a more holistic view of the economic situation after the bombings, encompassing tourism workers and suppliers of agricultural and manufactured products, as well as those who sell these products at market. The findings are based on data compiled during a seven-week period in mid-2004. The researcher lived in the Balinese village of Tegaljadi and conducted in-depth interviews in Indonesian with approximately 25 people, and gathered further data through participant observation (Tarplee 2004).

Tegaljadi and tourism

The village of Tegaljadi is located in the regency of Tabanan, and is approximately 25 kilometres from the provincial capital Denpasar and the nearby tourism resorts of Kuta, Sanur and Nusa Dua. Agriculture remains Tabanan's economic base, the area being known as the 'rice bowl of Bali'. However, in Tabanan, as in the rest of Bali, tourism has begun to encroach on agricultural land, and Tabanan has been earmarked as a potential tourism growth area. Tegaljadi maintains many pre-tourism characteristics, whilst becoming increasingly involved in the broader tourism industry. Forty per cent of the population are farmers, with the most common agricultural activity being rice farming. Along the edges of the rice fields that surround the village, crops of vegetables are grown and poultry and livestock are also raised by many people.

Despite the ongoing importance of agriculture for the village, residents are gradually becoming more bound up in the tourist economy. Although only a small number of residents are directly employed by hotels and other tourism enterprises or as guides at nearby tourist attractions such as the Alas Kedaton Monkey Forest in neighbouring Kukuh, several cottage industries have been established in the village, and a number of women make beaded sandals and other garments for sale to tourists and for export. Various residents also roast coffee and peanuts for sale at markets in nearby Tabanan as well as in Denpasar and Kuta. Finally there are many residents who are involved in tourism on a casual basis, especially on project work such as construction as opportunities arise. After the bombings, all these groups felt the effects of the decline of tourism arrivals throughout Bali, although to varying degrees depending on to the type and level of involvement of household numbers in tourism-related industries.

Individual experiences of the tourism decline in Tegaljadi

A number of individuals from Tegaljadi were particularly affected in the aftermath of the bombings. PT,[1] who worked as a bartender in a foreign-owned Nusa Dua hotel, had two components to his salary each month: a base salary and a 'service fee' that was dependent on the number of tourists staying in his hotel. After the bombings, the occupancy rate at his hotel fell to just over 8 per cent, forcing his take-home pay down to approximately A\$200, a third of its pre-bombing level. After the bombings, many employees of hotels and other large (usually foreign-owned) tourism enterprises, rather than being laid off, were forced to take unpaid leave. At PT's hotel, rotating rosters were imposed involving one month working and one month unpaid leave for the six months after the bombings.

MM is a supplier of building materials to both tourism enterprises and individuals and she was without work for almost six months following the bombings. Demand for her products fell dramatically and many existing orders were cancelled as tourism operators were too scared to build new, or expand existing, enterprises, and individuals could not afford to renovate their homes or build new ones. It is common practice in Indonesia for people to conduct business with personal friends, and employment often occurs along family and community lines in an attempt to assist family and friends. Thus, the impact of the bombings upon MM's business flowed on to others in Tegaljadi as she tended to hire labourers from amongst her friends and family in the village.

NM runs a business with her husband and wider family, roasting coffee and peanuts and packaging snacks for sale to tourism suppliers, restaurateurs and individuals within markets in Tabanan, as well as in markets in Badung, close to Kuta. Other suppliers of raw and processed goods in Tegaljadi, like NM, also found that beyond the decline in demand for their products they faced the challenge of increased competition as former tourism workers moved into the sector after losing their jobs in hotels and shops.

A final group within the village who experienced difficulties during the tourism downturn were young people who had recently graduated from high school and were unable to find employment after the bombings. AS, an 18-year-old man who was unmarried and still lived with his widowed mother, had attended a special Tourism High School, which had aimed to prepare students for employment in the tourism industry. Focus at these schools was placed on gaining skills useful for the tourism industry, especially English language ability, and providing hands-on work experience in the tourism sector prior to graduation. AS regretted his decision to attend such a school, which placed less emphasis on academic ability, because after the bombings his specialty knowledge of the hospitality industry was less useful than other skills which might have enabled him to secure work in a more stable sector.

Prior studies have found that responses to economic pressures tend to be very similar, despite varying circumstances (Moser and McIlwaine 1997). The particular experiences following the bombings outlined above were typical of many of the respondents who were interviewed in Tegaljadi, and responses to economic decline were also relatively uniform across households in the village.

As an immediate response to the economic effects of the bombings, households needed to reduce their spending, which they achieved through a prioritisation of expenses. Food was the most important household expense, yet even expenditure on food was reduced by eating less meat and fish and eating more vegetables, *tempeh* (fermented soya beans) and rice. In general people found that the variety in their diets tended to decrease and spending on luxuries such as treats for children, like biscuits, lollies and nuts, was reduced. Others reduced their expenditure on 'take-away' food from nearby night markets, and the variety of food consumption declined. One resident observed: 'whereas before we used to eat maybe chicken, now it is sufficient if we just eat tofu; what is important is that we eat enough to stay healthy for work.' Expenditure on other 'non-essential' items such as clothing, travel and entertainment was significantly reduced. This created flow-on effects for other community members who relied upon this expenditure for their own livelihoods.

Further, people had to be creative and versatile in 'making do' with what they already had. This was especially important when it came to special events and ceremonies, which can add significantly to household expenses. The Hindu-Balinese religion is an integral part of daily life in Bali, the practice of which is highly symbolic. People are expected to bring food and other offerings to religious ceremonies, which are held relatively frequently. All members of a community are expected to contribute towards such events; even if they are living elsewhere, they are expected to contribute financially to their *banjar* (village ward), which is a key community structure in Bali and the most basic administrative unit on the island, covering all aspects of life from civic and religious duties to agriculture (Warren 1993: 7).

Whereas experiences of the impact of tourism decline were relatively similar across households, within households the situation varied greatly. The influence of culture on gender roles and the linkage between this and economic vulnerability within society were brought to light in the wake of the bombings. Practice dictates that following marriage women cease paid employment; although the social norm is becoming less binding, this still influences household decisions regarding employment patterns of both men and women. As Long and Kindon (1997: 108) noted:

> Men are generally free to work wherever they need to and many migrate to tourism developments and live away from home . . . For women, they too are able to migrate to work until they get married. Then they are expected to stay closer to their family and be able to carry out their reproductive duties as wives and mothers.

The ongoing influence of New Order discourse regarding the role of women, which viewed them primarily as wives and mothers, continues to influence gender roles in Indonesian society despite the fall of Suharto's regime (see Suryakusuma 1996 for a discussion of the role of women in Suharto's New Order Indonesia). This added to the vulnerability of families following the bombings. Households with a number of adults of working age who can be mobilised during periods of financial difficulty are potentially better able to cope when faced with extreme economic circumstances. Prior to the bombings, many women had not been employed; rather they stayed home to look after children, carry out household duties and observe community obligations. Those households with dual incomes had more to fall back on following the bombings, but many only had a single source of income.

The decision-making process regarding household expenditure links into societal power structures and the differing perceptions of male and female roles within both family and society. Whereas both men and women faced uncertainty regarding employment security and future options, women had the added responsibility of 'making do with less'. Respondents stated that household expenditure was a communal decision; however, it became clear that, as women generally had the responsibility for shopping for the family and managing household finances, they therefore faced the most stress when trying to prioritise day-to-day expenses following the bombings. Many tourism workers stated that before the tourism downturn they had not had to think about what they would spend their money on. Following the bombings however, decision-makers (notably women) faced considerable stress determining their economic priorities, and both women and men experienced and expressed shame when they were unable to support their extended families.

In response to this added pressure, women tended not to spend money on themselves, attempting to fulfil the needs of the rest of their family before their own. Men, on the other hand, tended not to have much input at this level of household decision-making. Men were also observed to have more money to spend on personal expenses that did not necessarily contribute towards their family's collective wellbeing. Men commonly smoked cigarettes, gambled and utilised personal vehicles for leisure activities. Women, on the other hand, did not smoke or gamble and rarely left the village or used motorcycles other than to get to work or to run special errands that required them to travel to the city.

Safety nets

After the bombings a number of villagers had been forced to borrow money from family and friends or their employer if possible, or to take loans from institutions such as the LPD (Lembaga Perkreditan Desa), a credit institution which has been formed in most villages in Bali with the intention of helping villages to fund religious ceremonies and economic development. Following the bombings, any new loans were for necessities only, and usually for larger expenses than day-to-day expenditure on food and other household necessities. One villager explained that it had been common practice for him to take regular loans from his hotel's cooperative prior to the bombings; these loans were generally for luxuries such as new motorbikes, a car and house improvements. Following the bombings, as his income was no longer guaranteed, he ceased to borrow money for luxury items. Repayments of money borrowed before the bombings had also been delayed in many cases, with a certain amount of leniency on the part of banks and lending institutions following the bombings. Several people who had borrowed money before the bombings, when their wages or profits appeared to be guaranteed, found that afterwards they were unable to pay even the interest from their loan. Money was also borrowed from family, and less commonly from friends, but this was effective only if those affected by the economic downturn had family who did not also rely on tourism for their livelihoods, or if their family was wealthy enough to spare some of their own money.

The importance of the village itself in providing coping mechanisms was a reflection of the ongoing influence of 'traditional' culture, despite modernisation. Various traditional practices are still deeply entrenched in villages, and their significance has also extended to urban areas where many people had only relatively recently migrated (following the tourism boom in the 1960s). The concept of the village in Balinese society is closely linked to those of family and community, and each played a role in supporting tourism workers following the bombings. The importance of households and community at times of crisis has been identified in previous studies such as those into the impacts of the 1997 Asian economic crisis. As Setiawan (1999: 295) stated:

> For the poor, households are essentially social structures, which can range in size from one individual to a large, extended family, or (as in the case of street children) a fabricated family. The household economy is therefore made up of a network of social and economic relationships and interdependence among members. The members of a household depend on each other, and organise together to provide for the necessities for mutual survival. Households are therefore important adaptive institutions for the poor, providing mechanisms for pooling income and other resources and for sharing access to subsistence and other basic needs.

One of the most evident and widely practiced cultural norms is the grouping of families patrilineally, according to which women marry into their husband's family, unless they have only female siblings, in which case one of the sisters takes on the role of a man and takes her husband into her family compound. During tourism downturns this allowed families to join together to share resources and when some family members were affected by the bombings, they were able to turn to other family members for financial assistance and other material support.

The village formed a safety net for urban workers, provided that migrants maintained their linkages to the village and upheld their place in village structures (McKean 1989: 130). Although some residents who had moved away from the village and visited during the time of the field research stated that a permanent return to the village was an option for their family, there was no permanent return migration following the bombings, since by moving back to the village they would be moving further away from tourism destinations and employment. Most were waiting and hoping that, in the near future, tourism employment would again become viable, negating any need to return.

This unwillingness of urban residents to return to rural areas and work in agriculture reflects predominant attitudes toward tourism and agriculture. Whereas agriculture is widely regarded by Balinese as the basis of their economy and a fundamental part of traditional Balinese culture, the importance of both of which endures, no respondents who worked in tourism at the time of the bombing expressed any desire to work in agriculture. Their prevailing attitude was that, although agriculture was vital to Bali's economy and culture, and should be maintained in the face of pressures such as less available land, it was the way of the past: tourism represented the economic future of Bali. For some there was little choice. Access to land has been identified in previous studies as being of considerable importance in strengthening coping strategies following tourism decline, as it both diversifies livelihood activities and is an asset that provides financial security (Kareithi 2003: 18). Similarly in Bali, the environment of the village was a primary support mechanism, allowing villagers to fall back on farming to generate income, preventing complete reliance on tourism. Urban dwellers, on the other hand, had no other choice than to work in tourism given a shortage of arable land and higher land prices (Mitchell 1994: 191). Villagers were able to save money and rely on their surroundings for some of their daily provisions – including rice, vegetables, meat and the basic items necessary for religious offerings. This self-reliance enabled people to maintain a basic standard of living despite the economic effects of tourism downturn. One lady, IM, who was not involved in tourism herself, but a number of whose children had left the village to work in tourism, commented that urban dwellers have to 'pay for everything'. They were thus in a less fortunate position following the bombings with a lower capacity to cope, having lost the safety net that agriculture provided.

'Just waiting around for something to change' – economic vulnerability and tourism

After the bombings, the assumption of most respondents who had been affected by the tourism downturn was that the tourism industry would soon 'bounce back' to pre-bombing levels. Indeed, by the time that the study was carried out, more than a year after the bombings, tourism arrivals had already recovered considerably. For those respondents who had been working in tourism for at least a decade, their prediction was based on prior experiences of tourism downturns caused by global events, such as the Gulf War in 1991, regional downturns, such as the 1997 Asian economic crisis, and national events, including the fall of Suharto in 1998.

During previous periods of tourism downturn, the industry had been able to restore tourist confidence through concerted advertising campaigns and the impacts had been relatively short-lived. However, the Bali bombings differed from previous extreme occurrences in Indonesia, in that this crisis was the result of a direct attack against tourists, which had never before occurred in Bali and which presented a special challenge to tourist operators attempting to restore confidence in their destination (Hall and O'Sullivan 1996: 110). The international reaction was far more severe following the Bali bombings, as evidenced in the Australian government's issuing direct travel warnings against Bali, which, even during other extreme events such as the fall of the Suharto government, had been previously excluded from the list of no-go areas as 'Bali ha[d] always been considered the safest place in Indonesia' (Allard 2002). More recently, following further attacks, according to the Australian government, 'The 1 October 2005 Bali bombings and bomb attack outside the Australian Embassy in Jakarta in September 2004 underscore that the terrorist threat to Australians in Indonesia is real. Further terrorist attacks, including in Bali, cannot be ruled out' (Australian Government, Department of Foreign Affairs and Trade 2005). Bali had lost its distinctive and idyllic position.

In Tegaljadi people believed that the situation facing the tourism industry would inevitably improve, and thus so would their fortunes, and they could do nothing other than wait for this to occur. Tourism workers felt that they could neither control nor avoid the impact of extreme events upon their fortunes, and that they could only attempt to overcome crises as they occurred, rather than strengthen themselves against such vulnerability in the future. This inaction was attributed by some to the general culture of Bali, and perhaps Indonesia as a whole. Several respondents stated that the Balinese do not tend to think too much about the future and, while not apathetic towards it, they feel they have little control over it. Indeed, few respondents seemed to question their own circumstances, or the reasons for their economic difficulties.

Furthermore, many respondents believed that the Bali bombings were a 'one-off' event, which would never recur: a similar event that would destroy the tourism economy could not happen again as the Balinese people had 'already suffered enough'. Whether unaware of the risks they faced or incapable of remedying their circumstances, residents of Tegaljadi were making few attempts to limit their own exposure to similar events in the future, tending instead to focus on the macro level when questioned about their efforts to strengthen themselves against vulnerability. Many believed that the most successful methods of improving their economic situation were initiatives at the provincial and national government levels, such as increased security measures that were intended to protect tourists and strengthen the tourism economy, for example the particular focus on controlling domestic migration to Bali from Java, which was viewed by many respondents as the source of Islamic terrorism. The capture and trial of some of those responsible for organising the bombings was seen as further proof of government success at curbing terrorism and securing the future of tourism.

In urban areas, and increasingly in rural areas such as Tegaljadi, people are constrained in their employment options and, although they may prefer to work in another industry, they are prevented by the limitations of the Balinese economy, which has for so long been geared towards the development of tourism at the expense of other sectors. One woman, KD, stated 'At the moment, many people are telling me "look for work elsewhere, maybe as a public servant", but it's hard, especially now when there's so much unemployment.' The ability of people to get a job in another industry has been hindered by increased competition following the bombings, as well as the widespread corruption and nepotism within both public and private sectors, which requires either substantial sums of money in order to bribe potential employers, or powerful contacts within one's own family in order to gain a job.

Attitudes towards tourism (before and) after the bombings

Attitudes towards tourism within Tegaljadi were mixed. Although tourism has brought significant benefits to Bali, it was viewed by some as a 'double-edged sword', which had benefited many people economically yet had also resulted in cultural, social and environmental degradation (Wall 1996). Despite the increasing involvement of many of its residents in tourism, Tegaljadi is still on the margins of the mainstream tourism economy. There were two schools of thought within the village regarding residents' involvement in tourism. On the one hand, some were quite keen to become involved in tourism enterprises and wished to see increased tourism development within the village itself. These people often cited the economic benefits of the tourism industry as their primary rationale, and many stated that they would be happy to work in tourism themselves, if they had the skills, or for their children to become involved in tourism. On the other hand, many residents were far more wary of the benefits of tourism, and wished to limit the level of future tourism development within the village and its immediate surrounds. Those who were opposed to tourism in the village emphasised the need to ensure the security of the agricultural industry and prevent the economic dominance of tourism. They believed that the growth of the tourism industry could lead to a dangerous level of reliance, resulting in further vulnerability given any more tourism downturns in the future.

Bali's reliance on tourism was recognised by some as a significant problem that needed to be addressed; however, this attitude was more common amongst urban residents than those living in Tegaljadi. Within Tegaljadi, respondents who were not yet involved in tourism often stated that they would like to work in tourism in the future, despite seeing the impact of tourism downturns for others within the village, and more commonly for family and friends in urban areas. Household incomes prior to the bombings were much higher for those with some involvement in tourism than for people reliant on agriculture or other activities; thus many people already involved in tourism indicated that, once the crisis ended, they would be satisfied to return to a dependence on tourism, despite the ongoing threat of tourism downturns. The potentially high incomes outweighed the threat of periods of underemployment or unemployment. Those who were not yet involved in tourism similarly stated that they were attracted to the prospect of high wages in tourism and were not overly concerned by the threat of tourism downturns.

Owing to the linkages between all economic sectors in Bali even apparent alternatives to tourism were unreliable following the bombings, as they too were dependent upon the ongoing fortunes of the tourism economy. No realistic substitutes for tourism could be perceived, apart from agriculture. Agriculture was idealised by many, especially older respondents, as the way out of Bali's present

predicament; many suggested that agriculture's importance needed to be emphasised so that Bali would be less vulnerable to future tourism downturns if they
occurred. However, one tourism worker, MR, pointed out the flaws in the ideal-
istic notion of agriculture, noting that all Balinese live on the economic margins,
as agriculture was a similarly vulnerable industry in which 'harvests can still fail
because of plagues of mice'. Other young people, such as AR, a 26-year-old man
who worked in a non-tourism-related sector, but whose parents roasted peanuts
and sold them in Denpasar and tourism centres, saw a 'return to agriculture' as
impossible, on account of the shortage of land for people to return to agriculture, even if they so desired. Despite this, the ability to fall back on agriculture
had proved to be a vital coping mechanism for many respondents following the
bombings. Those who worked in tourism tended to combine income from tourism sources with agricultural activities, most commonly through the possession
of a small plot of rice paddy, or keeping cattle, pigs or poultry for household
consumption. All of these proved to be assets that could be released during times
of adversity.

Understandably, many residents of Tegaljadi hoped for the 'best of both
worlds': the ability to become involved in tourism and to take advantage of the
economic benefits it offers, yet to also be able to sustain traditional economic
activities through the maintenance of agricultural land. This is a potentially positive trend for, with an awareness of the need for proper management and control
over future developments in the area, it may be possible to maintain a level of
diversity, presenting a model for the future development of Bali that will decrease
vulnerability to extreme events.

The impacts of tourism extend into rural areas that are remote from the primary
tourism areas, and there is still scope for growth. In the case of Tegaljadi, the
direct involvement of residents was expected to escalate in the near future as one
man planned to construct tourist bungalows on the outskirts of the village. BS
intended to manage this enterprise himself, the income to be combined with that
from raising pigs and from his wife's construction business. Although perhaps not
intentionally, he was minimising the risk of joining the tourism economy through
the maintenance of a variety of income sources. A range of opinions, often negative, reflected on the potential benefits of this development, which was still in
the early planning stages. The proposed development was lamented by several
respondents in the village, who believed that it would inevitably be detrimental to
the village; it would encourage more people to become involved in tourism and
increase economic risk within the village. One respondent expressed her concern,
citing the experiences of a neighbouring village, where rice fields had been sold
to become a site for bungalows. She believed that, whereas the *sawah* (wet rice)
would have supported a whole family for generations to come, it had been sold
to meet the immediate needs of only a few people, and if the proceeds were not
invested wisely the sale could result in poverty in the future. Further, the employment opportunities created could not replace what was lost, as only a few people
could possibly be employed there. In other words, tourism might result in new
risk and new village-level inequality.

The importance of diversity

Those who were most vulnerable to the tourism downturn proved not to be employ-ees of large companies such as international hotel chains, which to some extent were able to absorb the economic shock following the tourism downturn and avoided retrenching employees until it became an economic imperative. Although employees of large-scale operations still felt the impacts of the decreased tourism numbers through reduced working hours and a consequent cut in take-home pay, many still maintained their jobs. Those most severely affected by the downturn tended to be informal workers largely 'on the fringes' of the mainstream tourism economy. Within Tegaljadi, most workers reliant on tourism were suppliers to the industry. Following the bombings they experienced a sudden decline in the demand for their produce as direct suppliers to tourists lost their customer base. Thus, even 25 kilometres from the core of the Balinese tourism industry, villag-ers who were not directly involved in tourism experienced the consequences of downturn, and the burden fell especially heavily on women. The poorest, and therefore the most vulnerable, lost most, but their losses were attenuated by local cultural circumstances.

Villagers generally recognised that tourism is an extremely sensitive industry, the ongoing success of which relies on a perception of political stability and the promise of personal safety for visitors; however, few could perceive a way to decrease the flow-on effects of instability for their households' economic security. Many viewed agriculture as the only reliable alternative to tourism. However, the effects of the Asian economic crisis were still a recent memory, and many agri-cultural workers in the village contrasted the hardship that they had experienced during that period with the relative good fortune of tourism workers, who had enjoyed greater tourism arrivals on account of very favourable exchange rates. Particular forms of instability had quite different impacts at village level.

Diversity of income sources was one of the most effective preventative mecha-nisms for reducing vulnerability. Those reliant on tourism in the village were more able than their urban counterparts to fall back on the incomes of family members involved in agriculture, or other industries not directly aligned with tourism. Similarly, suppliers to the tourist trade were able to turn to other (less lucrative) markets than tourism to sell their goods, as long as there was demand among the local population for their goods and they were not items – such as flowers – targeted too specifically at the tourist trade.

Because of the heavy reliance on tourism and the interdependence of all eco-nomic sectors in Bali, the impacts of the downturn affected not only those directly employed in tourism but also those involved in industries such as agriculture and manufacturing. This is not peculiar to Bali. As Robinson (1976: 120) points out, tourism generally has significant impacts on economies of host regions not only through creating a market for 'broad consumption items' manufactured within the region, but also through its linkages to industries that supply tourism opera-tors with goods and services. Thus, when demand for tourism lessens, effects are quickly felt in both other industries and other regions that may supply goods and services to tourism regions. Hence, access to alternate sources of income, and the ability to turn to activities not contingent on tourism to secure household income, are fundamental in decreasing economic vulnerability to external shocks.

Vulnerability to extreme events – economic or environmental – is compounded in a situation of limited resources and opportunities (Blaikie *et al.* 1994; Moser 1996). Similarly, the risk of tourism downturns can be mitigated by the maintenance of other existing economic resources, or the development of new ones that can act as a substitute. A primary limitation is the inability of those already reliant on tourism to develop such other industries themselves owing to a lack of the capital and skills required to establish new enterprises. Redirection of resources towards alternative markets allows people who are reliant on tourism operators and suppliers to overcome the economic pressures faced in extreme circumstances. However, tourism suppliers are constrained by the specialisation of many of their products for tourist consumption, as they are generally not suitable for local markets.

In the case of Bali, until the bombings occurred in 2002, apparent alternatives to tourism such as manufacturing (of handicrafts, textiles and food) and also agriculture were often geared towards sale to the tourism industry. As in the parallel case of Kenya's tourism industry (Kareithi 2003: 18) there was a lack of skill diversity among the residents of Tegaljadi, alongside few opportunities for change and response. Most had only a slight understanding of the broader tourism market and basic business management skills, which had been sufficient during times of economic prosperity, but which did not encompass long-term risk management strategies that could assist during a downturn. The adaptation to tourism from agriculture was a large step to take and many respondents had only recently become involved in tourism, hence they had not built formal networks, tending to take a more ad hoc approach to their business and capitalising on opportunities as they arose. Many primary producers had moved into tourism to take advantage of a new market; while they had experienced better economic prospects, they had not greatly changed their activities to adapt to the tourism industry. When losing access to direct tourism markets, Tegaljadi residents did not have the assets, networks or business skills to be able to redirect their goods to other markets further afield, despite some believing this was a potential 'way out' of their predicament.

It is necessary for the Balinese to look beyond local markets to ensure that there are alternatives to the tourism industry, which can support them in lean tourism periods. However, even after the bombings, most people working in tourism had scarcely looked beyond their current activities for new forms of income generation, mainly because of the lack of capital, skills and opportunities, as well as a continued desire to share in the economic prosperity offered by tourism during periods of normal operation. Upon 'discovering' the economic potential of tourism, many Balinese structured their household economies around tourism incomes, leaving behind more established livelihoods such as agriculture, while encouraged by a government that has relied heavily on external investment to develop the island. Tourist dreams are embedded in the lives and attitudes of even the poorest villagers. In recent years, much of Bali's development has been focused on the expansion of the tourism industry, which has presented an opportunity for the Balinese to exploit their culture and environment and earn money. The limitations of Indonesia's developing economy, the over-population of the island, the inability to rely on agriculture, and the lack of obvious income-generating alternatives have forced many people into tourism. Many more joined the tourism economy because of its obvious benefits. This was not questioned while tourism was viable; however, since the bombings the extent of the benefits of tourism has become a point of discussion and a clear cause for concern. The decline of the role of agriculture has been expedited by the sale of agricultural land for transformation into non-productive land and the expansion of urban and tourism areas (Mitchell 1994: 191). Although this trend was noted many years earlier (Noronha 1979: 183), it was only following the bombings that the implications became clear to many and the issues more widely debated.

Many rural workers now aspire to become involved in tourism, which is often equated with affluence and access to a Westernised, urban lifestyle and, prior to the bombings, incomes from tourism were substantially higher than those from agriculture. It is not intended here to romanticise notions of the agrarian lifestyle, especially in Bali, where intensive 'agricultural involution' (Geertz 1970) characterises agricultural labour; however, it is important to urge against the wholesale abandonment of alternatives to tourism. In the aftermath of the 2002 bombings there was widespread optimism, not only in Tegaljadi, that life would return to normal and that tourism would continue to grow. However, this hope for a more secure future and a return to 'normality' was shattered by further terrorist attacks in September 2005, when three more bombs were detonated in different popular tourist areas in Bali, killing 23 people. The macro-economic impacts of the continuing terrorist threat in Indonesia contributed to the collapse of Air Paradise, a small airline carrier that was established following the first Bali bombing, and the effects of cancelled bookings and diverted travel to safer destinations continue to filter down through all levels of Bali's economy (Burrell 2005). Although the downturn following the second Bali bombings was less severe than that in 2002, hotel occupancy rates two months after the attacks were only 36 per cent compared with 90 per cent prior to the bombings (Burrell 2005) and, at the end of 2006, tourist numbers were yet to return to pre-2002 levels. The desire of many to become involved in tourism had already been tempered by a healthy dose of cynicism regarding its benefits – whether economic, social or cultural – as, following the bombings, the effects of tourism on the wider economy became glaringly obvious, far beyond that on workers in tourism areas. Not only at times of crisis have economic diversity and social and economic safety nets been central to the livelihoods of the people of Tegaljadi.

Note

1 Throughout this chapter, pseudonyms have been used in place of participants' names.

References

Adger, W.N. and Kelly, P.M. (2001) 'Social Vulnerability and Resilience', in W.N. Adger, P.M. Kelly and N.H. Ninh (eds), *Living with Environmental Change: Social Vulnerability, Adaptation and Resilience in Vietnam*, London: Routledge.

Allard, T. (2002) 'Two Nations Caught Tragically Unprepared', *Sydney Morning Herald*, 14 October.

Allen, L. and Strutt, S. (2003) 'Travellers Cancel Holiday Plans', *Australian Financial Review*, 16 October.

Australian Government, Department of Foreign Affairs and Trade (2005) Travel Advice for Indonesia. Online. Available <www.smartraveller.gov.au/zw-cgi/view/Advice/Indonesia> (accessed 28 November 2005).

Beirman, D. (2003) *Restoring Tourism Destinations in Crisis: A Strategic Marketing Approach*, Wallingford: CABI Publishing.

Blaikie, P., Cannon, T., Davis, I. and Wisner, B. (1994) *At Risk: Natural Hazards, People's Vulnerability, and Disasters*, London: Routledge.

Burrell, A. (2005) 'Air Paradise Collapse Another Bali Blow', *Australian Financial Review*, 25 November.

Geertz, C. (1970) *Agricultural Involution*, Berkeley, CA: University of California Press.

Hall, C.M. and O'Sullivan, V. (1996) 'Tourism, Political Stability and Violence', in A. Pizam and Y. Mansfeld (eds), *Tourism, Crime and International Security Issues*, Chichester: John Wiley.

Hitchcock, M. and Darma Putra, M. (2005) 'The Bali Bombings: Tourism Crisis Management and Conflict Avoidance', *Current Issues in Tourism*, 8: 62–76.

Kareithi, S. (2003) *Coping with Declining Tourism, Examples from Communities in Kenya*, PPT Working Paper Number 13, London: Pro-Poor Tourism Partnership.

Long, V.H. and Kindon, S.L. (1997) 'Gender and Tourism Development in Balinese Villages', in M.T. Sinclair (ed.), *Gender, Work and Tourism*, London: Routledge.

McKean, P.F. (1989) 'Towards a Theoretical Analysis of Tourism: Economic Dualism and Cultural Involution in Bali', in V.L. Smith (ed.), *Hosts and Guests: The Anthropology of Tourism*, Philadelphia, PA: University of Pennsylvania Press.

Maskrey, A. (1989) *Disaster Mitigation: A Community Based Approach*, Oxford: Oxfam.

Mitchell, B. (1994) 'Sustainable Development at the Village Level in Bali, Indonesia', *Human Ecology*, 22: 189–211.

Moser, C. (1996) 'How Do the Poor Manage in an Economic Crisis?', *Finance and Development*, 33(4), 42–44.

Moser, C.O.N. and McIlwaine, C. (1997) *Household Responses to Poverty and Vulnerability, Volume 2: Confronting Crisis in Angyalfold, Budapest, Hungary*, Washington, DC: World Bank.

Noronha, R. (1979) 'Paradise Reviewed: Tourism in Bali', in E. De Kadt (ed.), *Tourism: Passport to Development?*, New York: Oxford University Press.

Picard, M. (1996) *Bali: Cultural Tourism and Touristic Culture*, Singapore: Archipelago Press.

Pizam, A. and Mansfeld, Y. (1996) *Tourism, Crime and International Security Issues*, Chichester: John Wiley.

Robinson, H. (1976) *A Geography of Tourism*, London: MacDonald and Evans.

Setiawan, B. (1999) 'Survival Strategies by the Poor in Yogyakarta, Indonesia: The Importance of "Social Capital"', in G.B. Hainsworth (ed.), *Globalization and the Asian Economic Crisis*, Vancouver: Centre for Southeast Asian Research.

Suryakusuma, J. (1996) 'The State and Sexuality in New Order Indonesia', in L. Sears (ed.), *Fantasizing the Feminine in Indonesia*, Durham, NC: Duke University Press.

Tarplee, S. (2004) *Postcards from the Edge: Vulnerability in the Balinese Tourism Industry*, unpublished honours thesis, University of Sydney.

UNDP, World Bank and USAID (2002) Bali Beyond the Tragedy: Impact and Challenges for Tourism-Led Development in Indonesia. Online. Available <www.undp.or.id/programme/conflict/documents/bali%20beyond%20the%20tragedy_en.pdf> (accessed 14 February 2004).

Wahab, S. (1996) 'Tourism and Terrorism: Synthesis of the Problem with Emphasis on Egypt', in A. Pizam and Y. Mansfeld (eds), *Tourism, Crime and International Security Issues*, Chichester: John Wiley.

Wall, G. (1996) 'Perspectives on Tourism in Selected Balinese Villages', *Annals of Tourism Research*, 23: 123–137.

Warren, C. (1993) *Adat and Dinas: Balinese Communities in the Indonesian State*, Kuala Lumpur: Oxford University Press.

9   Sustainability and security

Lombok hotels' link with local communities

Fleur Fallon

Tourism has become increasingly critical for the island of Lombok, located east of Bali, the primary leisure tourism destination in Indonesia. This chapter examines employment patterns and opportunities, both formal and informal, for indigenous Lombok people within the tourism industry. The employment status of local people by key tourism service providers is viewed as a prime indicator of sustainable tourism development. This study broke new ground as limited official government statistics are available for the indigenous Lombok people either engaged within the tourism industry or in related informal employment. An overview of the Lombok context and a discussion of key tourism employment issues here precede an analysis of formal employment for 15 hotels located in two of the principal designated tourism development areas on Lombok: Senggigi Beach, northwest Lombok, and Kute Beach in the south of Central Lombok. This is followed by a discussion of other links between hotels and the local community with particular reference to security issues and sustainability indicators.

Lombok in context

Being the next largest island east of Bali, Lombok is seen as the gateway to eastern Indonesia and attracts an overspill of visitors (approximately 20 per cent) from Bali. It is therefore the primary focus of this study, although the overall context of the province of Nusa Tenggara Barat (NTB), comprising Lombok and Sumbawa, needs to be considered. Sumbawa is twice as large as Lombok but, like the south and southeast of Lombok, is prone to long periods of drought, and supports just one-quarter of NTB's population of nearly 4.4 million (BPS *et al.* 2004). NTB is a relatively poor Indonesian province. Although it has shown signs of improvement across many economic and human development indicators in the last decade, NTB still ranks thirtieth and last of the Indonesian provinces in terms of the Human Development Indicator (HDI), which is a combination of life expectancy, adult literacy rate, years of schooling and per capita expenditure (BPS *et al.* 2004: 97). This has an impact on occupation choices and access to higher level jobs in all industries, including tourism. More than 50 per cent of the population is still engaged in agriculture or fishing. Other industries include cultured pearl, pumice stone and a gold and copper mine located in southwest Sumbawa.

The provincial capital of NTB is Mataram, on the west coast of Lombok. Mataram and the fertile central plains of Lombok have the highest population densities in NTB. The indigenous Lombok people, the Sasaks, who are nominally Muslim, make up approximately 90 per cent of Lombok's population. Sasaks in the north and northwest of Lombok still practise an indigenous faith that combines elements of Hinduism, animism and Islam, and experience some pressure and discrimination as they do not belong to the 'true' faith. Other Sasaks from East Lombok, impoverished by drought, famine, political neglect and past heavy taxation burdens, have engaged in periodic conflict with their more prosperous cousins in the west since the times of Balinese and Dutch colonial rule. More recently, many have been relocated through the government-supported transmigration programmes to Sulawesi, Kalimantan, Maluku and West Papua, not always successfully, and others have migrated to Saudi Arabia or Malaysia as contract workers, often illegally. The remaining population consists of Balinese born on Lombok (*Baloks*), about 10,000 ethnic Chinese, and other islanders, mostly from Bali, Java, Sulawesi and Sumbawa. The Balinese, Baloks and Chinese tend to be more materially successful than many of the local Sasaks, and this creates tensions that simmer below a surface harmony.

Lombok is traversed by a volcanic mountain range, dominated by Gunung Rinjani (3726 metres), a major tourism attraction. A number of smaller islands, in particular the Gilis, off the northwest coast, offer pristine waters and relaxed marine tourism attractions. Although there is one major hotel resort in Central Lombok, and part of the Rinjani tourism area covers all three Lombok districts, most tourism activities are concentrated in West Lombok (Map 9.1).

The first master tourism plan for Lombok was prepared in 1981 and reviewed in 1987. The key to attracting 'luxury' tourists was an integrated resort area similar to Bali's Nusa Dua region. However, political events, including the downfall of President Suharto in 1998 and four subsequent changes of president, and economic factors have conspired to limit the construction of such resorts, whose developers paid little attention to local environmental limitations or the needs of communities located in or near the designated tourism areas. Lombok's tourism industry officially commenced in 1989 and tourist numbers steadily increased until the monetary crisis (*krismon*) in 1997. Since then, the number of international visitors has been volatile, although domestic numbers have climbed. Whereas the emphasis of tourism in developing countries is often placed on attracting and serving the needs of international visitors, the mainstay of Indonesian tourism is the domestic sector, with 20 times the number of trips within Indonesia (over 100 million), compared with 4.9 million overseas arrivals in 2005 (Anon. 2005, 2006). The number of foreign arrivals to NTB plummeted from a peak of 245,000 in 1997 to a low of 80,000 in 2003 (NTB Department of Culture and Tourism 2004). The total number of arrivals for both domestic and foreign visitors was 327,000 in 2003, compared with the 1997 peak of over 400,000 (NTB Provincial Tourism Office 2000) (Table 9.1).

Many factors have combined to curtail NTB and Indonesian tourism, including the economic recession of 1997, political shifts in 1998, the East Timor

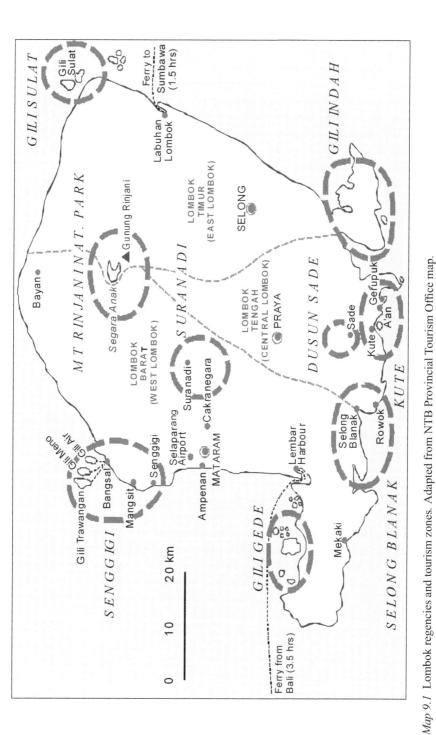

Map 9.1 Lombok regencies and tourism zones. Adapted from NTB Provincial Tourism Office map.

Table 9.1 Number of tourists to NTB for selected years from 1990 to 2003

Year	Foreign visitors	Domestic visitors	Total
1990	107,210	76,817	184,027
1995	167,267	140,940	308,207
1997	245,049	158,894	403,943
1998	211,812	168,727	380,539
1999	189,659	144,953	334,612
2000	107,286	126,364	233,650
2001	129,356	189,672	319,028
2002	120,637	226,635	347,272
2003	80,023	246,701	326,724

Source: NTB Provincial Tourism Office (2000).

independence conflict of 1999, 9/11 New York 2001, Bali bombs in both October 2002 and October 2005, additional bombs targeting the Australian Embassy and the Marriott Hotel in Jakarta in 2003 and 2004, the US–Iraq war and increased risk of terrorism, SARS in 2003, the tsunami in northwestern Indonesia of December 2004, and perhaps the risk of avian flu. These are all factors that have limited arrivals, alongside government travel warnings and media reports.

On 17 January 2000, Lombok itself was thrust onto the main news pages of the international media as riots broke out in Mataram. The rioters, including many youths from East Lombok, were provoked by charismatic outside leaders, at a mass rally protesting against the government's response to Christian–Muslim conflict in Maluku. Three days of riots resulted in damage to nine churches and 77 Christian and Chinese houses. Ironically, the same week the Indonesian government had revoked the 33-year-old Presidential Instruction that restricted the public observance of Chinese religious practices and traditions. Locals also used the opportunity of the riot to set fire to two nightclubs in central Senggigi, alleged centres of prostitution, drug dealing and constant nightly disturbances. A total of 6000 people, both residents and foreign tourists, left the island with over 2500 people evacuated by air, mostly to Bali. Over 1500 police, naval and military personnel from Java arrived to quell the violence. Orders to shoot rioters and looters were given. Five people were killed, and a total of 203 alleged rioters and provocateurs were detained. Fears were raised about unrest on Lombok sparking outbreaks of violence in other parts of the country.

Attempts since then have sought to engage local communities more in both dealing with conflict and developing appropriate tourism strategies. More recent attention worldwide, and locally on Lombok, has been given to the concept of sustainable tourism development. Of the three key pillars of the sustainable tourism development paradigm – environment, economy and local community – the environment and economic returns are often given precedence over community. It is argued that links with the local community, including direct and indirect employment opportunities, supplies of food, souvenirs, hotel equipment and decorations, as well as support for community cultural, educational and religious activities will engender strong support and local protection in times of security threats, and enhance both protection of the environment and economic returns. They will thus be critical to revitalising the struggling tourism sector and supporting the overall goal of sustainability.

Key issues in tourism employment

The World Travel and Tourism Council (WTTC 2005) estimated that one in twelve jobs are directly or indirectly related to tourism. The tourism and leisure sector is noted for beneficial local multiplier effects through the creation of indirect employment in both the formal and informal sectors. Tourism employment is thus one of the key indicators of sustainable tourism (Weaver 1998: 9), involving both the numbers of local people directly and indirectly employed within the industry, and job satisfaction within both the formal and informal sectors. The NTB Department of Culture and Tourism (2004: 37) estimated that the total number of people employed in the tourism sector for NTB was just 7490 in 2003, a 17.8 per cent decrease since 1999. This represents just 0.3 per cent of the employed labour force (BPS *et al.* 2004: 184). It falls far short of the estimation by WTTC (2004, cited in Thadani 2005) that 8.5 per cent of the Indonesian population is employed in tourism.

Although it is easier to quantify formal employment, informal indirect and induced employment 'cannot be underestimated or ignored in a consideration' (Baum 1993: 67) of tourism within the context of sustainability. Informal tourism workers tend to operate on the fringes of hotel and tourist beach areas, using their entrepreneurial and networking skills to directly tap into the wealth of tourists (Dahles and Bras 1999a,b). They differ from other informal workers, who tend to 'cater predominantly to other locals at a similar low economic level' (Cukier 1996a: 57). BPS *et al.* (2004: 184) indicate that for NTB the level of employment in the informal sector is as high as 78 per cent of the labour force, compared with 59 per cent for Bali, but not necessarily or even primarily in the tourist sector.

Although tourism is heavily influenced by perceived images and identity, limited emphasis has been placed on an analysis of ethnicity, gender or informal employment within the tourism industry (Cukier 2002; Richter 1995). Equity issues, including indigenous access to tourism opportunities, or the 'indigenisation' of the tourism industry, must be addressed in the interests of sustainable tourism (Din 1997: 76). Most tourism employment data available for Indonesia do not analyse employment in terms of local island populations, and local populations may miss out on many tourism jobs to a more highly educated, skilled and experienced migrant workforce from other islands. Referring to Eastern Indonesia, Tirtosudarmo (1996: 212) claims that 'local people . . . (are) most likely to be defeated by the more aggressive and highly skilled in-migrants . . . potential conflicts between the local population and migrants . . . affect social cohesion and in turn the political stability of the region'. Such conflicts may also arise between districts as demonstrated on Lombok in January 2000.

Gender analysis within the tourism sector in Indonesia is likewise limited. About 46 per cent of all jobs in tourism employment globally are held by women. However, women make up 90 per cent of jobs in the catering and accommodation sector, but many of these jobs are temporary, casual or part-time. Women constitute just 44 per cent of the total work force for NTB (BPS *et al.* 2004: 100), and special consideration needs to be given to their situation within the tourism sector.

Need for local community participation

Many constraints inhibit the involvement of local communities in tourism planning and development. These include the cultural and political legacy of decades of authoritarian rule; the lack of understanding by local communities; economic conditions; time factors; and the lack of collaboration skills of government officials (Timothy 1999). However, as Pearce *et al.* (1996: 211–12) stress:

> It is likely to be more productive (for tourism business interests) to view community involvement as good business. The process of effective community involvement in tourism can generate ideas from extensive local knowledge, it can develop partnerships in the purchasing and production of tourism related goods, and it can be a first plank in the total marketing plan for the business.

If local communities are not involved, a backlash against tourism may develop, with security and safety risks for tourists, who may be vulnerable to attack by local people, and protracted land disputes and conflict between local communities, developers and government (ibid.). These situations all contribute to reduced security for both local communities and tourists and combine to inhibit the growth of tourism. Local participation does not guarantee success, but its absence most certainly will create problems, leading to unsuccessful integration.

In addition to the jobs in the formal economy and in the informal sector, consisting mainly of guides and vendors, linkages with the local community for supply of food to hotels, construction and decorating materials, and craftwork need to be considered as tourism-linked employment opportunities for local people. The employment links of hotels with local communities on Lombok are examined below.

Tourism employment on Lombok

Surveys were conducted during a peak employment period in February 1999 and a post-riot period in June 2000, to obtain a more detailed analysis of island of origin and gender of employees in the hotel sector. They involved semi-structured interviews with managers of selected hotels in the Senggigi and Kute areas of Lombok, and supplementary interviews with staff and members of the local community. Hotels and guesthouses were selected on the basis of size, and then proximity to the larger hotels, and further analysis of eight of the hotels in Senggigi was undertaken in 2005. Two small guesthouses no longer existed and hotels at Kute were not revisited. Table 9.2 outlines the breakdown of staff, according to their island of origin, for the 15 hotels surveyed in February 1999 and June 2000 and for the eight hotels contacted in 2005.

In 1999, Sasaks, the local Lombok ethnic group, constituted the main group of employees (54.7 per cent). Baloks, Balinese and Sasaks composed 85.8 per cent of all staff. Balinese outnumbered Baloks by approximately 2:1. Javanese staff constituted 7.6 per cent of hotel staff; staff from other Indonesian islands, mainly the nearby islands of Sumbawa, Sumba, Flores and Sulawesi, made up 6.1 per cent. In February 1999, nine expatriate staff were working in only four of the fifteen surveyed hotels. This represented just 0.5 per cent of total staff in these hotels and this percentage was still evident in 2005.

After the January 2000 riots, the employment proportions changed, as most casual staff, who were mainly low-skilled, local Sasaks and women, were not retained. The staff numbers in Table 9.3 include both casual and contract staff. In 1999 the casual staff totalled 104, or 6.4 per cent of the total employees. By June 2000, this number had decreased to 39, just 2.7 per cent of total employees. Numbers of contract staff remained relatively static, representing 135, or 8.2 per cent of the total number of employees, in 1999, and 133, or 9.1 per cent of the total number of employees, in 2000. By 2005, contract and casual staff combined were virtually non-existent, just 2 per cent of the total staff numbers. One hotel, which had more than 200 staff, reduced its staff by 50 per cent following the Bali bombings in October 2002. The number of Javanese employees almost doubled between surveys in 1999 and 2000, hence the concern that more highly skilled employees from other islands would displace locals and lead to social unrest.

Table 9.2 Origin of hotel staff (contract and casual) from selected hotels at Senggigi and Kute, 1999 and 2000, and at Senggigi, 2005

Origin of staff	Number of staff, 1999	Percentage of total staff in 15 hotels, 1999	Number of staff, 2000	Percentage of total staff in 15 hotels, 2000	Percentage of total staff in 8 Senggigi hotels, 2005
Expatriates	9	0.5	8	0.6	0.5
Lombok Sasak	896	54.7	631	43.2	46.3
Baloks (Balinese born on Lombok), plus Balinese	508	31.1	502	34.4	40.0
Java	124	7.6	208	14.2	9.6
Other – Sumbawa, Sumba, Flores, Sulawesi	99	6.1	111	7.6	3.6
Total	1636	100.0	1460	100.0	100.0

Source: Author's survey of hotel managers.

Table 9.3 Percentage of positions held by Sasaks in selected hotels, 2000 and 2005

Position	Percentage of Sasaks in 15 hotels, 2000	Total positions in 15 hotels, 2000	Percentage of Sasaks in 8 hotels, 2005	Total positions in 8 hotels, 2005
Management	32.9	97	32.4	37
Accounting/ administration	31.0	219	20.0	127
Food/beverage	44.0	432	60.0	265
Housekeeping	37.3	324	52.6	154
Security	58.1	117	68.0	71
Gardening	82.2	73	83.0	52
Other	46.5	198	26.0	15
Total	43.2	1460	46.3	721

Source: Author's survey of hotel managers, 2000 and 2005.

However, the continuing volatility of tourism meant that by 2005 the Javanese and other islanders had declined to 9.6 per cent and 3.6 per cent of the total number of staff, and Lombok and Balinese/Balok staff climbed back to 86.3 per cent of the total staff in the eight hotels.

Sasaks are employed in a full range of positions, but are proportionately under-represented in the higher skilled positions of management and administration. To encourage a more equitable distribution of positions for the Sasak people within the context of sustainable tourism, training and education programmes would need to be designed to address the imbalance.

In 1999, the proportion of male to female employees in the hotels surveyed by the author was approximately 5:1. This did not change significantly following the January 2000 riots. In June 2000, women represented 17.6 per cent of total staff, in comparison with 17.2 per cent in 1999. However, with the steep reduction in casual and contract employees, women represented 15 per cent of the total number of staff in the eight hotels by 2005. The low ratio of women to men as hotel workers was also evident in Bali in its earlier stages of tourism development, as women represented 17 per cent of all hotel workers in 1974, whereas this had risen to 27 per cent by 1991 (Cukier 1996a: 64; UNDP 1992). This pattern may not be replicated so quickly on NTB. With low education levels of only 5.2 years of schooling for women in NTB, compared with 6.6 for men (BPS *et al.* 2004: 99), women have fewer opportunities for gaining positions within the formal tourism sector and since 2000 the opportunities for even casual work have been severely limited.

Managers indicated that they would like to employ more women, and they do not actively exclude women. However, cultural and religious beliefs in this predominantly Islamic community prevent many people, especially women, from working in the hotel industry. In 2005, 55 per cent of staff in the selected eight Senggigi hotels were Muslim, 41 per cent were Hindu (from Bali) and the rest were Christian. Most room service personnel on Lombok are men, unlike, say, in the Muslim state of Brunei, where the room service personnel are mainly women, but are migrants from the Philippines. Hiring beauty therapists and ethical massage therapists has proved difficult, despite offers of attractive salaries and even additional overseas training in France. Perceived links with prostitution have proven to be strong self-deterrents to hotel employment for local women. Muslim women who always wear a *jilbab*, the traditional Islamic headscarf, and those who wish to have no association with alcohol or pork products, are also self-restrained from applying for positions within hotels. The Islamic religion of the local people thus proves to be an impediment in moving towards gender equity in local employment. Moreover, those Sasak men who have a higher status, designated by *Lalu*, are reluctant to clean, prepare food or serve others. Thus, the government requirement for at least 50 per cent of hotel staff to be local is made difficult by cultural and religious issues, and lack of English language skills.

Despite these issues, competition for most hotel jobs is very strong, and voluntary turnover of hotel staff is minimal. Total salaries are comparatively high locally and permanent hotel employees in larger hotels receive benefits over and above their base 1999 salary of Rp135,000 (US$15) each month, including health insurance, uniforms, pension plan, one meal per day, sometimes travel support or accommodation on site, plus a share of the service charge. In 1999, a permanent employee in one of the larger hotels could often average Rp1 million (US$110) per month, including a share of the service charge.

Most jobs are advertised internally and by word of mouth, which is an advantage for the local community. Managers report that the main problems with recruitment are limited English language skills and limited education. Although experience at other hotels was not essential for employment, most of those employed at senior levels did have experience at other hotels. Staff were mainly aged between 25 and 40 years, about half had previous experience in other tourism establishments, and most had held only one position in their current hotel. Of the staff who supplied additional data, just 21 per cent had held other positions in the same hotel, that is, had been promoted. Those who aspired to promotion (69 per cent) did not perceive that there were significant barriers to promotion, given additional training opportunities, experience and the availability of vacancies. Opportunities for advancement are nonetheless limited and have been reduced further since the events of January 2000, given the downturn in tourist numbers. However, many staff did not aspire to work for the larger hotels, or had worked there and now preferred to work for smaller, 'more friendly' guesthouses, where there was less pressure. They were prepared to sacrifice pay and benefits for the more relaxed and informal atmosphere of a smaller establishment.

One hotel manager stated that a 'good' personality and English skills plus aptitude were more important than tourism training at an academy or another tourism establishment. Most hotels had their own training systems, with the international hotels providing extensive in-house training. Hotel training may include English language, computer skills, selling, complaint handling, train-the-trainer, fire safety, leadership, cost control and gardening skills, as well as specific room service and food and beverage skills. International hotels also supported sister hotel training within Indonesia and overseas for employees with management potential. Despite this, close and extended supervision is required for training, as staff often seem to lack discipline and initiative. Productivity is lower than in developed countries, and this problem is experienced throughout Indonesia.

Despite general management criticism of the quality of training provided by outside tourism training schools, most hotels participated in work experience programmes conducted by the tourism training schools and recruited the best of these students. Placement for work experience in the top international hotels is very competitive, as it gives students an edge for employment with the best monetary rewards and conditions. Whilst a student is on a work experience programme, which ranges from one to three months with one employer, that employer provides one meal per day, but does not pay the student any wages. Thus, the training schools are a source of cheap, trained labour, and a recruitment pool. Since the January 2000 riots, most hotels on Lombok stopped taking work experience students. With the general downturn in tourism across Indonesia, the opportunity to place students in work experience is more difficult.

When tourism again expands in Lombok, those who have experience in hotels, some tourism training or university education with good English language skills will be poised to compete effectively for available vacancies. Graduates from the Javanese and Balinese tourism training schools with international links may prove to be a threat. Thus, it is important that the training schools continue to operate and continue to train local people during the tourism downturn, so that these graduates can compete effectively with skilled personnel from other islands and thus involve local communities as beneficiaries of the tourism industry.

The informal sector

As tourism develops, the informal sector usually increases, diversifies and flourishes side by side with formal tourism enterprises (Cukier 1996a,b, 2002; Dahles 2000), but may be absorbed into the formal sector, or restricted by licensing regulations (Dahles 2000). The informal sector largely consists of hawking cheap goods on the beach, or on the street, or hustling for guide and transport business. These sellers are now officially known as vendors, 'the internationally preferred name for people who sell souvenirs, arts and crafts . . . as the image of "hawker" tends to be a negative one of aggressive selling of low quality goods' (Anon. 1999: 22–3). Vendors rely heavily on their entrepreneurial and networking skills to make contact with tourists. They therefore need to be creative, friendly and confident to hold the interest of a tourist until a sale, or series of sales, is made. Although income is erratic, it is seen as less arduous work than traditional agricultural labour. Most beach and street vendors on Lombok are local people, living adjacent to the main designated tourist areas, unlike on Bali, where many are migrants from Java (Cukier-Snow and Wall 1993; Timothy and Wall 1997). The Lombok vendors usually work together in groups and cooperate to defend their territory against outsiders from other areas. Beach vendors, by necessity, have learnt rudimentary English language skills, as well as Japanese, Dutch or German. Given their limited formal education and limited English skills, these vendors feel it is unlikely that they can compete for higher level tourism jobs. Some actually prefer the relaxed beach lifestyle and, on a good day, may sell between Rp200,000 and Rp300,000 (*c*. US$30) worth of goods. If the vendor is also the producer with low overheads and low costs of raw materials, most of the profit is retained. If they are selling on commission, the vendor retains 10 per cent of the sale price as commission.

Some groups of beach vendors are well organised. On the main beach frontage near the Senggigi Beach hotel, 150 vendors from one village are organised in a licensed cooperative association, ironically instigated after discussions with a foreigner. The street vendors are seen as a potential problem that hotel managers deal with in a variety of ways, from consultation with local village leaders, and advice to customers not to deal with them, to threats. At Kute, a local craftsman was nominated as the key organiser for a twice-a-week night market within the hotel grounds to reduce the number of itinerant vendors approaching customers near the hotel. In this way, hotels attempt inclusionary strategies to increase their profits, reduce economic resentment and limit personal security risks for tourists and tourism establishments. Large hotels, such as the Sheraton, may have good intentions of linking up with local businesses, but cannot associate with any guide or tour operator who is unlicensed, or who lacks the appropriate registration and insurance, in order to comply with international codes of conduct. Thus, company codes of conduct and strong adherence to company standards constrain deeper working relationships with some local groups. However, many hotel operators do attempt multi-pronged strategies of employing from the local community and developing positive relationships with the informal sector, whose members live side by side with those employed, while also contributing to the social welfare of the local community with health programmes and educational and nutrition support.

Links with the local community

In the context of sustainability, linkages between formal and informal tourism enterprises and the local community are advantageous for mutual benefit and support (Telfer and Wall 1996), as the following examples demonstrate. Tourist hotel operators first attempt to increase the numbers of tourists, their length of stay and their overall expenditure to maximise profits. By concurrently developing economic linkages with the host community, they both protect their profits and create a safe, secure environment that will continue to attract tourists. These economic linkages include employment of those who live adjacent to the hotel, supporting local handicrafts and setting up markets for beach and street vendors – all measures that retain tourist expenditure in the local community. Greater use of local food produce and materials means less reliance on imports, and stimulates local income growth.

Integration with the local community is dependent on the goodwill and commitment of senior staff, such as the general manager and the executive chefs. One of the biggest challenges, as Telfer and Wall (1996: 650) suggest, is to establish 'institutional commitments which transcend the interests and involvement of specific individuals' to create continuing backward linkages between tourism developers and operators with the local community for mutual benefit. Both the Sheraton Senggigi Resort (Telfer and Wall 1996) and the Holiday Inn Resort Lombok (now the Holiday Resort) achieved strong positive reputations in the community, through their commitment to employing local people and buying local products. The Holiday Resort provides an excellent example of how very close links and loyalty of the local community can provide protection in times of security threats.

The Holiday Resort opened in 1995. It is managed by Bass Hotels, a UK-based firm, which manages over 2700 hotels worldwide, and includes brand names such as Crowne Plaza, Intercontinental, and Holiday Inn. The Holiday Resort is located north of the main tourism area of Senggigi between Mangsit village and the beach. The land was not used for agriculture or housing when it was purchased. Access to the beach and their fishing boats has been maintained for the villagers, unlike in other areas in central Senggigi, and the main access road even cuts through the Holiday Resort property. The first general manager worked pro-actively with the local community from the very beginning of the plan approval to develop excellent local community relationships. It was difficult to employ local people, especially for food and beverage positions. By June 2000, the local Mangsit Sasak villagers represented just 3 per cent of the Holiday Inn staff, but Sasaks overall made up half the staff.

Extensive in-house training in English language skills and other standard hotel skills training are conducted. Staff meetings are conducted in English, in Indo-nesian and in the Sasak language. The hotel takes into account its employees' religious beliefs for rostering staff and allocating holiday periods. In addition to a pro-active policy to hire locals, the Holiday Resort also supports the local vil-lage by providing food, clothing, and medical assistance from its resident medical staff. Improved living conditions are evident throughout the village. The goodwill generated by Holiday Inn management secured loyalty of staff and protection of premises during the January 2000 riots. The Mangsit community erected large signs in Arabic – *Allah-u Akbar* ('God is Great') – and remained on close watch throughout the crisis. They also voluntarily guarded the house of the food and beverage manager after he was evacuated.

Tourists were slow to return to Lombok. Room occupancy rates for Lombok hotels plummeted from an average of 42 per cent to less than an average of 10 per cent in February 2000. Five months later, occupancy rates had risen to almost 40 per cent (NTB Provincial Tourism Office 2000), but consistent occupancy rates of 60 per cent are required to achieve reasonable profits. Hotel managers used the post-riot period to send employees on annual leave, organise room renovations, clean up hotel grounds and train retained staff. Permanent staff who decided to leave were not replaced and casual and contract employment almost ceased.

With an estimated 30,000 of Lombok's population directly or indirectly depend-ent on tourism (Segran 2000a,b), the economic impact of the riot was widespread. Without tourists, the guides, drivers, beach vendors and others working on the margins of formal tourism employment needed to look for other employment opportunities, even though these opportunities were scarce and possibly more arduous than tourism-related activities. As a response to the crisis, hotel management and other tourism businesses on Lombok made attempts to strengthen community links after the riot. The Starwood Hotel group, of which Senggigi Sheraton is a part, collected Rp42 million (US$4700) from its employees worldwide. This money was used to help staff rebuild houses and to supply basic food items to local communities. Suppliers to the hotels outside Lombok also provided some free services for up to three months after the riot (Segran 2000a). The big hotels and smaller tourism operators formed the Senggigi Business Association (SBA) to combine efforts and resources for the benefit of Lombok. However, after the initial enthusiasm and clean-up post-riot, much more work still needs to be done to ensure better working relationships between all tourism stakeholders, including the tourism service providers, government organisations and local community groups.

Conclusion

Despite constraints, the opportunities for training, promotion and secure employment for local Sasak people are better than in many other island tourism contexts (Kontogeorgopoulos 1998; Tirtosudarmo 1996; Weaver 1998). Employment prospects for local Sasak people with tertiary education, good English skills and a friendly personality, but no tourism school training, have been strong in the tourist hotel sector on Lombok. Yet few actually have tertiary education and English language skills. Employees from both small and large establishments in the hotel sector reported a strong degree of job satisfaction and were positive about their long-term prospects within the tourism industry on Lombok. However, the dominant Islamic culture places considerable constraints on the employment of women, hence employment levels for women on Lombok are lower than in neighbouring Bali, and also remain lower in 2005 than those reported in 1999 and 2000, on account of the lack of casual and contract positions.

Employment on the fringes of the formal tourism area may appear to be precarious, but many choose to be beach and street vendors as an alternative to labour in traditional agriculture, or to the constraints of formal tourism employment. When tourism flourishes, the informal sector also flourishes. When there is a downturn, or particular sections of the community are excluded from tourism, potential resentment may lead to a crisis.

Some hotel managers and owners and, more recently, the Indonesian government have recognised that efforts must be made at the grassroots level to develop and support community-based initiatives and small vendors, as well as to develop relationships with local communities to ensure mutually sustainable tourism activities. Some actions have reduced the gap between government rhetoric and help for the poor, by local sourcing of employment, food and handicraft vendors. Such practices provide reciprocal social and political support for the hotels. Employment opportunities and related training of local island people are basic to sustainable tourism development. When local people receive sufficient economic benefits from tourism both directly and indirectly, they are more likely to remain positive towards tourism operators and towards tourists.

The recent recognition of Lombok as a desirable destination has raised hopes that tourism will improve. Apart from the International Destination Stewardship Award, readers of *Condé Nast Traveler* magazine voted Lombok as the seventh best island destination in Asia and the Indian Ocean (Bali was the second best, after Phuket) in 2005. The construction of an international airport in Central Lombok has commenced after a ten-year delay. Local people have not opposed the airport itself, recognising that it may bring economic opportunities, but they have demonstrated against poor levels of compensation for their land. That opposition demonstrates both the sensitivity of all tourist ventures that require land alienation, and the need to work closely with local people. With the optimism that the tourists will return in larger numbers, tourism stakeholders have a renewed opportunity to demonstrate their commitment to sustainable tourism development goals by developing stronger links with the local community in the form of fair compensation, employment and local community management strategies.

References

Anon. (1999) 'Archaeological Wonders, the Past, Present and Future', *Travel Indonesia*, 21: 17–23.
—— (2005) Bali Update 505, 15 May. Online. Available <www.balidiscovery.com/update/default.asp> (accessed 18 June 2007).
—— (2006) Bali Update 491, 6 February. Online. Available < www.balidiscovery.com/update/default.asp> (accessed 18 June 2007).
Baum, T. (ed.) (1993) *Human Resource Issues in International Tourism*, Oxford: Butterworth-Heinemann.
BPS Statistics Indonesia, BAPPENAS and UNDP (2004) 'The Economics of Democracy: Financing Human Development in Indonesia', National Human Development Report. Online. Available <http://hdr.undp.org/docs/reports/national/INS_Indonesia/Indonesia_2004_en.pdf> (accessed 25 November 2005).
Cukier, J. (1996a) 'Tourism Employment in Bali: Trends and Implications', in R. Butler and T. Hinch (eds), *Tourism and Indigenous Peoples*, London: International Thomson Business Press.
—— (1996b) *Tourism Employment in Developing Countries: Analyzing an Alternative to Traditional Employment in Bali, Indonesia*, unpublished PhD thesis, University of Waterloo.
—— (2002) 'Tourism Employment Issues in Developing Countries: Examples from Indonesia', in R. Sharpley and D. Telfer (eds), *Tourism and Development*, Clevedon: Channel View.
Cukier-Snow, J. and Wall, G. (1993) 'Tourism Employment: Perspectives from Bali', *Tourism Management*, 14: 195–201.
Dahles, H. (2000) 'Tourism, Small Enterprises and Community Development', in G. Richards and D. Hall (eds), *Tourism and Sustainable Community Development*, London: Routledge.
Dahles, H. and Bras, K. (1999a) 'Entrepreneurs in Romance: Tourism in Indonesia', *Annals of Tourism Research*, 26: 267–93.

—— (eds) (1999b) *Tourism and Small Entrepreneurs: Development, National Policy and Entrepreneurial Culture: Indonesian Cases*, New York: Cognizant Communication Corporation.

Din, K. (1997) 'Indigenization of Tourism Development: Some Constraints and Possibilities', in M. Oppermann (ed.), *Pacific Rim Tourism*, London: CAB International.

Kontogeorgopoulos, N. (1998) 'Accommodation Employment Patterns and Opportunities', *Annals of Tourism Research*, 25: 314–39.

NTB Department of Culture and Tourism (2004) *NTB Tourism in Figures 2003* (*Pariwisata Nusa Tenggara Barat Dalam Angka Tahun 2003*), Mataram: NTB Department of Culture and Tourism.

NTB Provincial Tourism Office (2000) *Ten Years of Tourism Development* (*Sepuluh tahun pembangunan pariwisata Nusa Tenggara Barat*), Mataram: NTB Provincial Tourism Office.

Pearce, P., Moscardo, G. and Ross, G. (1996) *Tourism Community Relationships*, Oxford: Pergamon.

Richter, L. (1995) 'Gender and Race: Neglected Variables in Tourism Research', in R. Butler and D. Pearce (eds), *Change in Tourism: People, Places and Processes*, London: Routledge.

Segran, G. (2000a) 'After the Riots, Lombok Waits for the Tourists', *Jakarta Post*, 2 April. Online. Available <www.thejakartapost.com> (accessed April 2000).

—— (2000b) 'Lombok Waits for Tourists to Return', *Jakarta Post*, 15 April. Online. Available < www.thejakartapost.com> (accessed April 2000).

Telfer, D. and Wall, G. (1996) 'Linkages between Tourism and Food Production', *Annals of Tourism Research*, 23: 635–53.

Thadani, M. (2005) 'Tsunami Wave – An Economic Tourist Disaster', *Hospitality Net News*, 7 January. Online. Available <www.hospitalitynet.org/news/4021783> (accessed 29 November 2005).

Timothy, D. (1999) 'Participatory Planning: A View of Tourism in Indonesia', *Annals of Tourism Research*, 25: 371–91.

Timothy, D. and Wall, G. (1997) 'Selling to Tourists: Indonesian Street Vendors', *Annals of Tourism Research*, 24: 322–40.

Tirtosudarmo, R. (1996) 'Human Resources Development in Eastern Indonesia', in C. Barlow and J. Hardjono (eds), *Indonesia Assessment 1995: Development in Eastern Indonesia*, Singapore: Institute of Southeast Asian Studies.

United Nations Development Programme (1992) *Comprehensive Tourism Development Plan for Bali*, Vols 1 and 2, Annexes, Report by Hassall and Associates for the Government of Indonesia, Jakarta: UNDP.

Weaver, D. (1998) *Ecotourism in the Less Developed World*, New York: CAB International.

World Travel and Tourism Council (2005) 'Executive Summary: Travel and Tourism Sowing the Seeds for Growth', in *The 2005 Travel and Tourism Economic Research*, London: WTTC.

10 Priorities, people and preservation

Nature-based tourism at Cuc Phuong National Park, Vietnam

Barbara Rugendyke and Nguyen Thi Son

Priority given to environmental conservation has frequently been at cost to local communities, displaced from national parks to reduce the impacts of shifting subsistence agriculture (for example Buergin 2003; Schmidt-Soltau 2003). In turn, nature-based tourism has been encouraged in national parks. The hope has been that the economic benefits tourism is expected to bring will provide alternative livelihoods for relocated people, while being less environmentally destructive than traditional agricultural systems; this has been the case at Cuc Phuong National Park in northern Vietnam. This chapter explores the impacts of the nascent tourism industry at Cuc Phuong on the residents of nine villages, located within the park, or in resettlement villages in the buffer zone surrounding it, or in longer-established villages close to the park boundaries. Villagers' perspectives, alongside those of domestic and international visitors, provide understandings of the extent to which tourism has contributed to improved livelihoods, and of its wider social, economic and environmental consequences.

Prioritising conservation

Forests in Vietnam provide various timber and non-wood products to meet both domestic and export demands. As in most places, their importance in preserving biodiversity and various ecosystems has been recognised increasingly. Species conservation is also a serious issue in Vietnam: in most areas where hunting has occurred, the population of forest wildlife has been seriously reduced. Similarly, the number of endemic plants is steadily being reduced and many species are becoming extinct. In 1943, forest covered 43 per cent of total land area in Vietnam, but by 1993 natural and planted forests covered 9.3 million hectares and occupied only 28 per cent of the land (Le Trung Tien and Nguyen Cao Doanh 1995: 25). Remote sensing data revealed that, by 2000, only 2 million hectares of natural primary forests remained, and these were being reduced at a rate of 100,000 to 200,000 hectares each year (Smith 2000: 4). In response to the seriousness of this issue, reafforestation has been increasing each year so that, by 2005, total forest area was over 12 million hectares, occupying over 35 per cent of land area (GSO 2005).

However, pressure on forest resources continues – hardly surprising in a nation which had a population density of 245 people per square kilometre in 2005 (MSN Encarta Encyclopedia 2005: 1) and where nearly 57 per cent of the labour force is engaged in agriculture (CIA 2006: 9), and therefore directly dependent on the land or forests for survival. Reduction of forested land has resulted from their being continually subjected to logging, hunting, clearance for agriculture and settlement, and collection of medicinal and other useful plants. Vietnam's minority peoples, who mainly live in mountainous zones close to forests, still use forests to meet their basic needs. The low levels of remaining forest cover, and the rapidity of forest destruction in Vietnam, make forest conservation urgent, while the discovery of two previously unrecorded animals in 1992 and 1994 in Vietnamese forests has added impetus to the need to preserve the few remaining natural forest areas (WWF 1996).

Concerned to preserve primary forests, the Vietnamese Ministry of Forestry classified some areas as 'Special Use Forests', including national parks, nature reserves, historic/cultural sites of significance and environmental sites (Le Trung Tien and Nguyen Cao Doanh 1995: 29). These represent nearly 8 per cent of remaining natural forests, set aside for nature conservation and research purposes, to preserve historic and cultural relics and beauty spots, and for recreation and tourism. One of these primary forest areas, Cuc Phuong National Park, now one of the Vietnamese parks most popular with tourists, reflects the impacts of the new nature-based tourism development on local residents and on the park itself.

Cuc Phuong National Park

The first Vietnamese national park, declared a protected area in 1962 and granted national park status in 1966 (Vo Quy *et al.* 1996: 9), Cuc Phuong is an area of largely undisturbed forest about 120 km southwest of Hanoi. The park straddles Hoa Binh province to the north, Thanh Hoa to the south and Ninh Binh province to the east. In the foothills of the Annamite Mountains, it incorporates two limestone ridges, separated by a valley that is wide at the mouth but becomes a very narrow canyon as it proceeds eastwards. The altitude at the entrance to the park is about 100 m and the tallest peak is 650 m above sea level (Vo Quy *et al.* 1996: 11). In 1984 Pfeiffer claimed 'this very narrow canyon . . . prevented the mountain people from entering much of the forest, to destroy it in their slash and burn culture. That is why Cuc Phuong still has its unique and largely virgin flora and vegetation' (Pfeiffer 1984: 218). Fortunately, the park escaped the defoliation that decimated so much forest land in Vietnam during prolonged war (Constable 1982).

The only remaining primary forest in the northern delta in Vietnam (Le Trung Tien and Nguyen Cao Doanh 1995: 28), Cuc Phuong is known for its natural beauty and important biodiversity. It offers a range of attractions to the visitor. Covering 22,200 hectares, it contains diverse tropical flora and fauna, including some species now unique to the park. Brochures distributed to visitors to Cuc Phuong claim that, of the total number of plant families observed in Vietnam, 60 per cent are represented in the park, amongst them seven species unique to the park, including the yellow camellia. Over 250 species of vertebrate animals have been identified in the park (Le Trung Tien and Nguyen Cao Doanh 1995: 28) and, according to a park official, over 8000 species of insects.

Tourist attractions include the large and very scenic lake, Yen Quang, located in the buffer zone to the southeast of the park; a spectacular waterfall known as Giao Thuy; large limestone caves filled with ancient archaeological remains, including one popular with visitors known as the 'Cave of Early Man'; many ancient trees including the '1000-Year-Old Tree', a 45-metre high *Terminalia myriocarpa* specimen; and caves that are important roosting sites for bats. Within Cuc Phuong, visitors are able to visit a museum, botanical gardens and a rescue centre for endangered primates; the last houses a breeding programme for the endangered Delacour's langur, a primate found only in this park (Cao Van Sung n.d.). Publicity materials offer visitors the choice to trek to the village of Ban Khanh, home to local indigenous Muong, located approximately 15 kilometres from the Park Centre, and to stay overnight with the villagers. This adventure promises 'exposure to the culture and life of the Muong . . . a new and interesting experience for visitors' (Vo Quy *et al.* 1996: 38).

Organised package tours for overseas visitors to Vietnam frequently include a stopover at Cuc Phuong, which is conveniently located near the highway, en route to several major tourist attractions, including Bich Dong Pagoda, Hoa Lu and Sam Son Beach. According to the tourist office at the park, increasing numbers of visitors have been attracted to Cuc Phuong: 46,894 domestic and 2456 foreign visitors in 2000; 56,236 Vietnamese and 4227 foreign visitors in 2003; to nearly 70,000 Vietnamese and 6900 foreign visitors in 2006.

With an average annual growth rate of tourism in Vietnam of 9.2 per cent between 1995 and 2004 (WTO 2005), park managers hope to capitalise on Vietnam's emergence as a major tourist destination by attracting tourists to the park, and that the revenue earned will support preservation of the park. One park official also expressed the hope to 'create conditions for people in the buffer zone to be involved in tourism and get more income. When knowing that the park can provide good returns to their family, farmers will be much more active in its protection.' Thus, conservation could be enabled by tourism, as well as offering a possible source of income for local people.

The visitors

Domestic visitors

Of 100 domestic visitors interviewed at Cuc Phuong, the majority were from Hanoi (34 per cent) or from Hai Phong (15 per cent), so nearly half of Vietnamese visitors were from large urban centres in Northern Vietnam. A further 16 per cent of visitors were from the nearby town of Ninh Binh, or from other provinces close to the park. Thus, the park is primarily a regional attraction, ideally located to attract urban day trippers or visitors from the local region. Most (72 per cent) were there for a recreational day visit or on holidays at the park, 15 per cent combined a recreational trip to the park with business in the area, and the remainder were on a school excursion. For the majority (89 per cent), the primary reason to visit the park was simply to observe it; bushwalking and viewing plants and wildlife were also significant reasons for park visits. To see historical sites was a secondary reason for some, and 'escaping crowds' and 'seeing local culture' were mentioned by a few visitors.

The majority (53 per cent) visited Cuc Phuong for one day only; others stayed for no more than three days. For over half of them (53 per cent) this was their first visit to the area; a little over a quarter (26 per cent) were on the second visit to the area and some (13 per cent) were visiting the park for a third time. So, with sufficient appeal to encourage repeat visits, and the majority of domestic visitors drawn from the nearby region, Cuc Phuong is clearly used as a recreational facility by local people. This is reinforced by the fact that a further 35 per cent visited on the recommendation of friends or relatives, 9 per cent had learnt about Cuc Phuong from tourist brochures or advertisements, and a few others had learnt of it through TV, newspapers or local people. Almost all domestic tourists were either independent travellers or on school excursions. For most, the visit to Cuc Phuong was combined with other activities in the area, with 58 per cent of visitors stating that it was one of several destinations of equal importance visited, though for 40 per cent it was the major destination for their visit to the Ninh Binh area.

Of domestic visitors who spent more than one day at Cuc Phuong, 54 per cent stayed at the guesthouses in the park, a further 21 per cent stayed in local homes (probably with friends or relatives), while 9 per cent each stayed in hotels and campsites. A further 7 per cent of visitors trekked to stay at the homes of residents of Ban Khanh, located within the park. Remarkably, these few home-stay visits were the only mention of any contact between villagers and visitors suggestive of an income flow to villagers from tourists.

Most visitors enjoyed the pleasant natural environment and 'atmosphere' more than any other aspect of the park. Plants, animals and scenery featured strongly as enjoyable aspects of visits to Cuc Phuong, and a number of visitors nominated bushwalking and fresh air as important. However, advertised park attractions, such as the Big Tree, the Cave of Early Man and local people and their customs were not often identified as 'especially enjoyed' by visitors. For domestic visitors then, Cuc Phuong primarily represents a pleasant site for outdoor recreation, where wildlife, vegetation and scenery are general attractions.

When asked about what aspects of their visit were the least satisfactory, many visitors (17 per cent of total responses) commented that tourist services to the park, including guides, brochures, information and educational materials, needed improvement, while an equivalent percentage felt that food available at the park should be improved. The standard of walking trails was also of concern for 16 per cent of the visitors. The largest proportion of responses related to the need to protect Cuc Phuong better from vandalism and litter (19 per cent of responses); thus, the strongest concern of domestic visitors related to the need for better preservation of the park environment.

When asked if they would like to comment further about their experience of Cuc Phuong, of the 80 comments made by domestic visitors, 38 per cent related to poor provision of services for park visitors, including available park information, tours, tour guides and the activities they offered, accommodation, toilets and food. A further 20 per cent of responses suggested park management could be improved, better facilities provided, park access restricted and park 'rules' improved. Other comments related to the experiences of domestic visitors, including that roads were too narrow for buses, and that marked bushwalks were too long, too difficult and with poorly marked tracks. Of other specific comments, 15 per cent reiterated the need to better protect the park from crowding, litter, vandalism and environmental damage – again, the visitors themselves expressed concern that their activities were harming the very environment which had attracted them (Figure 10.1).

Figure 10.1 Tourists' impact on the environment at Cuc Phuong: soil compaction and litter at the 'Big Tree'.

International tourists

Of 28 international tourists interviewed, 50 per cent were visiting Vietnam for 14 days. It is therefore not surprising that visits to Cuc Phuong by overseas tourists were of limited duration, with 15 of 28 visiting for not more than one day, and a further 13 for not more than three days. Visitors were from Germany (17 per cent), Australia and France (10 per cent each), with one or two from each of 11 other 'developed' nations; a couple from each of Korea and Japan were the only visitors from elsewhere in the Asian region.

The most common reason given for the visit to Cuc Phuong, by 18 of 28 visitors, was to 'look at the national park', while bushwalking, viewing local plants and animals and getting away from crowds were all important to international visitors. Bushwalking and hiking were the activities most commonly mentioned (46 per cent), with taking photos and sightseeing, in conjunction with bushwalking, mentioned by many visitors. Far fewer international tourists visited the 'Big Tree' or 'Cave of Early Man' in comparison with domestic tourists, perhaps because visiting them requires a lengthy hike, and time available to foreign visitors is limited.

For 63 per cent of the foreign visitors, their visit was part of a holiday; others combined recreation with either business or education (17 per cent in each category). Ten of the foreign visitors learnt of the park from friends or relatives, seven from a travel guide book, and an equivalent number through their university. Only a few learnt of the park from travel brochures or advertisements, and only three were there as part of a package tour. Most were independent travellers, partly because marketing of the attractions of the park within Vietnam or incorporation of visits to Cuc Phuong into package tours is in its infancy. Only two had visited the park previously, both within the preceding year, and for three-quarters of the foreign visitors the visit to Cuc Phuong was to one of several destinations of equal importance in the area.

International tourists were asked to provide basic demographic data both about themselves and their travelling companions (about 78 persons in all). This revealed that, of 28 female visitors, 20 were students, 4 were teachers and 27 were aged between 16 and 39 years. Although some males in the older age group were amongst the visitors – with 18 of 50 male visitors over 40 – the majority were still in the younger age groups, with 32 visitors aged between 16 and 39. Of the male visitors, 20 were students and 20 were engaged in professional occupations. In all, this indicates that Cuc Phuong is attractive primarily to younger, fit and educated people – many likely to be student backpackers.

Only half of the 28 foreign visitors sought accommodation at or near Cuc Phuong, with eight staying in guesthouses within the park, three in hotels in nearby Ninh Binh township, two in a bungalow within the Park, and only one reporting having trekked to stay at Ban Khanh. So, only one village household – at Ban Khanh – earned income directly by accommodating an overseas visitor.

Of the foreign visitors, three-quarters brought their own food to the park, and 26 of the 28 people interviewed brought their own water to the park. The vast majority brought waste material to Cuc Phuong with them, yet every foreign tourist complained that they saw litter within the national park and 63 per cent of visitors complained of graffiti and vandalism. Trampled vegetation and toilet wastes were of concern to a quarter of respondents, and others observed soil compaction, visitors and local people collecting forest products, carving on tree trunks and what was described generally as 'pollution' – all were observed close to tracks, roads and the major tourist attractions.

For international visitors, the most important aspect of their visit was the pleasant atmosphere, 'nature', plants, animals and landforms. The numbers that were very satisfied or fairly satisfied with their visit were equal to those that were either neutral or dissatisfied. In the latter group, a quarter of respondents felt there was little to see and that they had not encountered any animals (perhaps suggesting potential for wildlife tours conducted by local guides), 17 per cent of responses here expressed concern about litter, waste and vandalism, and an equivalent number felt their visit had been too rushed for them to have sufficient time to appreciate the park. A smaller number, 10 per cent of total responses, found the park too noisy and crowded. For those who were satisfied with their visit, enjoyment of the natural environment was paramount.

When asked what they would like to see improved at the park, half the responses related to the need to improve tourist services, including guides, brochures and information, and that more walking trails be provided. One tourist suggested the friendly local Muong should be trained and used as tour guides. However, the overwhelming majority (26 of 28 foreign visitors) listed protection from crowding, litter and vandalism as their chief concern. So, foreign visitors to Cuc Phuong were already interested in and educated about conservation issues, expressed their concern more strongly than domestic visitors, and emphasised that greater priority should be given to preservation of the national park.

Villagers

Residents of villages within and near Cuc Phuong mostly belong to the Muong ethnic minority group. The second most populous ethnic minority group in northern Vietnam, the Muong compose almost 1.5 per cent of the total population of Vietnam, representing close to 1.1 million people at the time of the latest census (GSO 1999). The largest populations of Muong today live in Hoa Binh province and in the mountainous districts of north central Vietnam in Thanh Hoa (Rugendyke and Nguyen Thi Son 2007).

Historically, the survival of the Muong was based solely on subsistence agriculture, hunting and gathering. Produce from slash-and-burn agriculture in forest clearings was supplemented by hunting and by some irrigated rice farming in valleys. In the forest gardens, cotton, maize, some pumpkins and gourds and cotton were grown (Nguyen Tu Chi 1972: 80). Buffalo are raised as a meat source and as draught animals, and many Muong breed small livestock for consumption, including poultry and pigs. Most Muong have been sedentarised; however, those with access to forests supplement their diet with such products as bamboo shoots, mushrooms, wild tubers, vegetables and meat from a variety of forest animals, which were traditionally part of the Muong diet. As leisure time activities, women make baskets and weave cloth, known for its traditional geometric woven designs. However, production of handicrafts as an income-generating activity is limited – these activities are primarily to meet family needs. Wooden houses on stilts, grouped in hamlets called *quel* and in turn grouped into larger clusters called *muong*, historically provided homes for the Muong (Pfeiffer 1984; VWAM 2006).

In 1984, there were some 1000 Muong still living within Cuc Phuong. Pfeiffer described driving up the centre of the valley running through the park, passing many thatched houses on stilts and meeting 'a young Muong hunter with his cross-bow and quiver of bamboo arrows . . . hunting birds, which is illegal in Cuc Phuong' (Pfeiffer 1984: 219). There was then concern that the slash-and-burn subsistence agriculture would cause serious degradation of the park environment (Pfeiffer 1984). Although then, as now, it was illegal to hunt within the park boundaries, hunting of birds and animals continued, and chickens, pigs and cattle were still to be found in parts of the park. The extension of medical care had led to rapidly increasing birth rates among the Muong, and it was feared that increased population would place further strain on park resources.

The need to produce sufficient food for the growing population resulted in intensified swidden farming and associated forest destruction (Nguyen Ba Thu 1995: 199) so these early concerns were well-founded. Cuc Phuong Commune in 1995 had a population of 1500 and a 2.3 per cent growth rate, an increase from 500 people in 1962. According to national park authorities, about 2000 Muong have been resettled out of the park since then; given the limited size of the population within the park in 1984, this indicates both consistent population growth and the encroachment of Muong from surrounding areas into the park. Muong houses are no longer observable in the park centre, although the site of former settlements is obvious from changes in vegetation patterns. Eight hundred hectares of swidden land was freed for restoration as a result of the relocation of villagers (Nguyen Ba Thu 1995, 2000). In 1995, even after the resettlement programmes, about 2500 people were still living in the park, primarily close to its boundaries, and about 13 communes surrounding Cuc Phuong housed 50,000 settlers of Muong origin. At the turn of the century, about 2000 people were believed to still be living within the park boundaries. In conversation, national park authorities claimed that 1000 people, those described as having 'lived a traditional life in the park for many generations' could remain in the park, but said that they planned to continue to relocate more recent settlers into the buffer zones of the park.

Land available for resettlement in the buffer zone surrounding the park was observably poor, with thin, stony soil unsuitable for successful cultivation. With their skills based on use of forest resources and shifting subsistence agriculture, the Muong lacked skills for permanent settlement and sedentary agriculture. 'Voluntary' resettlement projects therefore have not been very successful, increasing the need to find alternative livelihood options for local people. Thus, the encouragement of nature-based tourism to Cuc Phuong was seen as a possible means of bringing new economic opportunities to the region.

The case study villages

To shed light on the extent to which the nascent tourism industry may be contributing to improved livelihoods for local people, data were collected from residents of eight small villages near, or within, Cuc Phuong National Park (Rugendyke and Nguyen Thi Son 2005; see Map 10.1). It was quickly evident that people living any distance from the roadside had no contact at all with tourists, so the survey was confined to villages within walking distance from the main entrance to the park, within the park, or on a less used access road at the northernmost part of the park.

Dong Quan, Dong Bot and Dong Tam are all located close to the access road that leads into Cuc Phuong. All villagers there were Muong (apart from a resident Viet (Kinh) school teacher), had been relocated from the park, and had lived in the park for at least three or four generations; most, however, had lived in their current location for over ten years.

Just inside the boundary of the national park Nga 1 and Nga 2 straddle the major access road to the park; the former about 1 kilometre east of the access road, whereas Nga 2 is about 1.5 kilometres west of the main road. Nga consists of Muong, living in the park as they have for generations, surviving from shifting subsistence agriculture and gathering resources like honey, bamboo shoots and herbs from the surrounding forest. To the observer, though, traditional aspects of Muong culture have been replaced by the majority Viet (Kinh) culture, with traditional housing no longer evident and customary attire rarely seen.

Located in the buffer zone between 1 and 2 kilometres south of the main entrance to the park, Bai Ca and Sam are not on the major access road into the park. Yen Nghiep Commune is also in the buffer zone, but is located to the north of the park. Of the 41 families supplying data from these villages, 39 reported that their family members were all born in the area and not resettled there. Sedentary rice production was the prime means of subsistence for these villagers and, as was the case in the Nga villages, villagers' attire and housing were not observably in the traditional style of the Muong.

Demographic and employment characteristics

A total of 78 persons resident in these villages were interviewed, providing demographic and employment data for 455 persons resident in their households, 223 of whom were females and the remainder males. All persons about whom data were collected were indigenous Muong, with the exception of two persons of Viet (Kinh) origin, one resident in Dong Quan and one in Nga.

A high proportion, nearly 43 per cent, of the total population were under the age of 16 (Rugendyke and Nguyen Thi Son 2007). A further 30 per cent of the population are in the 16–24 and 25–39 age groups, so the potential for further rapid population growth is high. With low education levels of all villagers – nearly half (49 per cent) had only completed primary education – there are few opportunities for local people to find any alternative means of survival in paid employment

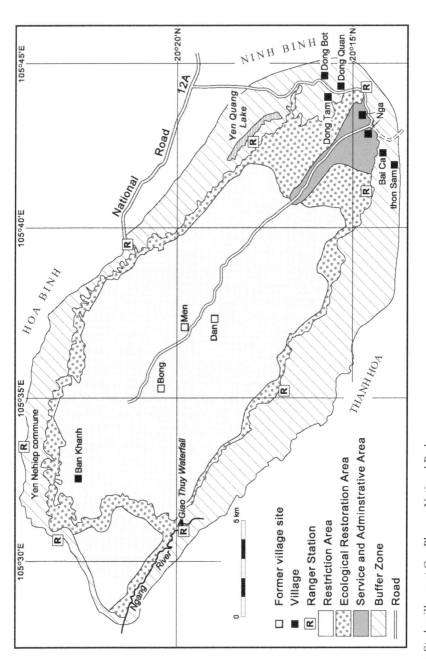

Map 10.1 Study villages at Cuc Phuong National Park.

elsewhere. If those in the younger age categories remain living in these villages, increased pressure on available resources is likely to be unavoidable.

Of the 222 male villagers, 103 (representing 46 per cent of the total males sampled or 85 per cent of the sampled male adult working population) were either totally or partially employed in agriculture (Table 10.1). Some worked in the local Commune Authority, as a nurse, social worker, brickworker or general 'worker', in addition to their agricultural activities. Very few (only 1 per cent of the sampled adult male working population) were employed entirely outside the agricultural sector, in occupations including park ranger, tour guide, engineer, nurse, motorbike driver, 'worker', or working at the Commune Authority. In Vietnam as a whole, just above 15 per cent (GSO 2006) of rural people are engaged in non-farm employment, so it is not surprising that few villagers located in marginal areas alongside or within the national park participate in off-farm wage employment. What is also evident from the survey findings is that very few villagers – probably only the ranger, tour guide and motorbike driver – directly derive income either from Cuc Phuong itself or from the emerging tourism industry associated with it.

The dominant activity of women within the villages was also in agriculture (54 per cent, or 120 women of the total 223 women in surveyed households, including three who also attended school and five aged under 16 years) (Table 10.2). Of

Table 10.1 Occupation by age of male villagers living near Cuc Phuong National Park

Occupation	Age group (years)					Total
	< 16	*16–24*	*25–39*	*40–55*	*> 55*	
Agriculture	2	17	21	33	9	82
Ag & bees		1	3	2	2	8
Ag & commune authority				1		1
Ag & husbandry			1	1	1	3
Ag & deer			2		1	3
Ag & cows			1	1		2
Ag & nurse			1			1
Ag & social work			1			1
Ag & brickworker				1		1
Ag & worker				1		1
School	72	6				78
At home	19				4	23
Ranger			1	1		2
Worker		3	1			4
Commune Authority			1	2		3
Tour guide at park			1			1
Engineer			1			1
Nurse		1				1
Motorbike driver			1			1
Total	93	28	36	43	17	222

Table 10.2 Occupation by age of female villagers living near Cuc Phuong National Park

| Occupation | Age group (years) | | | | | |
	< 16	16–24	25–39	40–55	> 55	TOTAL
Agriculture	4	20	31	30	13	98
Ag & weaving			3	2		5
Ag & bees			3	1	2	6
Ag & retail				3		3
Ag & school	3					3
Ag & husbandry		1				1
Ag & deer	1		2		1	4
School	63	5				68
At home	23	2			4	29
Retail					2	2
Teacher				1		1
Works in south		1				1
Restaurant owner				1		1
Park worker		1				1
Total	94	29	40	38	21	223

those women, 20 also engaged in other home or locally based income-generating activities, including weaving, keeping bees, raising deer, retailing (one operated a small roadside stall) and animal husbandry. This contrasted with the activities of males, for more men combined agricultural pursuits with formal employment outside the home. There were fewer full-time employment opportunities outside the agricultural sector for women than for men, with only six women engaged entirely in other occupations, including retailing, teaching (a Viet (Kinh) person originally from outside the locality), one who worked within Cuc Phuong National Park and one operating a small restaurant. Probably only the last two obtained income directly from the tourism industry.

For those villagers resident in and near Cuc Phuong National Park agriculture continues to be the chief source of subsistence for the majority, supplemented in the case of women with some home-based or farm-based income-earning activities. These activities were not, of themselves, able to provide an adequate livelihood for many local villagers.

Villagers resettled to Dong Bot, Dong Tam and Dong Quan had been provided with an initial grant (equivalent to US$2000) to establish agricultural activity and build a house, but subsistence is precarious. Villagers complained that soils were infertile and that there was no reliable source of water for agriculture. With establishment grants depleted, nearly all resettled villagers had difficulty surviving in their current location. Residents of all three villages preferred their former location in Cuc Phuong, for agricultural productivity was higher there, they could grow maize and cassava (formerly staple foods of the Muong) in the forest gardens, they could supplement their diets with forest products and they had better access to water. Thus, in the words of residents of Dong Bot: 'when living in [the] national park [we did] not worry about food', 'there is not enough food here for the whole year round' and agriculture is 'more difficult' in the present location.

Villagers who had not suffered the disruption of relocation still hoped for more income-earning opportunities, though none had memories of better living conditions elsewhere. Residents of Nga, located just inside Cuc Phuong, and able to engage in more productive traditional forms of agriculture, did not complain of food shortages, but hoped the growing tourism industry would provide opportunities for employment or income generation. Villagers located in Sam and Bai Ca, further from the road, did not regard tourism as a potential source of income generation.

The struggle of the villagers to eke out a precarious existence inevitably has implications for the surrounding biophysical environment. Villagers were understandably somewhat reluctant to comment about their use of resources from Cuc Phuong, for illegal use of park resources is very contentious, in the past resulting in violent conflict between park rangers and local people (Ray and Yanagihara 2005: 179); those caught gathering park resources are subject to hefty fines. However, residents were prepared to admit to observing others using forest resources, with nearly 50 per cent of respondents indicating there was much deforestation in and around the park, a further 12.5 per cent mentioning changes for the worse and local exploitation of the park, and nearly 50 per cent of respondents observing collection of timber from the park (Figure 10.2). The latter is hardly surprising, given that wood is needed for cooking and there is little local alternative. The researchers observed buffalo being herded from the park at sunset, and villagers leaving the park with sacks of plants, tubers, timber, bark and bamboo shoots, and

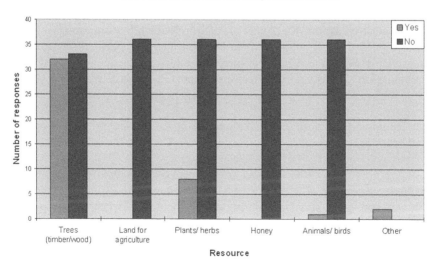

Figure 10.2 Villagers' observations of use of forest resources, Cuc Phuong National Park.

washing tubers collected in streams near the access road into the park. Although people did not admit to using the forest for agricultural purposes, they reported that plants, herbs, birds and animals were collected from the forest and many mentioned a reduction in forest animals. Given the illegality of such activities, it is highly likely there was significant under-reporting about the extent of use of forest resources. The collection of timber for firewood, herbs for medicinal purposes, bamboo for handicraft production, and tree bark (boiled to make a tea-like drink) was confirmed by park managers, and poaching of flora and fauna for sale is not unusual, the latter including monkeys, bears, pythons, gold tortoises and snakes (Cuc Phuong National Park 2000: 11; Nguyen Ba Thu 1995: 198).

Essential for the survival of villagers, park resources therefore are used to supplement agricultural production and very limited incomes. Albeit illegal, they form to some extent a 'safety net', providing access to additional food resources, a situation not unique to this region or nation (Baland and Platteau 1999). However, dependence on this 'safety net' is also environmentally destructive. With growing population, associated pressure on the biophysical environment and limited access to employment outside subsistence agriculture, the immediate future for villagers living in the vicinity of Cuc Phuong, and for their environment, seems to be bleak, unless alternative means of income generation can be found.

With local people unable to survive on agriculture alone, tourism would seem to offer a possible source of alternative livelihood. Yet the employment characteristics of the sample population demonstrate clearly that this is not the reality. Villagers were asked directly about the extent to which tourism impacted on their lives. Of the 78 villagers interviewed, 65 believed that Cuc Phuong would attract tourists. However, 62 had almost no contact with them, 22 said they had met some briefly (some nominated both answers) and only two had ever accommodated tourists at their homes. Only three people from all villages reported having gained any income directly from tourists, and very few were in employment generated by tourist activities at Cuc Phuong.

For most villagers (78 per cent of total responses), tourism brought no real change to their lives, although some believed there had been an improvement in service provision as a result of increased tourism, including improved roads and transport (76 per cent of responses), improved access to electricity (18 per cent) and better shopping opportunities (22 per cent), the latter facilitated by improved roads and transport. Whereas nearly 8 per cent thought access to jobs and employment opportunities had improved, others (6 per cent) were concerned at the increasing costs of goods and services resulting from the inflationary effects of tourism.

For most villagers, tourists were 'friendly and easy to talk to' or 'of no worry' to them (78 per cent of responses); the majority hoped more tourists would visit the area and to accommodate tourists at their home. Only one villager did not want to see more tourists and another commented that there should not be 'too many' more. Most, though, believed more visitors would make the villages 'more joyful', and that increased tourism might further develop the park, bring more attention to local people and improve their lives. One villager hoped increased tourism would provide the financial resources to enable better protection of the national park. Some did not believe that tourism would bring any economic benefits – perhaps the most realistic expectation given that villagers have had almost no contact with tourists.

When asked specifically to comment about tourism at Cuc Phuong, villagers, while advocating its expansion, felt that future planning for tourism development should involve local people and ensure they are the direct beneficiaries of any increases in tourist numbers. One respondent stressed that resettlement meant loss of land that had belonged to his family for generations, the forest could no longer be used for survival, and relocated villagers should be given direct financial support from tourism revenue earned by the national park. Another suggested they would like access to funds to enable them to invest in housing and facilities suitable to accommodate tourists. Such views were commonly expressed across all villages except by residents of Sam, whose 'We do not know much about tourism' reflected their location further from the major access road to the park than most of the other villages; its residents were less likely to see tourism as offering them income-earning potential. Conversely, Nga villagers, living inside Cuc Phuong National Park alongside the main central access road, were strongly in favour of investment in tourism, with 17 of a total of 24 respondents in Nga suggesting that the Park Board should invest in tourism activities that involved the villagers.

Ban Khanh

One other village, Ban Khanh, consisting of 20 families, is still located within Cuc Phuong, in a scenic part of the centre of the north of the park (Figure 10.3). A

Figure 10.3 Ban Khanh.

representative from each family living there was interviewed. Most villagers here were engaged in agriculture for survival, fared observably better than those inter- viewed elsewhere in and around Cuc Phuong, had better access to water and forest resources, and did not report food shortages. They expressed pride in being able to continue to live in their inherited environment. Their location allowed them to grow a wider range of agricultural produce, including the Muong staples maize and cassava, than was possible for those living in the buffer zone of the park. Also, development projects funded by the Vietnamese Ministry of Agriculture and Rural Development, with the Assistance of the United Nations Development Programme, have trained villagers in the protection of the park, assisted them with loans to establish bee keeping and deer breeding projects, allocated land for planting of lychees, provided a weaving loom and loan capital, and installed a micro-hydro electricity scheme, with a supply line for each household, and two water tanks and a filtration system. In return for loans, villagers were asked to reforest an area of land and return it to the national park. The villagers believed that their use of forest resources had declined as a result of these initiatives.

Continuing their traditional way of life, the Muong in Ban Khanh have been able to retain the locational advantages of an attractive physical and cultural envi- ronment, characteristics that give them the potential to capitalise on the 'cultural authenticity' sought by tourists. Tourist brochures and guidebooks encourage visi- tors to trek to this village to experience Muong culture and an overnight stay. The atmosphere of Ban Khanh was captured thus by one visitor:

> The village was near a river and had wooden stilt houses and lots of little gardens and palm trees . . . The village headman's house where we stayed is beautiful – a big airy room on stilts, with kittens, puppies and his eight kids running around the place . . . We walked to the river as the sun went down. It was beautiful . . . The men played instruments while the women did some different songs and dances.
>
> (Kerp 2004: 2)

Despite the attractiveness of the village, the villagers had very little contact with tourists – reflected also in the interviews conducted with tourists. Few visi- tors made time to trek through the forest to the village. However, tourists are able to accompany a park guide to the village as part of a package tour, where they may stay with a family for a night. The host family earns from 25,000 to 40,000 Vietnamese dong (about US$2–3) per group per night to host small groups of visitors. The fee is paid to the park, not directly from tourists to the villagers, so visitor 'supply' and payment for their stay is largely controlled by park managers. Most visitors are accommodated by the village head, so few other families benefit from the arrangement, unless they are able to sell some of the limited produce at their disposal – honey, weaving or crafts. All villagers said that they would like to have more tourists visit the village. Here, as the report from one visitor above indicates (a view shared by the researchers), is an environment which would be highly attractive to visitors, with the potential for income generation from 'cultural tourists'; to date, though, the daunting distance to the village has limited this. Ironically, the resettled villagers, dispossessed of their location within the park, lost similar advantages of location and tradition, which would have afforded them the greatest opportunity to gain from the growing tourism industry.

Prioritising people?

Although, in prioritising conservation, authorities have resettled villagers from Cuc Phuong National Park, for both resettled villagers and those living within Cuc Phuong and its buffer zone the park still provides resources for their survival. About 50 per cent of families in the resettlement villages, according to officials of Cuc Phuong Commune, were unable to grow or buy enough food, so that, for two or three months of every year, available food was inadequate. People living in the other villages also mentioned food shortages, which, despite its illegality, forced dependence on the resources of the national park for basic survival needs. Ironically, moreover, the anticipated source of alternative livelihoods for local people, tourism, is at risk if degradation of the park depreciates its value for visitors. One group of forest users has thus been removed in the interests of another – the tourists and conservationists – who are not dependent on the visited environment for survival, a situation not unique to Cuc Phuong (Honey 1999; Erb 2001). Paradoxically, the greatest concern expressed by visitors to the park was about the degradation of its environment, degradation caused by those same visitors.

The encouragement of tourism, it is believed by both villagers and park officials, will contribute to improving living standards for local people. However, despite the growth of tourism, any improvements in quality of life for villagers have been minimal, and primarily result indirectly from infrastructure and service provision designed to facilitate visits to the park. Tourism has not directly contributed to improving the quality of life of those living in the vicinity of Cuc Phuong or their access to food and employment, or to reducing levels of general poverty suffered by them, and very few even have any contact with visitors. The recent extension of the Ho Chi Minh Highway through the north of Cuc Phuong National Park, despite bitter opposition from environmentalists, is likely to increase accessibility of the region by visitors, and to bring major changes to the park and to local villagers. If those visitors are to contribute to improved livelihoods for local communities, it is important that those communities be encouraged to participate in tourism planning and activities (Din 1993; Hall 1998), to ensure they gain from tourism. Only with careful management of tourism development, and a 'pro-poor' emphasis designed to involve and benefit villagers, will tourism offer any hope of a more sustainable future for the villagers resident in the vicinity of Cuc Phuong National Park and, indeed, for the park itself.

This chapter has raised vexed issues about priorities in planning for environmental conservation and associated tourism development – the costs of environmental preservation for local people, the benefits for tourists and the paradox that attempts to preserve the environment and to encourage what is seen to be a more environmentally benign use than traditional agricultural systems, tourism, are at cost to both the environment and its displaced inhabitants. The World Tourism Organisation continues to suggest that 'As one of the most dynamic economic sectors, tourism has a key role to play among the instruments to fight against poverty, thus becoming a primary tool for sustainable development' (WTO 2007). Reflecting this view, park managers hope that tourism may provide a source of income for local villagers, while encouraging conservation of the park as a tourist destination; at Cuc Phuong, these hopes need to be realised. The tourist experience of Cuc Phuong's natural wonders has, thus far, contributed little to any improved quality of life for local villagers, many of whom struggle to meet their basic survival needs.

Note

This research would not have been possible without a University of New England Research Award (1999) and a Faculty of Arts Research Grant (2003), which funded fieldwork and collaborative writing processes. The authors are grateful both for that financial assistance and to Le Van Lanh of the Vietnam National Parks and Protected Areas Association for facilitating fieldwork at and around Cuc Phuong National Park. Finally, we are indebted to the many villagers and visitors who shared their stories – without their willingness to do so, this research would not have been possible.

References

Baland, J. and Platteau, P. (1999) 'The Ambiguous Impact of Inequality on Local Resource Management', *World Development*, 27: 773–788.

Buergin, R. (2003) 'Shifting Frames for Local People and Forests in a Global Heritage: The Thung Yai Naresuan Wildlife Sanctuary in the Context of Thailand's Globalization and Modernization', *Geoforum*, 34: 375–393.

Cao Van Sung (n.d.) Primate Conservation in Vietnam. Institute for Ecology and Biological Resources. Online. Available <www.coombs.anu.edu.au/~vern/iebr.html> (accessed February 2002).

CIA (2006) The World Factbook, Vietnam. Online. Available <www.cia.gov/cia/publications/factbook/geos/vm.html> (accessed 28 April 2006).

Constable, J. (1982) 'Visit to Vietnam', *ORYX*, 16: 249–254.

Cuc Phuong National Park (2000) *Cuc Phuong Report: News from the Cuc Phuong Conservation Project*, 3(1).

Din, K. (1993) 'Dialogue with the Hosts: An Educational Strategy towards Sustainable Tourism', in M. Hitchcock, V. King and M. Parnwell (eds), *Tourism in South-East Asia*, London: Routledge.

Erb, M. (2001) 'Ecotourism and Environmental Conservation in Western Flores: Who Benefits?', *Antropologi Indonesia*, 25: 72–78.

General Statistical Office of Vietnam (GSO) (1999) Structure of Population as of 1 April 1999 by Ethnic Group. Online. Available <www.gso.gov.vn/> (accessed 5 May 2006).

—— (2005) Area of Forest in 2005 by Province. Online. Available <www.gso.gov.vn/> (accessed 16 June 2007).

—— (2006) Structure of Population Age 15 and Over in Main Work over 12 Months by Type Area, Sex of Household, Region and 5 Income Quintile. Online. Available < www.gso.gov.vn/> (accessed 5 May 2006).

Hall, C.M. (1998) *Introduction to Tourism*, Melbourne: Longman.

Honey, M. (1999) *Ecotourism and Sustainable Development: Who Owns Paradise?*, Washington, DC: Island Press.

Kerp, N. (2004) 'Naomi Kerp's Diary: Restoring Vietnam's Forest', Alcoa Environment 2004 Earthwatch Diaries. Online. Available <www.alcoa/com/global/en/environment/ew/2004/diary_kerp.asp> (accessed 28 February 2005).

Le Trung Tien and Nguyen Cao Doanh (1995) *Vietnam Forestry*, Hanoi: Agricultural Publishing House.

MSN Encarta Encyclopedia (2005) Vietnam. Online. Available <www.encarta.msn.encyclopedia-761552648/vietnam.html> (accessed 28 February 2005).

Nguyen Ba Thu (1995) 'Measures to Build Up Buffer Zone in Cuc Phuong National Park', in *Proceedings of National Conference on National Parks and Protected Areas of Vietnam*, Hanoi: Agricultural Publishing House.

—— (2000) Opening speech, integrated conservation and development projects lessons learned workshop, Hanoi, 12–13 June.

Nguyen Tu Chi (1972) 'A Muong Sketch, Ethnographical Data', *Vietnamese Studies*, 32(1): 49–142.

Pfeiffer, E. (1984) 'The Conservation of Nature in Viet Nam', *Environmental Conservation*, 11: 217–221.

Ray, N. and Yanagihara, W. (2005) *Vietnam*, eighth edition, London: Lonely Planet.

Rugendyke, B. and Nguyen Thi Son (2005) 'Conservation Costs: Nature-based Tourism as Development at Cuc Phuong National Park, Vietnam', *Asia Pacific Viewpoint*, 46: 185–200.

Rugendyke, B. and Nguyen Thi Son (2007) 'Sustainable Futures? Displacement, Development and the Muong', in H. Cao, E. Morrell and S. Simon (eds), *Regional Minorities and Development: Challenges for the Future in South and Southeast Asia*, Ottawa: University of Ottawa Press.

Schmidt-Soltau, K. (2003) 'Conservation-related Resettlement in Central Africa: Environmental and Social Risks', *Development and Change*, 34: 525–551.

Smith, D. (2000) Paving Over the Ho Chi Minh Trail: National Geographic News. Online. Available <http://news.nationalgeographic.com/news/2000/11/1121_hochiminh.html> (accessed 28 February 2002).

Viets With a Mission (VWAM) (2006) Vietnam's Ethic Minorities. Online. Available <www.vwam.com/vets/trives/northern.html> (accessed 3 May 2006).

Vo Quy, Nguyen Ba Thu, Ha Dinh Duc and Le Van Tac (1996) *Cuc Phuong National Park* Hanoi: Agricultural Publishing House.

World Tourism Organisation (2005) World's Top Emerging Tourism Destinations in the Period 1995–2004. Online. Available <www.world-tourism.org/facts/menu/html> (accessed 7 July 2007).

—— (2007) 'Another Record Year for World Tourism', press release. Online. Available <www.world.tourism.org/facts/menu.html> (accessed 7 July 2007).

World Wide Fund for Nature (1996) Conservation Threats: Vietnam Country Profiles. Online. Available <www.panda.org/resources/countrypfiles/vietnam> (accessed August 1998).

11 Communities on edge

Conflicts over community tourism in Thailand

Tim Wong

Khao Yai National Park, located close to Thailand's capital city, Bangkok, has evolved into one of the country's premier ecotourism destinations based on its growing national and international status for conservation and recreation. The park is one of over 212 Protected Areas (PAs) in Thailand, whose management objectives are predominantly to protect biodiversity, facilitate scientific research and education, and promote recreation. As part of a nationwide tourism development initiative by the Thai government to promote economic development, Khao Yai is being developed as a 'showcase' site for tourism promotion, tapping lucrative domestic and international tourism markets (Wong 2005). The park's accessibility and proximity to Bangkok have made it one of the most popular tourist destinations in Thailand. Located less than three hours by road from Bangkok, it receives over 700,000 visitors per year, including large numbers of city dwellers, government officials, foreign dignitaries and tourists. The vast majority of visitors (80 per cent) are Thais visiting from a number of major urban centres nearby (NPWPCD 2004).

Although direct and indirect revenues from tourism continue to grow, their distribution has mostly failed to trickle down to rural communities living at the edge of the park, with benefits mostly accruing to the state and business interests. This case study explores the exclusionary nature of territorial resource management approaches of the state, and the inability of local communities to consolidate their involvement in, or benefits from, tourism development. Focusing primarily on one particular village at the edge of Khao Yai, and drawing on wider contexts of the evolution of the Thai Protected Area system, the chapter examines the micro-level politics and changing park–people relations surrounding local tourism development along the park boundary.

The business of parks and the role of local people

From its establishment in the early 1960s, the Thai Protected Area system has expanded to cover almost all of the nation's remaining forests (Carew-Reid 2004). Although these PAs make valuable contributions to the core functions of biodiversity and watershed protection, it would appear that the ongoing 'capitalisation of nature' in Thailand (Laungaramsri 2001), through the development of tourism, has become one of their most pervasive and defining features. Since 1960, development within national parks has been a common occurrence, with numerous government-approved roads, dams, mines and pharmaceutical research ventures by foreign companies allowed in these areas and, most prominent amongst these, tourism and tourist infrastructure.

The modern growth in tourism and ecotourism in Thailand is well recognised (Dearden and Hvenegaard 1998). By 1996, tourism revenues were generating US$11.25 billion, becoming the country's primary source of foreign revenue (Carew-Reid 2004). The popularity of visiting national parks in Thailand has grown significantly in recent years. Between 1995 and 1999, visitors to Thai national parks increased by 35 per cent (Carew-Reid 2004: 94), and an estimated 12.5 million tourists visited national parks in Thailand in 2003 (World Bank 2004: 31). Not surprisingly, PAs have the capacity to create enormous revenues through tourism. For instance, accommodation alone in Ko Samed National Park, in the Gulf of Thailand, has been estimated to generate in excess of US$47.5 million per annum (World Bank 2004: 31), and much of this is now coming from the domestic Thai tourism market.

The importance of tourism to the state, and of the role of PAs, have been clearly articulated in government development policies. As one senior conservationist in a Thai NGO explained of the government's policy: 'Thaksin's [the then Thai Prime Minister] plan is to "turn capital into assets", and tourism is now the biggest source of income revenue for Thailand' (personal communication 2003). The 'Visit National Parks Year 2000' promotion, for instance, encouraged huge increases in visitor numbers and development in national parks, but also drew significant criticism in its wake, dubbed by one organisation as 'The business of parks' (Third World Network 2002).

Yet, unfortunately for local people living in and around PAs, this economic development has largely failed to trickle down to the local level. Historically, community participation in protected areas has been exceptionally poor (Carew-Reid 2004: 14), and genuine community-based nature tourism in PAs is rare in Thailand. A number of national policies have explicit provisions for the involvement of communities in tourism. Some examples include the 8th and 9th National Economic and Social Development Plans (covering the period 1997–2006), which had specific provisions for the active participation of local people in natural resource management (IGES 2004). The National Ecotourism Policy (1997) also calls for involvement in ecotourism at the local level, and the provisions of the new people's constitution granting rights for the public to have a say in natural resource management are being slowly recognised. Yet, despite these and other policies calling for greater community involvement in PA tourism, there is still no formal mechanism for community participation.

Park ideology in Thailand: territorialisation and excluding local people

To understand the ways in which local communities are able to participate in tourism in and around Protected Areas in Thailand, it is necessary to first understand how these PA resources are exclusively managed and controlled by the state within PA boundaries, resulting in the almost complete exclusion of local interests, including involvement in management decisions or benefit sharing. Important too, are the essentialised negative images that have developed about local communities in the eyes of the state and the politically influential middle class in Thailand, which have damaged any sense of legitimacy that local people might have to participate in local-level natural resource management.

An expansionist, territorial state

The formal Thai PA system, established in 1962, and its rapid expansion over the last 40 years, has brought about wide-ranging curtailment of local people's access to forest resources, and almost exclusive control of these PA resources by the Royal Forest Department (RFD) (Buergin and Kessler 2000). Upon declaration of the National Park act in 1961, habitation or any kind of extractive use of resources within formally declared national parks became illegal. In subsequent years, millions of people living in, or reliant on, forests for their livelihoods and cultures have found themselves to be illegal residents and users of resources under this act. This history of strict enforcement of PA laws by the RFD, the impacts on local livelihoods, and the resultant conflict with communities around PAs is well documented in the literature (for example, Buergin and Kessler 2000; Roth 2004), and the outcome in many areas has been the exclusion of local communities from the PA resources on which they previously relied.

Many critics have been largely dismissive of 'conservation' as the major rationale for the state's expansion of PAs. Vandergeest (1996: 263) described the rapid expansion of PAs in Thailand over the last 40 years as the solution to a bureaucratic crisis. Until a nationwide logging ban in 1989 – prompted by disastrous floods in the south blamed on deforestation due to excessive logging – the RFD had been primarily responsible for managing logging concessions and timber extraction. The ban saw the RFD's legitimacy over forest management continue to decrease, prompting it to progressively 'reinvent' itself to focus on conservation, and through this process incorporate large areas of forest land and capital under its control. This process of 'territorialisation', by which through the power of mapping PA boundaries the state can effectively control land and resources within these 'strictly bounded and homogeneously defined space[s]' (Roth 2004: 23), excludes other stakeholders (such as local people) who might stake claims over these resources.

Local people as 'poor, greedy and ignorant'

The exclusion of local people from PA management – and thus the sharing of any benefits from PAs – has been legitimised in a number of ways. In earlier years, one major rationale was under the guise of 'national security concerns' in response to the growing 'communist threat' in southeast Asia in the twentieth century (which will be examined later in the chapter). The rationale used by the state to exclude local people from using PAs in more recent years concentrates more on the perceived nature of local people, and their apparent inability (in the eyes of the state and others) to manage natural resources in a sustainable manner.

A particular feature of park management in Thailand is that it allows certain kinds of uses, and not others, often termed consumptive and non-consumptive uses (Campbell 2002). Allowable uses within national parks are outlined under the National Park Act (1961), which permits recreation and tourism, but disallows activities that damage flora or fauna (Vandergeest 1996). Thus, wildlife viewing and ecotourism are classified as non-consumptive and, although potentially damaging to species and ecosystems, are considered allowable under this state management regime. Consumptive uses, on the other hand, are based strictly on 'deliberate removal of materials . . . direct harvesting of resources for food, materials or scientific testing [that are] deliberately consumptive, albeit with varying levels of impact' (Campbell 2002: 31). Within this conceptualisation, authors such as Neumann (1995) have termed this kind of use of PAs as a 'consumptive landscape', in which nature has been commodified and 'consumed' as part of a tourism experience. This ideology views tourism development as manageable and in character with the national park model, whereas use by local communities is perceived to be destructive. Through this logic, tourists and tourism practices were considered to be able to coexist with the forest, whereas local people's livelihood practices were not – a view still predominant in the present-day Thai PA system.

The typical understanding of local communities in Thailand is one of poverty and ignorance, and is a concept imported from the international conservation movement to Thailand since the 1970s (Ghimire 1994). A widespread belief exists that poverty necessitates the collection of forest products, simply in order to be able to subsist. This needs-driven, utilitarian-based concept of local communities' relations with the forests also labels local people as uneducated and ignorant of their own supposedly damaging impacts upon the forest. This seemingly predominant view in the RFD has dominated conventional approaches to PA management and policy, which still reflect these sentiments.

Local people's plight has been further damaged through their portrayal not just as poor and ignorant, but as greedy, often gun-toting 'nature malicious figures' (Laungaramsri 2001), who comb PAs for commercially valuable forest products. A great deal of sympathy has developed within the middle class over the plight of the RFD in combating the 'poaching threat', particularly in Khao Yai National Park, discussed below. Albeit no longer as frequently as in the early days of Khao Yai, ranger patrols still come into armed conflict with poachers while carrying out law enforcement operations, occasionally resulting in the death of both poachers and staff. Not surprisingly, this has further entrenched a sense of mistrust between the two groups.

Perhaps consequently, while locals have tried to establish their legitimacy to manage and share in the benefits from PAs in recent years, these efforts have achieved limited success. The Thailand Community Forestry Bill has been one of the main processes for communities to attempt to gain formal access to PAs. This bill has a long and controversial history, consisting of a multitude of drafts and counter-drafts from the government and people's organisations, arguing over a range of provisions about access and use, and in particular what kinds of forests could be 'community forests'. After 13 years of controversy, the bill has yet to be passed, mainly because of the government's rejection of community provisions in the draft bill that would give people access to particular types of protected forests – national parks, wildlife sanctuaries and watershed areas. Communities are thus left with little opportunity to share in any kind of benefits from PAs.

A history of Khao Yai – tourism and growing state control over local affairs

Khao Yai National Park was established in 1961 at the request of the then leader of Thailand, Field Marshal Sarit Thanarat, who viewed national parks as a way to modernise the country. Like those PAs in the United States on which Thai PAs were modelled, they were seen as an expression of civilisation, particularly under the utilitarian notion of parks as places of recreation and learning for the bourgeoisie in a more modern and civilised society (Laungaramsri 2001).

Early development in Khao Yai left no doubt about the utilitarian functions of a national park in the eyes of the government. In keeping with this thinking, a caveat covering 'development' was included in the National Parks Act of 1961, allowing development to take place so long as it supported tourism. Thanarat immediately began granting lucrative development opportunities to senior officials in his government. In 1961, before the park was even established, construction began on a nine-hole golf course in the centre of the park, and the head of the Tourism Authority of Thailand (TAT), a general in the military government, was allowed to build a range of tourist accommodation in the park's centre. The relationship with the TAT was formalised under the provisions of the National Park Act in 1964, and a period of almost 30 years of intensive tourist development in the park began. There is no indication that local people played any role in tourism in these earlier years. Local people's involvement with the park was largely in the form of continued subsistence collection, an activity that brought them into frequent conflict with heavily armed park authorities, and led to much violent confrontation and even deaths on both sides (Albers and Grinspoon 1997).

Increasing state control at the local level: security concerns and park laws

The process of increasing state control over resources at the local level around Khao Yai began before the park's establishment in 1962. This process is important not only because it has determined the ways in which benefits from tourism have been captured, but because it has also been highly influential in dictating the ways in which communities have reacted to and contested decisions about the use and management of park resources at the local level.

The first significant incursion of state control over local communities in and around the present-day park area came under the guise of security concerns over the spread of communism throughout rural Thailand. In the present-day Khao Yai area, the government relocated a number of forest-dwelling communities in the mid to late 1950s, over concerns that the insurgency would spread through these 'lawless' and 'uncontrolled' forest areas (though this was unrelated to the establishment of Khao Yai NP). The Thai military began nationwide counter-insurgency approaches, such as periodic crackdowns and subsequent large-scale military operations from the mid-1960s in forest areas suspected of harbouring insurgents (Bowie 1997).

The reaction of local people was one of relative compliance to government directives. Many villagers had become terrified of the communist threat based on propaganda from the government. One village resident recalled the time when her sister's village was ordered by the state to leave the forest in the early 1960s. She commented: 'If I was Yai Yim, I would also move, because we were all afraid of the government officers [more so than the communists]. Also, they came with official letters. So we thought we should agree.' Bowie (1997: 71) contends that this approach must be 'understood through an overview of the structure and rhetorical orientation of the Thai State', which used the fear of communism to enable it 'to suppress opposition to its policies' (Bowie 1997: 73–4), through increasing local control. Over time, the state's influence in forest areas was being increasingly legitimised under the label of security, allowing the government at all administrative levels to control local affairs and local communities more closely.

The second and more influential process of the state's control over local affairs (and local resources) was the enforcement of the park's authority through national park laws. Although Khao Yai was established in 1962, management capacity of the park was quite limited for the first couple of decades. Thus, in many places, particularly areas far away from the park headquarters, there was little or no effective local enforcement of NP laws until the late 1970s. The park then established a ring of field stations around the periphery of the park in an effort to expand jurisdiction into these areas. These stations were intended to play dual roles of local law enforcement and, later, promoting and developing tourism. For villagers at the local level, there was virtually no recognition of the park's establishment in many areas, such as the focal study area of this research – Baan Malangwanyurt – located on the southern edge of the park, far from the central headquarters.

Once the local field station was established, the park began to assert its authority. Enforcement of park laws was initially disorganised and arbitrary, and was made particularly difficult by the presence of armed insurgents in the local area. However, following the end of the insurgency in 1981, the park began to strictly enforce laws over collection of forest products, and impacts upon local livelihoods were drastic.

Reaction from locals was mixed. Some older residents were largely tolerant of the park's establishment, as one elderly resident explained: 'I didn't really feel like we'd lost any resources back then [in 1977 when they established the field station]. Some people complained, but then everyone accepted it.' When asked why villagers accepted the park's authority so easily, he replied: 'We just did what they told us to do. We had to accept it.' This attitude of compliance was commonplace in this generation, and can also be attributed to strong government control in forest areas during the anti-communist era.

Consolidating control of the park by the RFD

Rapid and relatively uncontrolled growth of tourism over the ensuing years led to increasing concern within the public, interest groups, government and the media. As a result, in 1992, the Prime Minister's office ordered a ban on overnight stays in the park, and removal of much of the tourist infrastructure including the golf course, most TAT accommodation and various buildings belonging to other government agencies. Although a significant amount of revenue was lost from removal of buildings, the RFD was left in the position of having sole agency over all management and development in the park, thus 'minimising meddling from other agencies' and reducing the likelihood of civil authorities assisting local people in their claims over resources within the PA (Vandergeest 1996: 264).

From a business perspective, the RFD has developed a monopoly on accommodation and business within the park. The visitor accommodation (all of which is owned by the park) provides various levels of comfort and expense, ranging from camping to more luxurious serviced cabins. At the centre of the park, visitors have access to a wide range of recreation options, including hikes along well-formed forest trails, wildlife spotting towers, birdwatching, whitewater rafting, canoeing, and mountain biking. The park offers guides for hire, mountain bike rentals, and numerous food stalls and souvenir shops, all owned by the park.

At the same time, big business was beginning to capitalise on growing economic opportunities from tourism around the park, and resort accommodation was beginning to proliferate outside the northern and western margins of Khao Yai. Once the accommodation ban came into force within the park, these resorts became even more popular, and led to a large windfall for landowners and investors, who suddenly saw land prices surge. Resort accommodation was squarely aimed at the affluent middle class – one popular resort near the main northern park entrance in the early 1990s was charging up to Bt17,000 (US$425 in today's terms) per night for accommodation (Vandergeest 1996). Resort growth has continued unabated up until the present. Today the park is ringed by resorts, particularly around the western margins of the park closest to Bangkok.

In Khao Yai, these tourism-focused policies have continued to dominate the park. The 'Visit Parks Year' policy has been strongly implemented in all levels of Khao Yai Park management, and increasing numbers has become the central management focus. As a senior park administrator commented during this research: 'The park is not necessarily important only because it is Thailand's first national park, but because it is one of the most interesting places for tourism in Thailand. We have to take good care of the environment so it will not affect tourism.' Much of this current tourism development is occurring at the margins of the park, and the impacts on community access to local tourism markets will be examined in the next section.

Livelihoods, transition and tourism at the local level

Although both the park and big business have been highly efficient in capturing and monopolising growing tourism markets in and around Khao Yai, it has yet to be established how the politics of this have played out over time at the local level. To better understand these processes, this chapter will examine the process of livelihood changes in a community at the southern edge of Khao Yai, called Baan Malangwanyurt (an alias provided for anonymity). This will concentrate on analysing the role of tourism, and how it has formed a key 'livelihood bridge' as part of a transition process in the village, and the local-level environmental politics surrounding the capture and consolidation of tourism benefits by local and non-local stakeholders.

Baan Malangwanyurt

One of over 100 villages ringing the park, Baan Malangwanyurt is located approximately 0.5 km from the RFD Field Station (here named 'Namtok'), outside the southern boundary of the park. The village has always been located outside the park boundary. Nestled in a 'sea of agriculture' ringing the 'green island' of Khao Yai, most of the 200 villagers today are engaged predominantly in wage labour, supplemented by paddy rice production and small market gardens on village lands. Today, nobody is reliant upon the forest for their livelihoods as in the past (though a handful of 'commercial' poachers hired by outside businessmen continue a lucrative trade), and the vast majority of forest use is for collection of small amounts of 'everyday' forest products for local use (Wong 2005).

It has only been since the mid-twentieth century that Baan Malangwanyurt has been integrated into the 'wider spatial economy and polity' of Thailand (Hirsch 1985: 3), as the state has slowly intruded and developed into these previously peripheral forest areas. The lives and livelihoods of villagers have experienced great changes over the last 50 years. External factors have increasingly brought about a quite rapid transition from an agrarian, forest-based existence to the complete incorporation of the village into the market economy of the modern Thai state.

Livelihoods in transition

Baan Malangwanyurt, old by the standards of a Thai village (Rigg 1997), has continuous lines of settlement that can be charted back at least 200 years, from when the first settlers migrated from northern Isan to settle in the area. Over time, various groups migrated to this village, and practised a form of mixed rotational and slash-and-burn agriculture in the forest, with supplementary collection of forest products for use and limited sale. For the majority of the village's history, until the early 1970s, villagers have been almost solely reliant on forest-based livelihoods.

Whereas the establishment of Khao Yai in 1962 had little bearing on local people's lives, it was the establishment of a determined local RFD presence through construction of the Field Station near the village in 1981 that led to severe restriction of access. These new law enforcement measures effectively stopped most local forest use in Baan Malangwanyurt. The ramifications for many in the village were significant. A high proportion of families still relied upon forest products for their livelihoods in the early 1980s, especially collection and sale of bamboo. A combination of decreased access to resources, coupled with falling prices for forest products (bamboo was being grown commercially or increasingly replaced by synthetic materials), meant many villagers could no longer rely on forests as their primary livelihood resource. Some attempted to persist with bamboo collection, and as one villager described: 'Those that stayed had to be like thieves and keep taking bamboo.' Yet the inevitable arrests, fines or bribes to local RFD officials, during these early years, made this uneconomic, and most had given up after two or three years. As one villager related: 'the first few years we were able to poach still but, after then, even the buyers were afraid to come here.'

As a result, many villagers related stories of having to move to Bangkok and other provinces throughout Thailand to find work, as they could no longer rely on the forest for a reliable livelihood. Total numbers who left the village at this time are unclear, but many, especially young women, had to migrate for a period of time in response to new hardships imposed by stricter law enforcement, and their integration into the emerging cash-based economy.

Tourism as a bridge to modern forms of livelihoods

Following the curtailment of forest use through stricter law enforcement, tourism became an increasingly important livelihood source for local people. The main road leading to Namtok waterfall through the village was sealed in 1983, and the waterfall began to receive its first tourists (from local areas and middle-class visitors from Bangkok) soon after. In 1985, a journalist came to write a story for a popular Bangkok magazine; soon university youth groups and increasing numbers of tourists were attracted. As tourist numbers continued to grow, local villagers began to establish roadside food stalls, selling local foods such as grilled pork, sticky rice and drinks, located just outside the front entrance of the park leading to the waterfall. Store owners relied upon both store income during busier times (public holidays) and income from bamboo collection during slower times.

According to local store owners, tourism numbers increased year by year. At the height of tourism flows in the late 1990s, villagers contend that hundreds of people would come each weekend, and thousands (mainly from the province) would visit on special holidays such as Songkran (Thai New Year). Villagers also say that they had more than 20 stores set up in front of the Field Station entrance. Although no official records of visitor numbers were ever kept by the village or the RFD, accounts of these large numbers are consistent, with one store owner telling stories of past Songkran holidays during which cars were parked back to the main road and further (a distance of at least 1.5 km from the waterfall).

The contribution of tourism to village livelihoods was extremely significant at this time. Some storeowners reported making profits of over Bt10,000 (over US$250), then a significant amount, on special holidays, with further significant profits during normal weekends. Many in the village helped in the stores of family members, and benefits were spread widely throughout the village, with most families becoming involved to various degrees in servicing tourism. The profits were seen to be so significant, that the local district (*tambon*)-level government made attempts to collect annual fees and taxes from these stores.

Despite a growth in tourism at this local level, real profits from tourism were being made in other places. In the centre of the park, tourism was continuing to expand. In 1983, the park received 275,000 visitors, and it expanded to over 470,000 visitors in 1990. The growth was supported by the TAT's expanding infrastructure, whose accommodation at its peak could house 2000 overnight visitors, and was regularly filled to capacity, especially on weekends as Bangkok residents flocked to the park. In 1992 accommodation costs ranged between Bt500 and Bt1200 (US$12–30) per night.

Tourism was not the only local wage labour opportunity that was becoming available for villagers, as they continued to reduce their reliance upon forests. Seasonal agricultural labouring in the new cash-cropping industry, and construction labouring on a new dam and irrigation development nearby, became available in the late 1980s. Another significant development that led to long-term employment of many in the village was the establishment of the local watershed management unit, which was established within the park just behind the Field Station in 1991, and has become a growing source of jobs for villagers over the years.

Work opportunities in local factories began to grow after about 1995, when factories were established in nearby towns. Previously, people needed to move to other distant towns or provinces to get this kind of work. Since then, and following occupational trends in villages more generally across Thailand (Rigg 2005), factory work has become an increasingly common occupation for villagers. The majority of young people who stay in the village now work in a local factory.

Entrance fees and souring of local relations

Around the same time as the 'Visit Parks Year' initiative discussed earlier in this chapter, park management in Khao Yai began an intensive programme of tourism infrastructure development, particularly at the periphery of the park. As part of this ongoing programme, Field Stations were encouraged to apply to the central RFD administration in Bangkok for approval and funding to upgrade tourism at each site 'once they are ready'.

This process occurred at Namtok Field Station in 2001. In addition to infrastructure development, one outcome was an increased park entry fee. According to villagers, this decision from the RFD came suddenly and with no warning. One day, a small boom-gate and sign was erected at the entrance to the Field Station, informing visitors of the immediate doubling of entry fees from Bt10 to Bt20. Villagers stressed that no consultation of any kind took place, a fact that was not disputed by the local Field Station head.

Villagers contend that impacts on visitor numbers and associated profits for businesses were dramatic, with estimates of a 70 per cent decrease in visitor numbers, and the number of stores falling from 20 before the fee rise to currently three to four permanent stalls. Moreover, owners said that they made little profit and commonly made no money on some weekdays. A number of RFD staff interviewed, including the Field Station head and Park Chief, did not think that the entry fees had a significant impact on visitor numbers, a fact that is impossible to verify without records. Despite a lack of documented evidence, it was evident that, over a period of several months, local villagers' businesses located outside the front entrance of the park had become marginal propositions at best, and that visiting the waterfall was never particularly busy. While many in the village were becoming increasingly engaged with wage labour (particularly young people working in factories), the apparent decrease in tourism revenues was perceived by many villagers to be significant, particularly for older residents working on these stalls who could not go back to agriculture and were too old for factory work.

A number of complaints from villagers were met with the evasive response from the local Field Station chief that this was not his decision, nor did he have the authority to change the situation. Villagers explained this as a common response when such complaints arise at the local level. Local officials in positions of authority, such as Field Station heads, often say that they have no power to influence decisions that come from central administrators in the park or located in the RFD headquarters in Bangkok. Villagers are instead told to direct their complaints to these central administration offices. Villagers expressed their dismay at this, as they felt it was impossible for them to be heard or taken seriously when they were in such a position of powerlessness, leaving them no grounds for negotiation. The sense of helplessness was accompanied by a great deal of anger, as all those interviewed thought that the RFD was seeking to capture all the benefits from tourism for themselves. The matter of the raised fees was taken to the authorities at park headquarters on the villagers' behalf by a member of the local *tambon* government administration, and subsequently referred to the central RFD administration in Bangkok. There were no replies from the central office, nor any further developments.

Analysing the politics at the local level

The consolidation of tourism opportunities in and around Khao Yai National Park by the state and business interests can be viewed as part of a wider set of 'environmental politics' in Thailand. These politics have arisen as a result of state and business attempts to centralise control over the nation's land, water and forests (Lohmann 1995), and concern the ways in which actors have been able to negotiate access to and control of these.

It has been earlier illustrated how this process of environmental politics has been played out on a local level at the park. At the park, the RFD has been increasingly successful in consolidating its control over park resources through the progressive monopolisation of tourism in Khao Yai, and exclusion of other stakeholders. At the local level, villagers were able to capture tourism benefits for a brief period, but then seemingly lost them through the actions of a park-wide push to vigorously develop tourism. This chapter has already touched upon some of the key reasons why local people have been both ineffective and relatively passive in consolidating and defending their benefits from tourism. Yet the details of these local-level environmental politics require some further analysis, particularly within the context of the broader changes in environmental politics in Thailand briefly described below.

During the last 30 years, a widespread environmental movement has arisen in relation to local resource conflicts between rural communities and state and business interests in rural Thailand. These processes encouraged the emergence and growth of civil society at both local and international levels, which has limited the power of the state and created some new political space for opposition (Pongpaichit 2000). The new features which have emerged through this context, 'the vital role of the media, the catalytic role of NGOs, the increasing importance of both people's movements . . . and the importance of the concept of *rights*' (Pongpaichit 2000: 26) amount to a radically 'new' political economy, in which local communities have begun to successfully contest local resource disputes with the state and other outside interests.

With Khao Yai, however, the environmental politics that have developed (or stagnated) over the park's history stand in clear contrast to these wider national environmental politics. While the park has steadily monopolised tourism opportunities, local people have been unable to effectively assert any kind of claim over park resources, and PA authorities have shown no willingness to recognise these claims when they have been forthcoming.

Environmentalism and environmental politics in Thailand are produced and received within particular historical and social conditions, and thus the reasons behind local communities' lack of a coordinated or effective resistance to the state (like those in the north of Thailand) become clearer. The particular stage of political and socioeconomic development in Thailand at the time Khao Yai was established precluded protest such as this from eventuating. Since then, the park has been dominated by powerful actors with tourism and conservation agendas (predominantly the state and urban elite), who have completely subjugated any rights of access of local communities to park resources (Wong 2005).

A former Khao Yai Park ranger, now senior project head for Wildlife Friends of Thailand, shared his point of view on the differences between Khao Yai and contemporary forest politics in the north and northwest of Thailand:

> Nothing like this [opposition to national parks] ever happened in Khao Yai, because at the time it was established, the power of the state was very strong. Back then, people did not know if they could fight or not. They had no community rights movement back then, and even up until now, the park has been full of soldiers and police. People were very scared of the government, and so NGOs didn't want to become involved in the area.

Ethnic communities living near protected forests in the north have been very successful in advocating their rights to use forest resources, and have been assisted by a range of external supporters including national and international NGOs, peoples' organisations, and academics, allowing conflicts over natural resources at the local level to be heard more widely. These local issues stay local and largely unresolved except where they are publicised and involve non-local actors in the role of 'catalysts, coordinators, or instigators' (Hirsch 1997: 26). Around Khao Yai, there are no recognised groups who have campaigned for the rights of local people's access to resources – either forest-based subsistence or tourism markets – around the park. Combined with the relatively passive reaction of local people towards encroachment on their resource domains by the state, owing partly to a long history of strong state control in these areas, this has left local people almost completely excluded from tourism or other benefits from the park.

New forms of resource conflicts

Though local communities around Khao Yai have remained relatively quiet and passive in their opposition to the establishment of the park, the status quo is beginning to change. While the park is trying to deal with increasing numbers of tourists each year, and overcrowding and related environmental problems, which are regularly criticised in the Thai media, management must also increasingly deal with encroachment into the park of business people and locals, all of whom are hoping to share in the profits from the growing tourism market. Senior park managers stated that encroachment along the lengthy, and poorly defined, park boundary continued to be one of their key management issues.

In Baan Malangwanyurt, the topic that elicited most heated comments about current relations with the park remained the issue of entry fees to the waterfall, and the impacts of this on local revenues. Particularly prevalent was the perception that the RFD was trying to monopolise tourism at Field Station Namtok, with obvious side effects on the profits of local vendors. Around the boundary of Khao Yai, the issue of gaining access to, or controlling benefits from, tourism was a common theme coming from local people. New forms of resource dispute were thus emerging as a result of the RFD's emphasis on promoting tourism in the park. Replacing previous conflicts over rights to use forest products for subsistence or sale, the RFD and local communities are now increasingly competing for tourism revenues generated by the new 'commodified' form of nature.

This new form of resource conflict is redefining the relations of the park with local people in more popular tourist areas. Near the waterfall site, the Field Station chief related his experience:

> Over time, tourism increased. Local people and local, district and provincial officials became increasingly resentful that the RFD would not allow them to develop tourism inside the Field Station. When I first came here, shopkeepers were trying to drag people into their shops on the way to see the waterfall. Everyone is trying to set up business here. We remove them, but they come back. Some people who own land next to the park are even filling in the creek to extend their land and get access.

The implications of these new forms of resource conflicts for park management are significant, not just because of the potentially damaging impacts upon the park, but because the park has little experience in the kind of community outreach and consultative expertise required to solve such complex resource disputes (see Albers and Grinspoon 1997; Wong 2005).

However, the rigid, law enforcement based management approaches in the park are beginning to shift in response to changing socioeconomic realities. In the north of the park, a new initiative is being tested, with the basic and important aim of establishing better relationships, and thus better communication, with local communities. This was begun by the park's preparing to work with local communities through the creation of 'Community Forest Committees' and 'National Parks Committees' at different administrative unit levels. These 'grassroots problem-solving committees' aim to help in the resolution of local conflicts such as land encroachment, poaching, and tourism opportunities, with benefit sharing as a key component. Although these committees are still in very early negotiation and planning stage, the ideas underpinning this initiative offer hope for significant positive change.

Conclusion

Despite immense tourism growth in Khao Yai over the last few decades, the state has allowed little formal opportunity for surrounding communities to share in the significant benefits being generated by increasing numbers of visitors at local and wider scales. Local people, unable to effectively protest over their exclusion, have been 'shut out' and silenced, creating significant resentment towards park managers. Whereas tourism seems to be replacing forest product collection as the new form of resource conflict between the park and local communities, basic issues of conflict have remained the same. Whether it is collection of forest products, or selling food and drinks to tourists, local people are pursuing what they perceive as their right to share in the benefits from local park resources, and to participate in the management decisions that directly affect the ways in which resources are used and benefits distributed.

This analysis of tourism in Khao Yai National Park can thus be seen as a subset of issues within a wider debate over rights and equitable benefit sharing of PA resources both in Thailand and more widely. Involving local people in conservation would be an equitable and productive step forward in reconciling both local livelihoods and wider conservation objectives, yet this remains difficult given the current political and legal structure in which park managers in Thailand must work. The current PA legal framework is one of the major hurdles in undermining the resolution of resource conflicts, while unclear policies for community-based tourism discourage its implementation in protected areas (IGES 2004). In Khao Yai, regardless of the overall national legal framework, moves have been made to improve the situation regarding benefit sharing and inclusion in some aspects of decision-making for local people. The head of the park community outreach unit outlined his vision for new park–people relations, that the park strive to make sure that 'Villagers should feel that the forest directly and indirectly benefits them.' To turn such words into effective actions on the ground will require not just commitment at the park level, but also fundamental change at central administration levels in regards to benefit sharing to allow these changes to take place. Until then small-scale tourism will be unnecessarily constrained.

References

Albers, H. and Grinspoon, E. (1997) 'A Comparison of Enforcement of Access Restrictions between Xishuangbanna Nature Reserve (China) and Khao Yai National Park (Thailand)', *Environmental Conservation*, 24: 351–62.

Bowie, K. (1997) *Rituals of National Loyalty: An Anthropology of the State and Village Scout Movement in Thailand*, New York: Columbia University Press.

Buergin, R. and Kessler, C. (2000) 'Intrusions and Exclusions: Democratization in Thailand in the Context of Environmental Discourses and Resource Conflicts', *GeoJournal*, 52: 71–80.

Campbell, L. (2002) 'Conservation Narratives in Costa Rica: Conflict and Co-Existence', *Development and Change*, 33: 29–56.

Carew-Reid, J. (2004) *Lessons Learned from Protected Areas Management Experience in Thailand: A Critical Review of Protected Areas and their Role in the Socio-economic Development of the Four Countries of the Lower Mekong River Region*, Brisbane: ICEM.

Dearden, P. and Hvenegaard, G. (1998) 'Ecotourism versus Tourism in a Thai National Park', *Annals of Tourism Research*, 25: 700–20.

Ghimire, K. (1994) 'Parks and People: Livelihood Issues in National Parks Management in Thailand and Madagascar', in D. Ghai (ed.), *Development and Environment: Sustaining People and Nature*, Oxford: Blackwell.

Hirsch, P. (1985) *Participation, Rural Development, and Changing Production Relations in Recently Settled Forest Areas of Thailand*, unpublished doctoral thesis, University of London.

—— (1997) 'Environment and Environmentalism in Thailand: Material and Ideological Bases', in P. Hirsch (ed.), *Seeing Forest for Trees: Environment and Environmentalism in Thailand*, Chiang Mai: Silkworm Books.

Institute for Global Environmental Strategies (IGES) (2004) *Research on Innovative and Strategic Policy Options (RISPO): 2nd Progress Report*, Asia-Pacific Environmental Innovation Strategy Project, Tokyo: IGES.

Laungaramsri, P. (2001) *Redefining Nature: Karen Ecological Knowledge and the Challenge to the Modern Conservation Paradigm*, Chiang Mai: Earthworm.

Lohmann, L. (1995) 'No Rules of Engagement: Interest Groups, Centralization and the Creative Politics of "Environment" in Thailand', in J. Rigg (ed.), *Counting the Costs: Economic Growth and Environmental Change in Thailand*, Singapore: Institute of Southeast Asian Studies.

Neumann, R. (1995) 'Local Challenges to Global Agendas: Conservation, Economic Liberalization and the Pastoralists' Rights Movement in Tanzania', *Antipode*, 27: 363–82.

NPWPCD (2004) *Management Plan for the Dong Phayayen – Khao Yai Forest Complex (Draft January 2004)*, Bangkok: National Park, Wildlife and Plant Conservation Department, and Department of Forestry, Kaesetsart University.

Pongpaichit, P. (2000) 'Civilising the State: State, Civil Society and Politics in Thailand', *Watershed*, 5: 20–7.

Rigg, J. (1997) *Southeast Asia: The Human Landscapes of Modernization and Development*, London: Routledge.

—— (2005) 'Poverty and Livelihoods after Full-Time Farming: A South East Asian View', *Asia Pacific Viewpoint*, 46: 173–84.

Roth, R. (2004) 'On the Colonial Margins and in the Global Hotspot: Park–People Conflicts in Highland Thailand', *Asia Pacific Viewpoint*, 45: 13–32.

Third World Network (2002) International Year of Reviewing Tourism: Thailand's Case. Bangkok: Tourism Investigation and Monitoring Team. Online. Available <www.twnside.org.sg/title/iye7.htm> (accessed 10 June 2007).

Vandergeest, P. (1996) 'Property Rights in Protected Areas: Obstacles to Community Involvement as a Solution in Thailand', *Environmental Conservation*, 23: 259–68.

World Bank (2004) *Thailand Environment Monitor 2004: Biodiversity*, Bangkok: The World Bank and Thailand Ministry of Natural Resources and Environment.

Wong, T. (2005) *Environmentalism, Ideology and Stereotypes – A Community Ethnography in a Thai National Park*, unpublished master's thesis, University of Sydney.

12 Community-based ecotourism in Thailand

Anucha Leksakundilok and Philip Hirsch

Mainstream tourism has had well-documented social and environmental impacts throughout the world. In developing countries in particular, such impacts are generally assumed to result from the scale and the cultural impositions of large-scale package tourism. In some cases, responses to impacts include social movements that campaign against the worst excesses. In other cases, the response has been to experiment with alternative forms of tourism that seek to spread the benefits more widely, to limit the social and cultural costs, to make the tourism industry more environmentally sustainable and low-impact and even to use it as a means to build cultural and environmental awareness.

Thailand has been one of the Asian countries at the forefront of international tourism, attracting visitors from wealthier countries in search of low-cost, 'exotic' destinations. Thailand has also borne the impacts of tourism, and achieved a reputation for hosting some of the seedier and more environmentally destructive sides of the international tourism industry. Non-governmental organisations (NGOs) and others have established campaigns against the more damaging aspects of international tourism in Thailand. However, the industry's economic significance in terms of employment and its contribution to the country's foreign exchange earnings have meant that such campaigns have not posed a serious threat to the structure of tourism. An alternative response has been a set of experiments with more socially and environmentally responsible tourism approaches that seek to spread the benefits of tourism more widely and to practise tourism development in a more environmentally sustainable manner. A key question that arises, with initiatives falling under the banner of 'community-based ecotourism' (CBET) in Thailand, is whether they are able to deliver in their own terms. In other words, is CBET economically and environmentally sustainable, and is it socially more equitable than other forms of tourism? Is CBET an alternative to mainstream tourism, or should it be seen as a supplementary niche?

In this chapter, concepts of ecotourism as an alternative are considered, then the growth of mainstream tourism in Thailand and the emergence of community-based ecotourism as a response described. The broad themes are illustrated empirically in four case studies of CBET in different regions of Thailand. These cases are considered in terms of their employment of natural and cultural resources, their experience of the social distribution of benefits from tourism at the community level, and the significance of access to capital and marketing expertise in setting potentials and constraints for CBET.

Ecotourism as an alternative and the move toward CBET

Ecotourism has its origins in a number of responses to mainstream tourism. In part, it emerged in Western societal contexts to deal with problems created by mainstream tourism. As such, it can be seen as a measure for conservation of the environment and natural resources. More positively, ecotourism was initiated to promote and establish responsibility in nature-based tourism development. From a social perspective, it is also an alternative approach to tourism that covers the rights and wellbeing of the host community. Ecotourism has sometimes been used to raise environmental awareness and increase natural and cultural heritage conservation among local people by using people's participation, but the processes of management and the outcomes can differ widely. Moreover, ecotourism itself can be seen as a contradiction in terms (Higham and Luck 2002; Weaver 2001a), and it is debatable whether it is a viable alternative for local communities in developing countries or is in fact, as some critics claim, a form of neo-colonialism (see Britton 1982; Cohen 1996; Mowforth and Munt 1998; Nash 1989; Pleumarom 2002).

The variety of experiences and objectives of ecotourism that has emerged over a couple of decades has produced typologies of ecotourism and ecotourists. Differentiation includes 'hard and soft', conservation (saving for use) and preservation (saving from use) (Fennell 1999), 'risk and safe', 'active and passive' (Orams 1995), 'deep and shallow' (Acott *et al.* 1998), 'rigid and liberal' (Weaver 2001b), 'Sound ET and ET Lite' (Honey 1999) and 'nature and social activities' ecotourism (Chapman 1995). These categories also show some overlap with non-ecotourism forms or mainstream tourism (Leksakundilok 2004). Therefore, to talk of the application of ecotourism is problematic. It is not a single form of tourism, but is closely related to the specifics of place, time, culture and acceptance by the local community.

Ecotourism has adopted a sub-discourse of local, small-scale orientation – in contrast to mainstream tourism – through local participation, the involvement of people who live in the locality of the ecotourism activity ('local people'), local benefits, producing economic opportunities for local people, and the sustained wellbeing of local people. To promote rural development, ecotourism has included a wide range of cultural attractions, no matter whether related to nature or not. The local experience and the implications of ecotourism at the community level in many countries including Thailand have shown the transformation of ecotourism into what has become termed 'community-based ecotourism' (CBET), and sometimes simply 'community-based tourism' (CBT) – both based on attempts to redirect or capture the flows of benefits hitherto largely accruing to outsiders.

As a major tourist destination, Thailand has gone a long way in mainstream tourism. But as a country with a strong awareness of the social and environmental impacts of mainstream economic development, and with an increasingly robust civil society movement, Thailand also has experience of ecotourism and most recently CBET. The seeking of such alternatives was boosted in response to the economic crisis after 1997. Further, the so-called 'People's Constitution' of 1997 supported trends toward 'people-centred' development, a 'self-sufficient economy' and people's involvement in natural resource management.

Many ecotourism development projects are promoted as alternatives to past forms of tourism development, which contributed little income to local communities and at the same time had adverse social, environmental and natural resources impacts in surrounding areas. Are such alternative projects viable local-level alternatives? This chapter addresses this question through detailed case studies from different regions of Thailand, which demonstrate how the constraints of the state (through limited local access to 'natural' eco-resources) and capital (through the limitations of local marketing expertise and the difficulty of community-run initiatives competing with go-it-alone entrepreneurs) tend to direct or divert CBET in particular directions. In particular, cooperative initiatives tend towards tourism based around cultural resources over which villagers have greater control, but with a packaging that often involves a degree of 'invention of tradition'. At another level, such packaging and assertion of control is in part an expression of cultural and resource claims vis-à-vis the state.

From mainstream to community-based ecotourism in Thailand

Tourism is one of the major economic sectors of Thailand and has played a crucial role in its national development for at least four decades. The number of international tourists in Thailand grew from 44,000 in 1957 to 81,340 in 1960 and to 5,298,860 in 1990 (Amnuay 1996: 187–8). Numbers accelerated their growth to 11.65 million in 2004 (TAT 2005). Tourism income has also increased from Bt196 million in 1960 to become the country's largest source of foreign exchange income in 1982 (Bt23.8 billion or US$1.19 billion). It climbed to Bt127.8 billion (US$5.1 billion) in 1993 (EIU 1995: 67) and to Bt384 billion (US$9.6 billion) in 2004 (TAT 2005), and has been estimated to provide around 3 million jobs (WTTC 2005).

The rapid growth of the tourism industry has generated a huge amount of income for the country, but income distribution has been very unequal, whereas cultural impacts have been substantial. The gross foreign exchange revenues from tourism must be set against leakage of tourism income overseas through the import of goods to serve the tourist industry and expatriation of foreign investment profits. Less tangible losses include changes in values, norms and identities caused by commodification of culture, including traditions, events, ceremonies, arts and festivals; social differentiation within destination communities, both among families and among villagers; social problems such as crime, drugs and prostitution, including increased sexually transmitted diseases; exploitation of natural resources, polluting of destination areas by waste water and garbage dumping, water shortages, soil degradation and erosion, and depletion of native flora and fauna. Some of these impacts are part and parcel of the nature of large-scale tourism. Others may be the result of poor understanding and planning as well as failure to incorporate measures to counteract foreseeable problems into tourism management initiatives. Furthermore, domination of decisions on tourism development by bureaucrats and a small group of private investors means that adverse impacts of tourism on the environment and local communities have received insufficient attention.

This adverse situation has led to widespread debate on tourism development within Thailand, with interest and input from many sectors of the community. Understanding the negative impacts of the tourism industry, coupled with pressures from affected communities and society in general, has led to some change within Thailand's tourism management policies. Two types of response are evident. First, a 'top-down' approach to planning and managing the tourism industry in a more responsible way has been adopted by government agencies in conjunction with entrepreneurs. This has meant more attention to the community and more environment-oriented development of the tourism industry than previously. Second, various other tourism approaches have been developed as an alternative to conventional tourism. Alternative tourism was initiated by some small operators, NGOs and villagers, and later on accepted and practised by related governmental agencies and tour operators. Ecotourism is one such alternative and is promoted by many stakeholders as a new and workable form of tourism and component of sustainable development.

Ecotourism was first adopted in Thailand in the early 1990s, only a few years after its emergence in Western countries. It was known in Thailand initially as *Kaanthongthiao Choeng Anurak* (conservation tourism), but later the term *Kaanthongthiao Choeng Niwet* (ecological-based tourism or ecotourism) became more common. Ecotourism is defined in the Thai National Ecotourism Policy as:

> responsible travel in areas containing natural resources that possess endemic characteristics and cultural or historical resources that are integrated into the area's ecological system. Its purpose is to create an awareness among all concerned parties of the need for and the measures used to conserve ecosystems and as such is oriented towards community participation as well as the provision of a joint learning experience in sustainable tourism and environmental management.
>
> (TISTR 1997: 8)

Although strictly speaking the two Thai terms refer to different forms of tourism, they are often used interchangeably. The new development paradigm and the political reforms introduced in Thailand in the 1990s, both of which recognise and accept the necessity for environmental conservation and the rights and responsibilities of local people in the management of cultural and natural resources, support the adoption of ecotourism.

The emphases on people's participation in tourism and the use of ecotourism as an economic tool to distribute income and relieve poverty in the local economy have become the main attractions of ecotourism within Thai communities. With an economic, social and environmental rationale, ecotourism has spread rapidly across the tourism industry and has been incorporated into community development practice in many rural areas. Despite its promotion by various groups, however, most of those involved have very little experience in managing this form of tourism. This has created confusion and conflict in its implementation, with very different practices and outcomes. Sometimes the concept is interpreted strictly, but sometimes it is simply used to promote conventional mass tourism. To some extent such practices reflect the Thai experience of integrating and adapting introduced practices to suit particular local ways of doing things, but it can also lead to the concept being distorted with adverse effects. One of the reasons for the distortion, or even failure, of implementing ecotourism may be the different understanding of ecotourism's components, such as 'nature', 'culture', 'local', 'participation' and 'community'. Community participation so far has proved difficult to achieve.

Community-based ecotourism in principle responds to the issue of control by outsiders, and it has become widespread in Thailand. Figure 12.1 summarises the transformation of tourism in the community to ecotourism and then what is specifically termed community-based tourism (CBT), whereby tourism is managed and controlled by the community. Ecotourism, as a tool for conserving the natural and biological environment worldwide, supports this concept effectively.

The government's development policy of using tourism to raise people's income and conserve their culture and identity has brought tourism into the community arena. From the local level, community-based initiatives have involved tourism activities for economic, social, cultural and political purposes. Village communities are also a focus of tourists' visits if their intention is to experience local culture. However, promotion of village visits in itself simply introduces mainstream tourism into rural communities with a range of accompanying impacts. Most local people receive few benefits from these visits because their involvement is usually limited to selling a small quantity of low-priced local crafts and services. The major share of any profit from these visits goes to various middlemen, from within or outside the community, especially tour operators and souvenir shop owners.

In response to the limitations of mainstream tourism targeted at rural ways of life, ecotourism has been promoted in order to allow local people to control tourism and its impacts, and earn the largest share of the tourists' expenditures, and thus to change the community from being the 'target' of tourism to its 'manager'. Many of these communities are applying the concept of ecotourism to the process of community development but have adapted it into CBT and CBET.

Thai rural communities are involved in CBT with support from outsiders – government agencies, academics, private operators, NGOs and local administration, mainly the Tambon Administrative Organisations (TAOs). Spontaneous community initiation of CBT is rare, particularly in the initial stage of CBT development. Nevertheless, almost all outsiders and other stakeholders accept that local people should play a crucial role at all stages of the development process. A broad survey

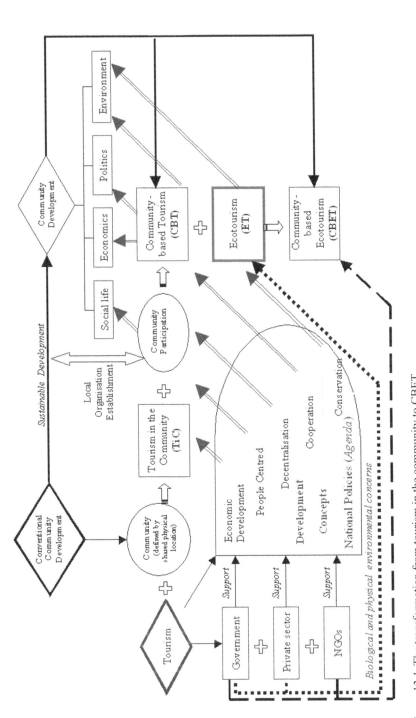

Figure 12.1 The transformation from tourism in the community to CBET.

of community initiatives in ecotourism conducted in 2000–2 outlines the general characteristics of CBT across rural Thailand:

1   CBT in rural areas occurs in all parts of the country, not limited by size or type of community, geography or administrative structure.
2   Communities, or their members, provide basic tourism services such as accommodation, local food, tours, cultural performances and transportation, which are often supported by outsiders.
3   Although people's participation in community and tourism management is often mentioned, there is no clear evidence about how they manage participation in the community, how many people are involved and what level of participation is occurring in the communities. Some respondents refer to what is called 'pseudo-participation' in the development process that relies heavily on the support or control of outsiders, and thus ignores the main idea of participation (Deshler and Sock 1989; Midgley 1986; Selener 1997).
4   Although problems in communities are mentioned, including some adverse impacts of tourism, the survey does not show any serious impacts. This may be because CBT is in its initial stages and the number of tourists is still low. Nevertheless, the main purpose of CBT was concerned with earning extra income, and therefore some impacts may be overlooked. To deal with problems, almost all communities need more support from the government or outsiders, particularly for finance and environmental management, but people have nonetheless risen to the challenge of establishing management tools for CBT and ecotourism (Leksakundilok 2004).

Local communities across Thailand are involved in several forms of sustainable tourism development. Responses to the general survey demonstrate both similarities in overall development patterns and also diversity in the specific processes of tourism development at the community level. However, in all cases CBET faces questions of viability, based not only on local constraints but also on those associated with state (for example bureaucratic procedure) and capital (for example the difficulty for community-run operations in competing with commercial enterprises). These issues can be examined through four village studies, where communities obtained support from different outside supporters.

Four case studies of community-based ecotourism

Four communities were selected for study from different regions of the country (Map 12.1). All four communities were considered to be best practice exemplars of good community-based management and a relatively high degree of local participation in development activities. Their organisational characteristics differ and illustrate the range of forms that CBET can take in Thailand. The selected communities are Umphang district, Tak province in the lower north; Ban (village) Khiriwong, Lansaka district, Nakhon Sri Thammarat province in the south; Ban Sasom, Khong Chiam district, Ubon Ratchathani province in the northeast; and

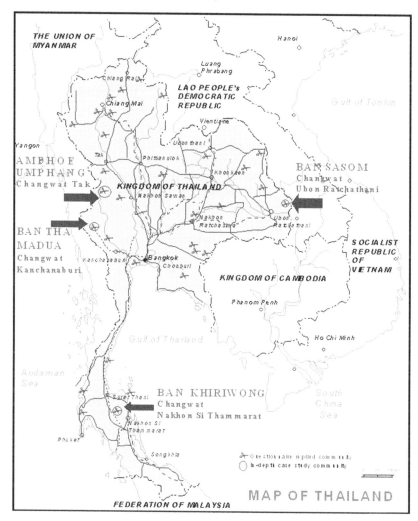

Map 12.1 Case study villages, Thailand.

Ban Tha Madua, Thong Pha Phum district, Kanchanaburi province in the western part of central Thailand. They all faced some difficulties and constraints, but their experiences of self-development provide lessons for others.

The key ecotourism characteristics of the four places were as follows.

• Umphang District was until recently one of the most isolated areas of Thailand, and remains heavily forested and lightly populated with mainly ethnic minority villages. Umphang is involved in ecotourism at district (*amphoe*) level, with the local administration seeking involvement of rural villagers in management. Government agencies, particularly TAT and the province, have tried to help local residents solve previous problems related to conventional tourism. Local operators set up the Umphang Tourism Promotion and Conservation Club (UTPC) and are taking responsibility for ecotourism management in this area. They received two international awards, from the Pacific and Asia Travel Association (PATA) and the ASEAN Tourism Association (ASEANTA) for ecotourism destination management in 1998 and 1999. Natural attractions in the wildlife sanctuary and Karen villages have attracted both domestic and international tourists. The Umphang ecotourism management arrangements can be characterised as an example of local private and public collaboration in ecotourism management. As this district-wide 'community' is rather large compared to other cases, the study focused on two villages: Ban Umphang as a town centre and Ban Khotha (of the Karen ethnic group) as an isolated destination village. Ban Umphang has 618 households and 1403 inhabitants (1998), while Ban Khotha has only 28 households and 136 inhabitants.

- Ban Khiriwong is a group of villages that received national attention in 1988 when several hundred villagers were swept away or buried by floods and mudslides caused in part by forest destruction on upper hillsides. This event triggered a national logging ban. Since then, and partly as a result of reconstruction efforts, Khiriwong villages have become well-known nationally as a self-reliant community. In 2001 there were 670 households and 2774 inhabitants. They established subsistence occupation groups to strengthen their community unity and increase the income of their members. Khiriwong Ecotourism Club (KEC) is a group that deals with tourism in the village, particularly trekking tours to the top of the mountain in Khao Luang National Park. The villages began ecotourism in order to raise the awareness of villagers and outsiders immediately after the devastation caused by environmental degradation, and they have nearly two decades of experience in ecotourism and village tourism. They also cooperate with the national park and the TAO. Khiriwong can be characterised as a self-mobilising CBET community.

- Ban Sasom is on the edge of Pha Taem National Park in northeastern Thailand. It had a population in 2000 of 474 living in 61 households. Ban Sasom villagers organise tours to Pa Dong Natham (dense, dry evergreen forest areas) in the national park. Ecotourism has been promoted since 1994 by several outsiders, including an NGO (the Native Care Foundation), the military and TAT, to conserve the forest and increase local incomes. Ecotourism was first promoted at *tambon* (sub-district) level, in tandem with community forestry. The Royal Forestry Department (RFD) selected Sasom Community Forest as one of its pilot project areas. Because of the scattered location of the villages and the location of Ban Sasom at the gateway to Pa Dong Natham, a Sasom Ecotourism Club (SEC) was established to serve tourists visiting the community forest area and Pa Dong Natham. The SEC employs cultural resources associated with Isan (northeast Thailand) tradition to welcome visitors in a set of staged performances including a procession as tourists arrive in the village, long drums, and local dancing and music.

The Pa Dong Natham Ecotourism Cooperative (DNT-EC) runs CBET and a community business, which can be characterised as CBET integrated into community forest management.

- Ban Tha Madua is a small village that was evacuated and resettled from the area flooded by the reservoir of the Khao Laem Dam project. The village had 115 households with a population of 658 in 2000, including about 100 Burmese temporary workers. As part of a home-stay ecotourism programme, villagers demonstrate the natural features around the village, and particularly the three-colour crab, alongside their culture and way of life, to tourists who are brought in by an international tour operator. They established the Ecotourism Village Group (EVG) and managed the community services with support from the International Volunteer Association Inc. (IVI) and the International Volunteers Thailand (IVTH). They host international group tours at least once per week. Attempts to 'revive' tradition as a tourist attraction have been modified somewhat to suit the preferences of tourists; for example, transporting foreign tourists in ox-carts was discontinued after some visitors expressed concern over the welfare of the animals. This village can be characterised as having internationally market-oriented CBET.

More than 10 per cent of the local population in each case study area reported being involved directly with tourism (18 per cent for Khotha, 13 per cent for Khiriwong, 12 per cent for Sasom and 14 per cent for Tha Madua). Other occupations and sources of income included agriculture, local business and non-local employment, and also remittances. A small village may have a high percentage of involvement, whereas a larger village may have a higher number of people involved.

Development objectives and management in the case study villages

Community-initiated tourism development in the four villages had several objectives. The first was economic development. Ban Tha Madua earned a relatively small income from tourism because of its small market, but receiving a tour group once a week on a regular basis nevertheless provided a useful income supplement. Umphang focused mainly on increasing the number of tourists, to increase job opportunities for the community's members and the incomes of tour operators. Khiriwong's main objective was to increase local people's awareness of environmental conservation, reforestation, and the need to coordinate with outsiders. Meanwhile, Ban Sasom used ecotourism to support the community's struggle to claim a community forest that overlapped with the national park, for conservation and to retain a relatively traditional way of life. Some minor objectives that influenced the development of CBET in the four cases were the promotion and demonstration of other development activities such as the subsistence farmers of Khiriwong, community forestry and crafts activities at Ban Sasom, and agroforestry at Tha Madua.

The collective management that is the basis of community development requires a degree of unity and solidarity among community members. Community solidarity differed from one community to another, reflected in part by the number of local organisations, their activities and the level of community participation in the tourism development process. Khiriwong had a very strong and united community that had almost all the villagers as members, and its management was achieved through local effort rather than the cooperation and support of outsiders. Umphang was also effective, maintaining one main ecotourism business organisation, which other agencies had to recognise and with which they had to cooperate, and which extended membership to all stakeholders in the community. Ban Tha Madua and Ban Sasom used ecotourism as a tool to strengthen their solidarity in struggles over resource management problems. The level of involvement with government agencies was low. On the one hand, this was beneficial for the communities in giving them a sense of rights and responsibility to manage resources independently. On the other hand, they experienced some lack of support or ignorance on the part of responsible agencies, particularly regarding natural resources utilisation and market operations.

Tourism resources, services and market

A key element of building and maintaining ecotourism is the management of an ecotourism image that attracts a particular type of tourist, particularly those who want to see and learn about the local community's efforts to manage ecotourism. The terms 'primitive', 'remote' and 'poor' communities appear to be used as a theme to promote ecotourism destinations, especially for those tourists who are seeking 'primitive' undeveloped areas and the authentic 'other'. Such perceptions create controversy over the strategies used for ecotourism and community development. Moreover the actual attractions in local sites, whether perceived by local people or tourists (Table 12.1), vary considerably in the manner in which they incorporate, human activities and cultural and natural features, including undisturbed natural resources, into tourist attractions.

Tourism services provided in the communities again vary considerably (Table 12.2). Home-stay, a popular service in CBT, has been introduced in most areas to extend the opportunity for villagers to gain benefits from tourism and increase contact between guests and hosts. Home-stay is a problematic activity in terms of security, language and other barriers to interaction, and with regard to the fair distribution of benefits within the community. In the case of Tha Madua, village-stay, whereby tourists stay at a purpose-built facility in the village but not with individual families, has been provided instead with the belief that it is safer and fairer for both tourist and villager.

The number of tourists visiting the different communities varied according to their attractions, services and promotion, and their accessibility. Umphang's visitors (25,389 in 1998) were mostly nature lovers or seeking adventure. Although the majority were domestic tourists (about 90 per cent), international tourists were also numerous relative to other case studies. Khiriwong's visitors (3320 day trippers in 2000) were mainly domestic tourists, who visited the village to

Table 12.1 Tourism resources in the case study areas, Thailand

Tourism attractions	Umphang	Khiriwong	Sasom	Tha Madua
Natural attractions	Umphang Wildlife Sanctuary (UWS) with rivers, waterfalls, caves and forest atmosphere	Khao Luang National Park (KLNP) with mountaintop, flora and fauna, waterfalls and creeks, wildlife. Natural area around the villages as part of mountainous forest area such as falls and creeks	Pa Dong Natham (rainforest in sandstone mountainous area). Sandstone and Mekong scenery, including sunrise. Community forest	Three-colour crab, common forest and backyards. Big old-growth trees in the forest plantation areas. Small hill with cave and trees
Human-related with nature activities	Karen villages living in the forest area	Integrated orchard plantations	Community Forest Group. Wild products. Forest monasteries (Wat Tham Patihan and Wat Phu Anon)	Forest plantation (by villagers)
Cultural attractions	Folk houses. Karen lives	Saving Group (local activity). Subsistence and craft groups. Historic buildings. Rural life and hospitality	Village area. Housewives Group (craft). Agriculture. Sacred places	Agriculture. Village tour and performance
Rough proportion of nature-based tourist destinations	90% Most of the destinations are natural places. The Karen villages are also located in the UWS area	40% There are many mountainous and waterfall areas around the village. These are less popular than the village's interesting places, because of their remoteness and inaccessibility	70% Main activity is taking tourists to the forest. However, tourists may spend some time in the village to observe local way of life and buy some local products	20% The natural destinations are forest plantation, three-colour crab swamp, but tourists spend most time in and around the village observing local way of life, including farming practices
Undisturbed, authentic areas	In the UWS, where Karen people are practising their traditional lifestyles	The natural resources and the rainforest in KLNP are undisturbed and authentic	Local people keep the community relatively traditional, but may seasonally utilise forests for their survival	All three-colour crab habitats are already preserved. There is no impact because of a low level of tourism

Source: Leksakundilok (2004: 218).

Table 12.2 Tourism services and activities in the case study areas, Thailand

Activities	Umphang	Khiriwong	Sasom	Tha Madua
Services	Resorts and home-stay. Trekking tour and camping. Rafting, canoeing and whitewater rafting. Adventure tour. Transportation (minibus, elephant riding and porter)	Home-stay. Tour guide (village tour, plantation tour, trekking tour and camping). Local products (demonstration and shop)	Home-stay. Tour guide (trekking tour and camping). Local products (craft demonstration and shop). Performance	Village-stay. Thai dancing performance. Tour (ox-cart riding and bushwalking)
Education and interpretation programmes	All tour operators provide guides for tourists, but most of them offer no educational programmes. Only some operators provide interpretation programmes	Many guides can explain and interpret forest environment. All guides can give information about integrated orchard and local crafts	A small number of guides can explain and interpret the nature around them. Some guides have knowledge but lack interpretation skill	There are no local guides
Specific or local guides	All guides were trained and registered as local guides for trekking. Few guides could interpret local knowledge. Many local people have knowledge and can interpret	There are no registered guides. Many guides were trained. Local people can interpret their local knowledge	All guides are trained and registered as local guides. Few guides have good local knowledge. Many local elders have good knowledge	There is no trained guide. All adult villagers can interpret their knowledge in agriculture

Source: Leksakundilok (2004: 220).

learn about self-reliant ways of life in line with the King's 'sufficiency economy' idea (*sethakit phorphiang*) and NGO and Buddhist ideals of limited consumption. Some were nature lovers, who took trekking tours to the top of Khao Luang Mountain in the summer. Few international tourists visited Khiriwong, although this village has been included in the Lonely Planet guide to Thailand since 1997 (Cummings 1999: 877–8). This is partly on account of difficulty in communicating with local people, and no tour operators offered trips there. The 'enclosure' practice, whereby local people control and collect a fee for entrance to the national park, may deter tour operators. Most of Tha Madua's visitors (478 in 2000) are weekly foreign tourists from Intrepid tour groups, and some are volunteer visitors. There are relatively few visitors to Sasom (365 in 2000–1), and most were urban students who came to learn about local culture as part of their educational programmes.

Environment, society and economy in community-based ecotourism

In principle, ecotourism brings sustainability, social equity and cultural concerns to the practice of tourism as an activity that has otherwise hitherto been overly concerned with the economic bottom-line. Yet ecotourism also needs an economic base. A key question that arises, and is examined below through the experience of the case study communities, is therefore whether these multiple objectives are achievable, manageable or realistic at the local level.

Natural and cultural resources

Like other forms of tourism, community-based ecotourism relies on specific attractions as resources for the sustained viability of the industry. In few cases in Thailand is ecotourism practised or practicable in a 'pure' sense of visitation of nature separate from its human context. Community-based ecotourism in particular tends to focus on culture and local ways of life; local practices interact with the environment and natural resource use and management, and are part of a discourse on rurality in urban Thailand. However, access to such tourism resources is governed by a range of constraints that take CBET in particular directions.

Human–nature relations as an ecotourism attraction

Rural Thais have long maintained a close relationship with nature, especially the forest, and this influences their lives and beliefs. They have used natural resources in their agro-ecosytems and changed the nature of the landscape, which can be seen as a cost to the environment and/or as a benefit to people. Some also gain benefits from ecotourism, which can draw on the uniqueness and diversity of human culture, particularly in terms of natural resources management (Yos Santasombat 2001).

Ways of using natural resources differ from place to place, but a common feature is the challenge of attaining balance between use and conservation. This challenge is illustrated at Suan Somrom, one of Khiriwong's areas of attraction, and an agro-forest area. People plant diverse fruit crops (such as mangosteen, rambutan, durian, papaya, banana and betel palm) in the old-growth forest, without clearing any trees and without using fertilisers and pesticides. Practised for a long time, this helps to conserve the environment, although it is under threat because of soil erosion and illegal encroachment. The Karen people at Ban Khotha use the natural space around them after consultation with a relevant spirit or god, particularly the 'Lord of land and water' who owns all jungle, fields, water and land in the vicinity. The spirits in the forest help Karen people to zone the conservation area and prohibit the cutting of some types of trees.

Ecotourism supports this kind of natural conservation as it enables people to change from intensive consumption of natural resources to non-consumptive and sustainable utilisation as the latter becomes a marketable attraction in its own right. It thus provides an economic incentive for people to conserve natural resources. Sasom people successfully manage the community forest and gradually gain benefits from ecotourism, just as the Tha Madua people stopped catching the three-colour crab and began using it as an ecotourism attraction. This creates further awareness among local people of the benefits of natural-resource conservation.

Cultural and social structure

Relative cohesion and unity within the community is an ideal – if not always a feature – of rural Thai community culture. Close relationships and kinship-based networking (*khwam samphan choeng kruea-yat*) at village level has been influenced by both ideology and pragmatic management. Unity and cooperation, including the selection of leaders, often start at the level of relatives and friends. In all the case studies, although conflicts may occur, ideals of social structure and harmony remain intact, but are influenced by change, such as an education system that focuses students' aspirations away from their local rural ways of life, or more individualistic forms of production as new market-oriented economic activities are introduced at the village level.

This kind of social structure has influenced community development and ecotourism. In the case of Khiriwong, living in a remote area with little external contact has forced local communities to be self-reliant, despite progressively more contact with outsiders in recent decades. Seven main clan groups trace their settlement at Khiriwong back for over 200 years, and close relationships among the villagers have been maintained, although intra-clan marriages and the number of migrants have increased. Conflicts within the community exist, but can be limited and controlled by kinship-based networking, as they are in Tha Madua.

Culture and social structure influence ecotourism development in every village, particularly in terms of management. Although projects may be initiated by a few individuals, they rely on support from wider groups of relatives. Development and management, but also various types of conflict, are organised within groups.

However, as opportunities for making money out of tourism have emerged, some groups have become more self-serving and seek to control or monopolise benefit distribution within the communities, as was the case in Khiriwong and Tha Madua in 2000–1.

Politics of management

Although national policy is to decentralise roles and responsibilities, most decision-making, budgeting and authority continues to be strongly reliant on the central government. Local administration, particularly at the Tambon Administrative Organisation (TAO) level, can regulate development, but must base this on national laws or regulations. Natural resources, particularly national parks, remain mostly under state control, and their management has frequently resulted in conflict with TAOs or community groups, as has occurred in Ban Sasom and Khiriwong (see also Chapter 11). These conflicts enter the arena of CBET in that communities whose main ecotourism attractions are natural find that their main resource is off-limits and beyond their control or right to manage. The increasing rights and expanded roles of TAOs positively affect CBET development, but also obstruct the role of the community, particularly at the grassroots, in managing CBET. In many cases, this difficulty of access has encouraged local groups to turn more to resources that remain under their control or ownership, which tend to be more culturally based.

Power relations within the community are a crucial issue. The establishment of TAOs according to principles of decentralisation has changed the power structure at the local level. The role and power of the village headman (*phuyai-ban*) and sub-district (*tambon*) chief (*kamnan*) have been reduced. Meanwhile, the TAO management team and the council members (as the people's representative) have more responsibility and power over development than the *kamnan* and *phuyai-ban*. This has created new nodes of power in the community. TAOs can support CBET well when they have good links with the local ecotourism group, as seen in Khiriwong in the early period (1994–9), but were ineffective when the management team did not represent Khiriwong during 1999–2002. In Sasom the Pa Dong Natham Ecotourism Cooperative management, and the Umphang Tourism Promotion and Conservation Club in Umphang, are both engaged in the TAOs as members or part of the management team, giving the local political support needed to sustain ecotourism development. However CBET is also a platform for tension and power-play in relations among formal leaders *(kamnan, phuyai-ban* and the representatives to TAO) and informal clan leaders.

In Thai society, the rise and fall of leaders not only depends on who can play the political game well, but also on their intelligence, sacrifices for and relations with the people. The *kamnan* and *phuyai-ban* as formal leaders can be seen locally as either government officers or informal leaders, or both. Many of them used to be 'natural leaders' (*phu-nam thammachat*) in the communities and, traditionally, formal and informal leaders were not always different. However, within the last two decades, the influences of liberal–capitalist politics at the lowest community level have changed local people's perception of the leaders. The benefits of new businesses, including tourism, have encouraged some leaders to use their positions of power to gain individual economic advantage, thereby undermining community initiative.

Economic aspects of CBET

Communities are generally involved with ecotourism in two ways: first as destinations or attractions in terms of place, and second as operators or actors in terms of management by people or local organisations. The opportunity to operate and develop ecotourism depends on the extent to which people have rights to manage resources. Such rights of local people to control or participate in ecotourism cannot be taken for granted where governments have proclaimed lands to be protected areas. Conflicts related to protecting and managing natural resources often result.

The concept of community-run tourism emphasises the cooperation of stakeholders and the role of local groups in managing tourism in the community, in order to limit the influence of individual entrepreneurs, both the local elite and outsiders, in controlling such economic activity. Therefore, most CBET concerns small-scale tourism, local services and integration with natural and cultural resources, involving the management of strategies and actions of local people and community organisations, through the adoption of a 'grassroots' approach (Boyd and Butler 1996) rather than an 'agencies-driven' approach. Investment costs are small, often based on collecting contributions from the members and supported by outside funds.

Marketing is the weak point of ecotourism at the community level; hence communities usually try to create relationships with outsiders. Each of the four village communities has relied on help and support from outside, especially from government agencies and sometimes from commercial operators and NGOs, though this necessitates some external dependence. It is also hard for communities to generate substantial incomes through offering basic services. Moreover many tourists engage in ecotourism as only part of their recreational experience, and there are now many CBETs competing for similar markets. CBET services provision and marketing are rather new activities for many rural communities, and cooperative enterprise sits uneasily with the private enterprise character of the tourism industry.

Community-run initiatives find it difficult to compete with private entrepreneurs. Every villager incurs some expenditure to provide services, mostly in labour costs and the cost of raw materials (in the case of home-stays). The operational expenditure that villagers incur to prepare tourism services is usually presumed to be lower than the income earned, or the price set by the group, but some villagers have to pay for the renovation of toilets and amenities, bedding, kitchenware, and enlarging the space for visitors before any services can begin. For example, the Sasom group calculated this cost to be about Bt1000 (US$22) per household or Bt3000 (US$66) per household when this included a new toilet. Consequently, to be able to invest in services is one of the criteria set by the local group for entry into ecotourism. This can be a barrier for the poorest in the village, who cannot afford the cost of entry or participation in some activities. This investment can also be wasted if there are not enough visitors.

Despite these constraints, ecotourism has overall positively affected the economy of the villagers. In each of the four case studies there were similar proportions of people (around 14–18 per cent) involved in CBET, and ecotourism income was mainly supplementary rather than primary income for households. Table 12.3 compares the income from tourism with household incomes from agriculture and other income of sample families. In Umphang, where tourist numbers were very much larger than elsewhere and some individuals played multiple roles in the CBET, people earned a relatively large share from tourism – often more than half their overall incomes, but in the other three cases people mostly earned around 10 per cent from ecotourism. Most were happy with this additional income and positive about its social and cultural benefits.

The analysis of case studies shows both positive and negative impacts of CBET development. Positively, the local communities earn income from their efforts and distribute it directly to the ecotourism members, although the distribution among them is inequitable and the sum distributed to other community members is minimal. The process can enhance nature conservation and environmental awareness among villagers. Negatively, ecotourism management in the communities creates and expands conflicts among stakeholders due to mismanagement, struggles for power, competitiveness and different approaches to development. It can also increase social tensions in the communities if some are seen to benefit more than others thanks to patronage by the village leader, or if some are seen to go it alone in competition with and/or at the expense of the community initiative. The pratice of 'enclosure' of resources is evident in the cases of Khiriwong and Sasom, and accepted among locals and their neighbours but not by outside stakeholders, particularly government and business groups. This situation presents key challenges to the empowerment of local people and cooperation with others on resources management.

Conclusion: inventing tradition and claiming resources

Ecotourism at the community level is dynamic and takes diverse forms. Each community described here differs in history, size, location, major source of initiation of ecotourism, and type of supporters, but the development of community tourism is similar in terms of aims, strategies and outcomes. They all aspire to achieve local benefits without incurring the adverse impacts on society, culture and environment commonly associated with large-scale mainstream tourism.

CBET has become an arena for asserting control of local natural resources in the form of tourist attractions and the rights to manage them. Community management is based on the principle that collective concepts of resource control are preferred over individually run enterprises. Ecotourism clubs, consisting of village-level associations of service providers, are established as collective

Table 12.3 Ratio of income from tourism to total income of some involved families, Thailand

Household[a]		Income (baht)[b]			Main occupation of household	Tourism role/ occupation
Respondent	Case	Total	Tourism	Share (%)		
Sayan	Khiriwong	60,000	7,000	11.7	Orchard plantation and craft employment	ET manager
Kanya	Khiriwong	74,000	24,000	32.4	Orchard plantation	Local products worker
Nipat	Khiriwong	115,000	35,000	30.4	Resort owner, TAO and orchard plantation	Leader and operators
Nu-soem	Khiriwong	115,000	15,000	13.0	Orchard plantation and construction/	Home-stay
Somboon	Khiriwong	203,000	3,000	1.5	Orchard plantation	Home-stay
Urai	Khiriwong	242,000	21,000	8.7	Orchard plantation	Local craft producers
Prachueb	Khiriwong	287,600	18,000	6.3	Orchard plantation, TAO/café and ET group	Leaders
Liang	Sasom	7,330	980	13.4	Paddy and gardening	Leader, home-stay and guide
Bubpha	Sasom	15,050	1,050	7.0	Paddy	Guide and home-stay
Aed	Sasom	24,000	3,000	12.5	Paddy	Crafts and home-stay
Khieo	Sasom	36,000	1,800	5.0	Paddy	Local guide
Sawaeng	Sasom	48,260	2,000	4.1	Shop owner and farmers	Grocery shop

Panya	Tha Madua	123,000	9,000	7.3	Shop, transport	Leader and home-stay
Niam	Tha Madua	19,900	2,000	10.1	Village headman, plantation	Leader and home-stay
Chanpen	Tha Madua	259,000	19,000	12.0	Teacher and ET members	ET coordinator
Johaejae	Umphang (Khotha)	36,500	31,500	86.3	Paddy, shop, house and elephant owner	Grocery shop; house and elephant for hire
Phanom	Umphang (Khotha)	55,992	36,000	64.3	Paddy, elephant owner	Headman, elephant for hire, home-stay
Assawin	Umphang (Khotha)	9,996	0	0	Paddy	Not involved
Niphaporn	Umphang (town)	300,000	300,000	100.0	Tour operators	Ecotour and tour guide
Khanuengha	Umphang (town)	660,000	420,000	63.6	Teachers	Food stall and chef
Kanchana	Umphang (town)	126,240	54,000	42.8	Officer	Shop
Maitri	Umphang (town)	67,240	7,000	10.4	Farmer, employee	Punter and guide
Rattikorn	Umphang (town)	420,000	300,000	71.4	Business	Shop and adventure gear for hire
Somkit	Umphang (town)	133,200	60,000	45.1	Farmer and guide	Guide
Average of 29 households	Umphang (township)	187,913	62,944	33.5	Officers, students, shop owners, employees and farmers	Grocery and food shops, tours, guides, punters, chefs and general employees

Source: Leksakundilok (2004), field surveys in 2001.

Notes

a Selected households involved with tourism, showing their income from tourism compared with their total income.

b Total income is estimated without subtracting expenditures.

managers, and work with other village groups to share and assign responsibility for self-sufficient, local economic development. Of the four cases, Khiriwong exemplifies the most advanced integration of ecotourism and community development through the establishment of a cooperative centre, and it has also been well supported by local state authorities, who have allowed the ecotourism activities access to forest areas nominally under state jurisdiction.

Despite its partial economic successes and its role in asserting rights to use and manage local resources, CBET continues to face multiple challenges. As these studies have shown, limitations need to be understood in terms of sociocultural, political and environmental conditions of the community. Internal constraints are governed by power relationships, management culture and the levels of knowledge required to cope with tourism and sequester its benefits. External factors include cooperation with private and state actors, and especially the ability to find a place in the wider, competitive tourism market, as well as securing rights in resource access and management normally reserved as the prerogative of state agencies. This challenges local people to develop CBET in harmony with both the conditions of Thai society in general and the specific local conditions.

Not surprisingly, villagers have tended to turn to those resources over which they have greatest control, namely cultural resources. Yet they have also secured or maintained access to state resources, through the process of ecotourism itself, which has made local authorities more aware and understanding of community claims on their local environments. Inevitably, the packaging of cultural resources has involved a degree of inventing tradition both to attract tourists and to reify community ownership. Whether it is the staged cultural traditions of Sasom village or the toned-down traditional transport of Ban Tha Madua, culture is presented, and thereby transformed, with the wants of tourists in mind.

References

Acott, T.G., La Trobe, H.L. and Howard, S.H. (1998) 'An Evaluation of Deep Ecotourism and Shallow Ecotourism', *Journal of Sustainable Tourism*, 6: 238–53.

Amnuay T. (1996) 'Promoting Tourism in Thailand Worldwide', in *36th Anniversary of Tourism Authority of Thailand*, Bangkok: Tourism Authority of Thailand (TAT).

Boyd, S.W. and Butler, R.W. (1996) 'Managing Ecotourism: An Opportunity Spectrum Approach', *Tourism Management*, 17: 557–66.

Britton, S.G. (1982) 'The Political Economy of Tourism in the Third World', *Annals of Tourism Research*, 9: 331–58.

Chapman, D.M. (1995) *Ecotourism in State Forests of New South Wales: Who Visits and Why?*, Sydney: State Forests of New South Wales.

Cohen, E. (1996) *Thai Tourism: Hill Tribes, Islands and Open-ended Prostitution*, Bangkok: White Lotus.

Cummings, J. (1999) *Thailand*, Melbourne: Lonely Planet.

Deshler, D. and Sock, D. (1989). *Community Development Participation: A Concept Review of the International Literature*, Ithaca, NY: Cornell University.

EIU (1995) *EIU International Tourism Reports No 3*, London: Economist Intelligence Unit.

Fennell, D.A. (1999) *Ecotourism: An Introduction*, London: Routledge.

Higham, J.E.S. and Luck, M. (2002) 'Urban Ecotourism: A Contradiction in Terms?', *Journal of Ecotourism*, 1: 36–51.

Honey, M. (1999) *Ecotourism and Sustainable Development: Who Owns Paradise?*, Washington, DC: Island Press.

Leksakundilok, A. (2004) *Community Participation in Ecotourism Management in Thailand*, unpublished thesis, School of Geosciences, University of Sydney.

Midgley, J. (1986) *Community Participation, Social Development, and the State*, London: Methuen.

Mowforth, M. and Munt, I. (1998) *Tourism and Sustainability: New Tourism in the Third World*, London: Routledge.

Nash, D. (1989) 'Tourism as a Form of Imperialism', in V.L. Smith (ed.), *Hosts and Guests: The Anthropology of Tourism*, second edition, Philadelphia, PA: University of Pennsylvania Press.

Orams, M.B. (1995) 'Towards a More Desirable Form of Ecotourism', *Tourism Management*, 16: 3–8.

Pleumarom, A. (2002) Eco-tourism or Eco-terrorism? Global Development Research Center (GDRC). Online. Available <www.gdrc.org/uem/eco-tour/negative.html> (accessed 10 March 2003).

Selener, D. (1997) *Participatory Action Research and Social Change*, Ithaca, NY: Cornell Participatory Action Research Network, Cornell University.

TAT (2005). Target of Tourism in Thailand 1997–2006, Tourism Authority of Thailand. Online. Available <www2.tat.or.th/stat/web/static_index.php> (accessed 25 December 2005).

Thailand Institute of Scientific and Technological Research (TISTR) (1997) *An Operational Study Project to Determine Ecotourism Policy* [in Thai], Bangkok: TISTR.

Weaver, D.B. (2001a) *Ecotourism*, Milton, Qld: John Wiley.

—— (2001b) 'Ecotourism as Mass Tourism: Contradiction or Reality?', *Cornell Hotel and Restaurant Administration Quarterly*, 42: 104–12.

WTTC (2005) *Thailand: Travel and Tourism: Sowing the Seeds of Growth*, London: World Travel and Tourism Council (WTTC).

Yos Santasombat (2001) *Kanthongthiao choeng niwet khwamlaklai thang watthanatham lae kanchatkan sabphayakon* [Ecotourism, Cultural Diversity and Natural Resource Management], Chiang Mai, Thailand: Khrongkan Patthana Ongkhwamru lae Sueksa Nayobai Kanchatkan Sabphayakon Chiwaphab nai Prathetthai (BRT).

13 Ecotourism and indigenous communities

The Lower Kinabatangan experience

Rajanathan Rajaratnam, Caroline Pang and Isabelle Lackman-Ancrenaz

Protected areas represent the heart of the world's political and economic commitment to conserve biodiversity and other natural and cultural resources (Borrini-Feyerabend *et al.* 2004). Covering more than 10 per cent of the world's land surface, their success as a tool for conservation largely depends on whether they are effectively managed to protect the values they contain (Hockings *et al.* 2006). The management of protected areas has been predominantly the responsibility of various local, state and/or national government agencies, often with influence from other interested parties, including ecologists, social scientists, conservation and human-rights advocates, legislators, policy makers and NGOs. In the last decade or so, protected area management has seen a dramatic shift from the conventional 'exclude people to protect nature' approach (Dowie 2005) to that of addressing the issues and concerns of people living within or adjacent to protected areas, particularly those of indigenous communities (Borrini-Feyerabend *et al.* 2004). As such, it is increasingly acknowledged that natural resources, people and culture are fundamentally interlinked (Wilson 2003). Since indigenous communities are often reliant on local resources for their livelihood (clean water, food, minor forest produce) and socioeconomy (Gadgil 1990), the stage is further set for loss of traditional access to resources (Eagles *et al.* 2002), potential wildlife–human conflicts (Wang *et al.* 2006) and loss of income from crop and property damage (Nepal and Weber 1995), as an indirect consequence of the protected area. The promotion of tourism, particularly nature-based tourism and/or ecotourism, is often highlighted as a strategy to address these conflicts by injecting so-called 'tourist dollars' into the socioeconomy of the indigenous community in question (see Ceballos-Lascuráin 1996; Eagles *et al.* 2002). However, a high proportion of tourism in or around protected areas is often commercially operated, resulting in indirect and sometimes negligible 'flow-on' benefits to indigenous communities, despite providing employment as guides, boatmen or porters and/or contributing to the local economy through the purchase of traditional handicrafts. This chapter explores and addresses the positive socioeconomic and conservation outcomes from the integration of ecotourism into the economy of a local indigenous community adjacent to a totally protected area on the island of Borneo.

The Lower Kinabatangan floodplain, located in the Malaysian state of Sabah on the island of Borneo, is an important wetland in Malaysia and much focus has been given to its rich wildlife (Sebastian 2000). Its principal river, the 560km-long Kinabatangan, is the largest river in Sabah and the second largest in the nation. Resident, charismatic wildlife in the floodplain includes the Bornean orang-utan (*Pongo pygmaeus morio*), Bornean elephant (*Elephas maximus borneensis*) and the unique proboscis monkey (*Nasalis larvatus*), all of which are endemic to Borneo. The floodplain is also one of the few places in the world that is home to ten sympatric primate species. In addition, all eight species of Borneo's spectacular hornbill, the oriental darter (*Anhinga melanogaster*) and the rare Storm's stork (*Ciconia stormii*) are present here, along with a myriad of other species of birds, amphibians and reptiles including large water monitors (*Varanus salvator*) and the estuarine crocodile (*Crocodylus porosus*). Lackman-Ancrenaz and Ancrenaz (1997) recorded approximately 95 species of mammals and 238 species of birds in the Lower Kinabatangan.

However, like the vast majority of unprotected forested areas in Borneo, the Kinabatangan floodplain has not been spared from exploitation for highly prized tropical hardwood and the development of agriculture. Over 40 years of selective timber extraction, followed closely by the opening up of large privately owned agricultural plantations of oil palm (the current mainstay of Sabah's economy), has resulted in substantial forest fragmentation. Nevertheless, the rich and varied wildlife, including the wide-ranging orang-utan and Bornean elephant, continue to persist in these fragmented and disturbed habitats. After a long history of requests to protect the area by various bodies, both governmental (Sabah Wildlife Department) and non-governmental (World Wide Fund for Nature Malaysia, HUTAN, The Sabah Society and others) alike, the state government gazetted approximately 26,000 ha of land in the Lower Kinabatangan as a wildlife sanctuary in August 2005 (Map 13.1).

Long before the sanctuary was gazetted, tourism in the highly biodiverse Lower Kinabatangan was realised as an economically viable venture that was, in the beginning, mainly undertaken by enterprising metropolitan entrepreneurs as nature tourism. In the last decade, nature tourism here has rapidly evolved from a fledgling industry to one of international significance and reputation. The Lower Kinabatangan has been acknowledged as a prominent tourism destination in the National Ecotourism Plan (Ministry of Culture, Arts and Tourism 1996), which was launched in 1998. The area has also been recognised for its nature tourism value in the state-based Sabah Tourism Master Plan. Readily observable, captivating wildlife along the Kinabatangan River and its tributaries, such as the unusual proboscis monkey and majestic hornbill, plus the occasional opportunity to observe orang-utans and Bornean elephants, have been actively promoted as a tourist attraction by both the state tourism board and private tour operators. These particular species have been marketing icons that have created wide publicity for the area. As aptly stated by Houston Zoo's Director of Science and Conservation, who recently visited the area, 'this is the only place in the world where orang-utans and elephants coexist in close proximity and where we are privileged enough to view them easily in their natural environment.'

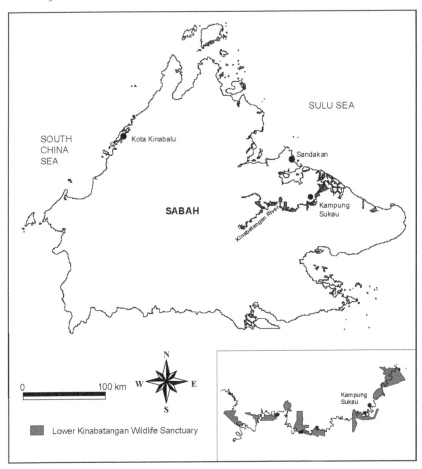

Map 13.1 Lower Kinabatangan Wildlife Sanctuary, Sabah, Malaysian Borneo.

The booming nature tourism industry in the Lower Kinabatangan is mainly centred on Kampung Sukau (Map 13.1), home to the indigenous Orang Sungai (River People) community. For such a healthy and thriving industry, there is negligible socioeconomic return to the local Orang Sungai, who have traditionally relied on the Kinabatangan and its environs for important resources like freshwater fish, prawns and forest produce. This chapter specifically focuses on Red Ape Encounters and Adventures (RAE), a community-based ecotourism project owned and managed by the villagers of Sukau, one of the many stakeholders in the Kinabatangan Wildlife Sanctuary. This model project is a collaborative effort between the Sabah Wildlife Department (SWD) (the official government agency with full jurisdiction over the sanctuary), the Kinabatangan Orang-Utan Conservation Project (KOCP, run by HUTAN, a French NGO) and the local community. The underlying concept behind RAE epitomises ecotourism, with its focus on integrating wildlife and habitat conservation with tourism for sustainable socioeconomic development of the community. This chapter also addresses important issues in the setting up of RAE, such as cultural and land use challenges, as well as strategies employed to maximise economic benefits to the community. It also looks at how RAE has incorporated cultural experience into ecotourism programmes, a concept not fully emphasised by other commercial nature tourism operators in the area. Most importantly, it highlights the successful partnership between all relevant stakeholders in integrating sustainable tourism into the traditional economy.

The sanctuary and its environs

The Kinabatangan floodplain is the largest in Malaysia, comprising a diverse array of habitats that range from lowland dipterocarp forest, freshwater swamp forest, riverine forest and limestone caves, to oxbow lakes and coastal mangrove forests (Boonratana 1994). Large-scale timber exploitation of its forests commenced as early as the 1930s, using traditional methods (Hutton 2004), before rapidly progressing from the 1950s, with the introduction of bulldozers, tracked skidders and the chainsaw (Payne *et al*. 1994). At the peak of the local logging industry in the Lower Kinabatangan during the 1970s and 1980s, the Kinabatangan River was the main conduit for the transportation of logs for export from the large coastal town of Sandakan (Map 13.1). With timber resources rapidly dwindling soon after, permanent agriculture in the form of large oil palm plantations gained prominence, to become the current mainstay of Sabah's economy (Payne *et al*. 1994; Rajaratnam and Ancrenaz 2004). Large areas of forest in the Kinabatangan floodplain have been converted to this monocultural crop, with some single estates covering as much as several thousand hectares. Kinabatangan District Office records in 2005 reveal that a total area of 268,358 ha within the district has been planted with oil palm, resulting in the remaining forests here existing as small, isolated fragments in an agricultural landscape.

Despite the long history of exploitation and development, conservation efforts in the Lower Kinabatangan date back to the 1980s, during which scientific research (Payne 1988, 1990; Dawson *et al*. 1993) began to show significant evidence for wildlife conservation, focusing on big, wide-ranging 'umbrella' species like the Sumatran rhinoceros, elephant and orang-utan, as well as unique species such as the proboscis monkey. The proboscis monkey population here is estimated to be the largest in the state (Boonratana 1994), and the orang-utan has achieved a significant population size of more than 1000 individuals (Ancrenaz *et al*. 2005). Payne (1989) first proposed the creation of a wildlife sanctuary. Progress towards it being gazetted was painstakingly slow, despite repeated follow-ups by relevant governmental and non-governmental agencies. In 1999, the state government made a commitment that the proposed sanctuary would be 'Malaysia's gift to the earth' before giving it some status as a bird sanctuary in 2001. The approximately 26,000 ha sanctuary was eventually permanently gazetted in August 2005. The sanctuary comprises several fragmented forest remnants or 'lots' (Map 13.1) interspaced with private oil palm plantations, linking the mangrove areas near the coast with existing forest reserves further inland.

The people of the Kinabatangan

The Orang Sungai (River People) is the local indigenous group living in the Lower Kinabatangan area. According to Hutton (2004), their name originated during colonial rule by the British, who collectively named all communities living along the Kinabatangan as the River People. Over the years, they have progressively intermarried with other indigenous races and migrant groups such as the early Chinese immigrants, Bugis from Indonesia, and Suluks and Cagayans from the Philippines. They now live largely in scattered settlements along the Kinabatangan from the upper to the lower reaches. The main community of focus in this chapter consists of the people of Kampung Sukau near the Kinabatangan Wildlife Sanctuary.

Kampung Sukau is a community of approximately 1200 people made up of 150 families who are all of the Muslim faith. It is accessible by unsealed road branching from the main Sandakan–Lahad Datu road or, alternatively, by boat along the eastern coastline of Sabah (Map 13.1). Basic infrastructure exists in the form of a small government clinic, a primary school and a larger secondary school (up to Year 10) 3 kilometres away. Electricity is available in the village but the supply is not consistent. There is still no clean running water and villagers rely on rainwater and river water from the Kinabatangan.

As their indigenous name suggests, the main economy of the residents of Kampung Sukau revolves around the Kinabatangan River and its forests. Fishing for freshwater fish and prawns, plus the collection of forest produce such as rattan, forest fruits and vegetables, have been their traditional activities (Lackman-Ancrenaz and Ancrenaz 1997). Fishing is still carried out using traditional methods such as cast nets (*rambat*), trammel nets (*pukat*) and fish traps (*bubu*). Some villagers practise some subsistence farming, mainly for fruits and vegetables. With the advent of oil palm cultivation in recent times, most of the villagers have opened up small-scale oil palm holdings on their individual lands. A few enterprising individuals have also opened small provision shops, and another villager has set up a bed-and-breakfast lodging. Other residents supplement their income with seasonal work including forest clearance for oil palm companies, collection of edible birds' nests from surrounding limestone caves and building small-scale village infrastructure (Lackman-Ancrenaz and Ancrenaz 1997). Overall, the economic existence of the village, although generally stable, can be precarious at times, especially during the rainy monsoon season between November and March when periodic floods occasionally cut off the village.

Nature tourism in the Lower Kinabatangan

With its rich and abundant wildlife, the tourism potential of the Lower Kinabatangan was first highlighted by Payne (1989). It is important to distinguish here between ecotourism and nature/wildlife-based tourism. As outlined by Ceballos-Lascuráin (1996), the essential difference is that ecotourism signifies the involvement and provision of socioeconomic benefits to local communities. A common misconception is to equate nature tourism in the Lower Kinabatangan with ecotourism. This is not the case because no significant and consistent socioeconomic benefits to the local Orang Sungai have been achieved since the inception of the tourism industry here, a fact observed by the authors, who are associated with the area through wildlife and human socioeconomic research, and echoed by various residents of the community.

The first tourist lodge was opened in 1991 in Kampung Sukau, taking advantage of the abundant proboscis monkeys and associated riverine wildlife along Sungai Menanggul, a nearby tributary of the Kinabatangan River. News of the rich and easily viewed wildlife spread, encouraging more metropolitan-based tour companies to build their lodges. Today, at least six different metropolitan tour companies operate lodges in Sukau, making it the hub of the nature tourism industry in the Lower Kinabatangan. For example, in 2001, the lodges received almost 18,000 tourists of which more than 80 per cent were foreign (Yoneda 2003), a big jump from the 13,000 tourists received in 2000 (Red Ape Encounters, unpublished data).

The majority, if not all, of the commercial wildlife tour operators out of Sukau offer a standard two-day/one-night package, which includes airport transfers to and from Sandakan, comfortable en-suite lodging, all meals and a morning and evening guided boat cruise. Boat cruises concentrate mainly on the Menanggul River but also cover the main Kinabatangan River. The main attractions in these boat cruises are undoubtedly the unusual proboscis monkey, other primate species such as long-tailed macaques (*Macaca fascicularis*) and silver langurs (*Presbytis cristata*), and birdlife such as hornbills, egrets, darters and kingfishers. Tourists also experience the ambience associated with a boat cruise, namely the misty mornings and sunsets, not to mention views of the picturesque Sukau village and the everyday riverside activities of its human inhabitants. Occasionally tourists might see a wild orang-utan by the river or a herd of elephants if they happen to be in the vicinity. These minimum packages vary in cost but generally are in the RM450–650 (US$118–171) range per person (twin share) depending on the comfort of the lodging. Optional trips such as a night cruise to view buffy fish owls (*Ketupa ketupu*) and black-crowned night herons (*Nycticorax nycticorax*), or a trip to a nearby oxbow lake, cost extra.

There are no official estimates of the tourism revenue being generated in the Lower Kinabatangan, particularly that revolving around Sukau. However, available data on tourism arrivals in 2000 and on projected expenditure were analysed by the authors to reveal that the total revenue may have achieved a figure in the region of RM5.3 million. For a thriving industry presumably generating substantial revenue, there is, however, negligible socioeconomic return to the indigenous people of Sukau. The metropolitan-based tour operators have the financial capital, marketing inroads into the domestic and international tourist market, and expertise to operate tourism facilities, but employment in and participation by the local community is low, a view echoed by Yoneda (2003). Personal observations by the authors reveal that tour operators provide their own boats instead of hiring boats from the community. Similarly, all tour guides and the majority of the boat drivers are from either the state capital, Kota Kinabalu, or nearby Sandakan. Boat fuel and all supplies for the lodges are also not purchased within the village, resulting in high economic leakage. What little employment there is for the locals exists mainly in housekeeping duties and cleaning of the lodges, most often at meagre monthly wages. The lone village bed-and-breakfast does provide optional boat cruises for its visitors, but with little economic return for the community.

The sanctuary: friend or foe

Conservationists, both local and international, unanimously supported the creation of the Kinabatangan Wildlife Sanctuary for its intrinsic value in protecting the rich wildlife still persisting in the remaining forest fragments of the Lower Kinabatangan floodplain. Furthermore, the extensive conversion of forests to oil palm and the resulting siltation and agricultural chemical runoff have polluted the waters of the Kinabatangan River, with serious repercussions for fisheries, an important livelihood for the Orang Sungai (Sebastian 2000). Again, conservationists argued that the creation of the sanctuary prevented further development, thereby affording increased protection of Kinabatangan's fresh water and its resources for the locals (SWD 1994; Sebastian 2000).

Oil palm plantations have also fragmented the rainforest in the region, which has posed a major obstacle to wildlife movement, especially for wide-ranging species such as elephants and orang-utans (Dawson *et al.* 1993; Lackman-Ancrenaz and Ancrenaz 1997). These animals have, in turn, been forced to encroach into plantations and orchards owned by villagers, leading to the emergence of wildlife–human conflict mainly through crop-raiding, which has threatened the ability of the locals to secure sustainable livelihoods (Nepal and Weber 1995; Wang *et al.* 2006). To most of the villagers of Sukau, the sanctuary is seen as a source of crop pests. They also see its creation as preventing access to traditional resources from the forest such as timber for construction, rattan and medicinal plants. With some resentment, they question the priority of the sanctuary towards wildlife conservation over their traditional use of the forest. This, in conjunction with the lack of direct economic return from metropolitan-based nature tourism in the area, created a situation whereby the local Orang Sungai, who are integral stakeholders, do not appear to directly benefit from the creation of the sanctuary. The orang-utan symbolises this conflict of interest; while the conservation of the highly endangered red ape is evoking great concern throughout the world (Ancrenaz *et al.* 2005), the Orang Sungai in the Lower Kinabatangan regard the orang-utan primarily as a pest to their crops and a possible cause of their loss of access to resources (Figure 13.1). It was therefore critical that the SWD, the government agency responsible for managing the sanctuary, took into account the local community's concerns as well as those of other stakeholders including adjacent oil palm plantations, commercial tour operators and the wider conservation community (SWD 1994). In line with Brechin *et al.* (2003), in the interest of advocacy of biodiversity conservation with social justice, it was imperative to work with, rather than against, indigenous communities, the private sector and NGOs, with the proviso that these parties were committed to basic conservation goals.

Figure 13.1 The orang-utan.

Red Ape Encounters and Adventures: a community-based orang-utan ecotourism project

In 2001, in response to those concerns, a community-based orang-utan ecotourism project called Red Ape Encounters and Adventure (RAE) was launched in Kampung Sukau. It had preliminary funding for two years from the Danish International Development Agency (DANIDA) and was initiated by SWD and HUTAN. HUTAN is a French NGO, which set up the Kinabatangan Orang-Utan Conservation Project (KOCP) in 1998 in collaboration with SWD. The objective of KOCP was to achieve long-term viability of orang-utan populations in Sabah by restoring harmonious relationships between humans and the orang-utan. This was carried out through rigorous scientific research on the ecology and behaviour of orang-utans in the Lower Kinabatangan, which made use of local research assistants scientifically trained by HUTAN. An important goal was to integrate habitat and wildlife conservation with the socioeconomic development of communities who are stakeholders in the Kinabatangan Wildlife Sanctuary. Since the inception of KOCP, HUTAN has trained almost 50 villagers from Sukau and other surrounding villages to assist and participate in the scientific research on orang-utans within the sanctuary.

RAE was set up after much consultation between SWD, HUTAN and the villagers of Kampung Sukau. The consultations created a platform for the local community to voice their opinions and concerns. Thoughts, ideas and opinions from relevant parties were heard about how to forge this ecotourism project. Although there was some minor scepticism and apprehension during this phase, parties were able to engage in constructive debates to arrive at workable community decisions by focusing on processes that could economically benefit the village, while at the same time applying conservation measures through community-based initiatives.

This initial consultative phase proved integral, since many ecotourism development projects geared toward communities have a tendency to impose the idea of tourism development onto individuals whose main interest is not tourism. Projects with this philosophy are often unsuccessful because their top-down nature ignores the very culture of the industry they are trying to develop. Instead of top-down approaches, grassroots approaches to ecotourism development that place local entrepreneurs with a passion for conservation at the centre of the process can have a greater probability of success (Borrini-Feyerabend *et al.* (2004) provide a comprehensive review of such ventures elsewhere in the world). The next phase in the formulation of RAE was to develop the organisational structure for this community-based ecotourism project. Eight keen villagers who were scientifically trained by KOCP formed the nucleus for this project. Their local knowledge combined with scientific training and understanding of the ecology and behaviour of orang-utans were invaluable as a starting platform for RAE.

Experts in business management were also engaged to identify opportunities for income generation, resulting in the formulation of a comprehensive business plan. The final step was to legally register the project as a company in order to acquire an official licence to operate as a tour operator in accordance with government requirements. This resulted in RAE being the first community-based ecotourism pilot project in Malaysia focusing on orang-utans in the wild, with the potential to be replicated on Borneo or applied globally in areas that combine ape conservation with ecotourism.

RAE's ecotourism approach

RAE models its programmes on the four basic principles of ecotourism outlined by Epler Wood (1996) as follows:

- avoids negative impacts that can damage or destroy the integrity or character of the natural or cultural environments being visited;

- educates the traveller about the importance of conservation;
- directs revenues to the conservation of natural areas and the management of protected areas;
- brings economic benefits to local communities and directs revenues to local people living adjacent to protected areas.

This is echoed in the words of RAE's Managing Director:

> While we see this venture as a means to inject more economy into our community, we have not lost sight of the overall picture which is the conservation of orang-utans and other wildlife in the sanctuary. RAE is a means whereby we can showcase our recognition of this invaluable resource by educating our visitors and other villagers alike on orang-utan ecology, behaviour and the conservation activities actively carried out by HUTAN and the Sabah Wildlife Department in the Kinabatangan Wildlife Sanctuary.
> (Sahdin Lias, personal communication 14 November 2006)

RAE is unique when compared with existing commercial tour companies operating in Kampung Sukau. Its uniqueness stems from the following factors:

- In line with traditional native customary laws, RAE has been awarded the exclusive right by the SWD to bring visitors to a well-planned and controlled ecotourism site in the Lower Kinabatangan Wildlife Sanctuary. It offers the best possible chance for a visitor to experience a true 'encounter' with an orang-utan in its natural environment.
- RAE's licensed nature guides are all scientifically trained research assistants for HUTAN. They are highly knowledgeable locals who have an intimate understanding of the forest and its diverse inhabitants. Being conversant in English puts them in the best possible position to effectively impart their knowledge to international visitors. Besides orang-utan research, they also have a sound knowledge of other taxonomic groups such as birds, reptiles and plants. This all-round ability enables RAE to cater flexibly to special-interest visitors such as botanists, ornithologists and herpetologists.
- RAE's ecotourism programmes are not merely nature sightseeing tours but also enhance knowledge about the local wildlife and the environment. They are carefully designed to ensure that conservation messages about orang-utans and other wildlife, plus the importance of the Kinabatangan floodplain, are clearly communicated to the target audience. These programmes also provide a unique 'hands-on' opportunity for visitors to experience and participate in activities directly related to conservation and research initiatives on orang-utans, such as reforestation, collecting behavioural data on orang-utans through observation, and tree species data collection. In the reforestation activity, visitors participate in the replanting of primary tree species in a highly degraded area in the sanctuary that has been specifically earmarked for forest recovery.

RAE is also committed to promoting and showcasing responsible ecotourism practices. Its operation is based on strict guidelines to reduce impact on the environment and the orang-utan. RAE's specific ecotourism programme incorporates trips to observe orang-utans in the sanctuary in ten months of a year. For the remaining two months, the orang-utan study area is left to rejuvenate naturally. Similarly, the carrying capacity of the forest with regard to visitors is carefully respected, with no more than five visitors allowed per trip. Based on its experience thus far, RAE is developing a set of guidelines for responsible viewing of wild orang-utans for eventual endorsement and adoption by the World Conservation Union (IUCN) for future use in areas involving orang-utans and ecotourism. These guidelines cover, among other things, important protocols such as length of time allowed per viewing, noise levels during viewing, minimum distance allowed between observers and orang-utans, and hygiene issues, such as not allowing unwell visitors to view orang-utans to prevent the risk of transmitting pathogens.

RAE's package

In line with international ecotourism packaging standards, RAE charges approximately RM1115 (approximately US$290) per night per person (twin share) for a basic three-day/two-night discovery programme. The programme includes transfers to and from Sandakan airport, accommodation with villagers under the 'home-stay' concept, all meals, two afternoon boat cruises along the Kinabatangan and the smaller Menanggul, a morning boat cruise to a nearby oxbow lake, and a visit to the adjacent Gomantong Caves to experience a cave ecosystem with its associated fauna, including bats and swiftlets. The highlight of this programme is undoubtedly the all-important trip into the sanctuary to 'encounter' an orang-utan and learn more about this flagship species and its conservation. If unsuccessful here, visitors have at additional cost the option of joining a five-hour jungle trek to locate an orang-utan. Other optional tours include boat cruises at night to view buffy fish owls, black-crowned night herons and crocodiles. RAE is also flexible and can organise specially catered programmes for up to two weeks. The 'home-stay' style accommodation, although spartan, provides the visitor with a 'personal' experience of the local culture. If numbers suffice, visitors are often entertained by the local village *warisan* or cultural troupe with traditional songs and music, using traditional musical instruments.

RAE and the community

Although still relatively recent, RAE has begun to contribute appreciably to the socioeconomy of Kampung Sukau. It was initially launched as a pilot endeavour in 2003 to test its viability and was an astounding success in its first two years of trial between 2003 and 2004. During this period, RAE received more than 270 visitors to generate an income of approximately RM157,423 (approximately US$41,427). The majority of these visitors were sourced through network contacts (supporting agencies like HUTAN and DANIDA) and subsequent word of mouth. RAE was made official in 2005 with its own business plan, marketing strategy and website at www.redapeencounters.com. In its first year of official operation, RAE received 109 visitors to generate an income of approximately RM78,215 (approximately US$20,582), a modest start but hopefully a prelude to higher numbers.

A benefit-sharing mechanism has been designed to ensure that opportunities from tourism have direct benefits for the local community. RAE supports only local community services such as the provision of boats and their drivers, 'home-stay programmes' as mentioned above, guides, hiring of the *warisan*, purchase of locally made handicrafts and other forms of transportation. Thus, the majority of tourist dollars generated by RAE are transparently distributed within the local community to enhance the local economy, except for a small percentage that may be paid to outside contractors for services that are not available in the village.

Figure 13.2 shows the profit allocation (in percentage terms) from the ecotourism project to the various sections of the community. The major allocation (44 per cent) is towards RAE's operational costs, including salaries of staff, administrative costs and other associated costs in running a licensed business. RAE maintains a permanent full-time staff of ten, thus directly contributing to their personal household incomes and that of extended families as well. The home-stay programme attracts 11 per cent of funds, which indirectly supplements the household income of the enrolled families. When this programme was initiated as part of RAE's tourism package, five families applied and were officially licensed by the State Ministry of Tourism, thus ensuring an acceptable standard of accommodation in line with ministry guidelines. Irrespective of tourist numbers, RAE employs a rotation system among the families, thus guaranteeing a fair and equitable allocation of income. In addition to the direct income from food and accommodation, the host families are also able to generate additional income through the sale of local handicrafts to tourists. Some home-stay families have ingeniously used their weaving skills to produce attractive baskets and vases out of newspapers and coloured plastic bags, further promoting and showcasing environmental consciousness and friendliness. The home-stay programme, although relatively small, has been an astounding success in supplementing individual household income in the community. At the time of writing, a further seven families had applied to be enrolled into the programme and were awaiting permits from the State Ministry of Tourism. Although RAE has its own boat and permanent boatman, 11 per cent of its profits supplements the income of at least seven households who own boats. These village boats are used for river cruises and trips to RAE's study site for groups of more than two tourists. As with the home-stay programme, the use of village boats is rotated to ensure an impartial distribution of income. RAE anticipates that the number of village boats and boatmen on the books will increase to at least 15 in the future.

RAE also regularly employs guides from the village to complement its own resident guides for large groups of tourists. Priority for the hiring of village guides is given to those households that do not have a regular stream of income such as from cash crops or fishing, thereby improving the overall economic wellbeing

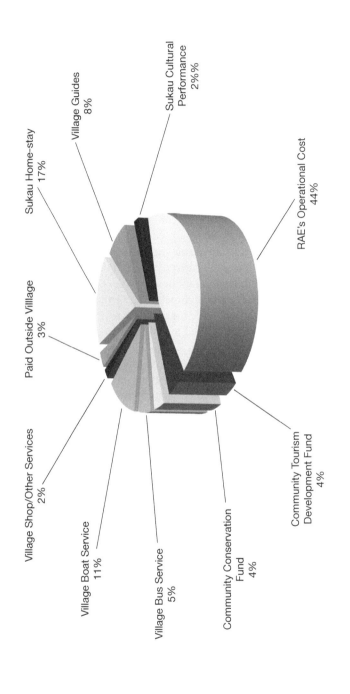

Figure 13.2 Village benefit-sharing, community-based ecotourism project, Kampung Sukau, Sabah, Malaysia.

of the community. The village bus service, made up of three minibuses and two four-wheel-drive (4WD) vehicles, is additionally used to transfer large groups of tourists from the Sandakan airport. The 4WDs are especially useful during the wet season when the road into Sukau can be treacherous. The village shop is occasionally used to procure supplies like mineral water and snacks for tourists. RAE, in conjunction with HUTAN, was also an important player behind the formation of the local *warisan* or cultural troupe, which consists of five widows who are skilled in traditional musical instruments (Figure 13.3). The entertainment by the troupe is offered as an additional paying option to tourists at a rate of RM300 (approximately US$80) per show. Although all of these widows are from households with some form of income stream emanating through sons and/or daughters (from employment as HUTAN research assistants, or through cash crops, agriculture or fishing), performing with the troupe nevertheless supplements household incomes and provides them with occasional economic independence.

Most importantly, two special village funds have been set up as part of RAE's economic integration with the local community. These are the Community Tourism Development Fund and the Community Conservation Fund, into which a proportion (4 per cent) of RAE's tourism revenue is channelled (see Figure 13.2). The Tourism Development Fund is managed by a village committee with a memorandum of understanding with RAE. The funds are utilised for education of nature guides, support of ongoing research and development into best practices for ecotourism, upgrading of tourism facilities, and capacity building for villagers, including training of new, interested villagers as tour guides. The Community Conservation Fund supports ongoing research by RAE (in conjunction with HUTAN) to continuously increase its knowledge of wildlife in the sanctuary

Figure 13.3 Warisan cultural troupe.

and its environs. The fund is also used to support wildlife law enforcement work by RAE and conservation education awareness programmes for the villagers. The majority of RAE's permanent staff are also honorary wildlife wardens, an official designation bestowed by the SWD in a statewide programme aimed at facilitating SWD's law enforcement operations. In line with this, RAE's honorary wildlife wardens regularly patrol the sanctuary to prevent illegal activities such as logging and poaching, using the Community Conservation Fund to support the associated costs of boats, maintenance of outboard engines and fuel. This shows the value of the fund in protecting the resources that are being used to generate income for the community. Probably the most significant use of the Community Conservation Fund is the alleviation of wildlife–human conflicts, one of the main reasons behind the community's apprehension about the Lower Kinabatangan Wildlife Sanctuary. One such successful application of this fund was in 2006 when it was used to pay RM4500 (approximately US$1185) for an electric fence around the village's communal graveyard to stop repeated encroachment by elephants. This attempt to alleviate wildlife–human conflict was an important initial step towards integrating wildlife conservation into the culture of the local community.

The response

Despite its being a relatively new community-based ecotourism venture, there has been positive response from both tourists and locals alike, based on interviews by the authors. Initial feedback from tourists has been good, especially from the conservation education and cultural viewpoints. A tourist from the United States used RAE twice and was highly impressed with its professionalism and sense of pride in wanting to integrate ecotourism, culture, wildlife conservation and education into some form of sustainable source of socioeconomic revenue for the community. This was mirrored in her statement:

> I learned a great deal about the rich, unique wildlife of the Lower Kinabatangan River and the Wildlife Sanctuary. RAE keeps virtually all the visitor dollars in the community, making it clear to the community that their wildlife are critical to their own prosperity. The community thus becomes the front line in wildlife conservation, as it certainly has with the village in which RAE works.

A Japanese tourist found that the opportunity to replant a primary forest tree (Figure 13.4) was the highlight of her trip: 'Although I did manage to see a wild orang-utan in addition to other wildlife like proboscis monkeys, silvered langurs and egrets, the tree replanting left me with a sense of fulfillment in being able to give something back to the environment of the orang-utan.' Another tourist from Switzerland felt that, while he learnt a lot about the wildlife and conservation, he was especially fortunate to experience and, more importantly, be a part of the local culture through the home-stay experience and entertainment by the local *warisan*. Thus:

The people are so friendly and hospitable and the food simple but tasty. It was always fun to walk through the village with a lot of young children trying their best to practice their English on you which is not even their first or second language. A highlight of my cultural experience was when I was cajoled into emulating the traditional dance during the cultural performance.

The home-stay in particular not only facilitates 'cultural experience' for the visitor but provides villagers participating in the programme an opportunity to meet people of other nationalities and cultures, thereby generating cross-cultural experience and awareness, something not readily available in packages offered by commercial tour operators in the area. Thus for Ibu (not her real name), a home-stay host:

I was apprehensive at first, always worried that I would make a mistake that might offend my guest. However, we always inevitably break the ice and warm up to each other. As much as I have enjoyed showing them our culture and way of life, I in turn have benefited from knowing theirs as well.

This has applied not only to the home-stay hosts but to RAE's guides as well. Several guides who were interviewed felt that interaction with tourists from different nationalities enabled them to improve their guiding skills, especially when faced with the same nationality again. Jai (not his real name), a regular tour guide with RAE, was understandably nervous at first:

Figure 13.4 Reforestation in action.

I was worried that I would not be able to adequately answer a visitor's query. However, almost all visitors from different cultures were understanding and, thanks to my scientific training with HUTAN, I became more and more confident in my answers.

The programme has brought socioeconomic benefits as well. Home-stay hosts who were interviewed all believed that that their individual household incomes had been boosted since the incorporation of this option into RAE's operations. Some felt that they were accumulating enough savings to complete extra renovations to improve the comfort levels for their guests. As one commented, 'I have been meaning to get a new mattress for the guest room and properly tile the bathroom floor. Hopefully with a few more tourist visits, I will be able to do that.' A few others felt that the extra income would prove useful in allowing their children to complete their education in boarding schools in Sandakan, important given the low rate of completion of secondary education in the community. Although it was not substantial, village boatmen were glad for the extra income facilitated by RAE as it assisted with the cost of boat and engine maintenance. One was even contemplating saving up any leftover disposable income towards the cost of purchasing a second boat:

My main income to support my family is fishing from the river and the nearby oxbow lake. I also fish for freshwater prawns, which are highly prized as far away as Sandakan but tend to be seasonal especially towards the end of the wet season. Some years you get a good bountiful catch and some years you don't. Getting extra work through RAE has helped my household income towards supporting a wife and four children, two of whom are at school. Maybe I could save up now for a second boat and get my nephew, who is single and jobless, to run it. I can still get a share of his profit since it will be my boat.

The potential for the creation of extra employment thus exists, albeit through a flow-on effect from RAE's contribution to the community.

The future

Despite being only in its third year of operation, RAE hopes to substantially increase community participation and support for their ecotourism activities. This can occur only once the local community of Kampung Sukau is fully convinced that ecotourism can provide an alternative, viable revenue and, at the same time, protect their natural surroundings. To this effect, RAE regularly organises conservation education programmes for the residents of Kampung Sukau, focusing on its activities and trying to instil in the community a strong sense of stewardship over the Lower Kinabatangan Wildlife Sanctuary and orang-utans in particular. When it comes to competition with commercial tour operators in the area, RAE is confident that it will achieve financial viability by drawing on its advantage in offering an ecotourism alternative that attains conservation outcomes while simultaneously providing sustainable development to Kampung Sukau.

Conclusion

From the conservation management and socioeconomic viewpoints, RAE is unique. Although there are several indigenous community-based ecotourism ventures throughout the world, such as the Quechua people's Potato Park in Peru (Alejandro Argumedo, personal communication, cited in Borrini-Feyerabend *et al.* 2004), Baghmara village and the Gurung people's Annapurna Conservation Areas in Nepal (Krishna *et al.* 1999; Kothari *et al.* 2000), the Innu people's Kamestastin Lake in Labrador, Canada (Brown *et al.* 2006), and the several landowning clans (*mataqali*) of Tavoro Forest Park in Fiji (Young 1992), they are almost all modelled on Community Conserved Areas (CCAs). Kothari (2006) presents a comprehensive review of the ecological and socioeconomic benefits of CCAs, namely, areas actually owned by the community in question. RAE, on the other hand, operates in a protected area that is under the full jurisdiction of the state government (Borrini-Feyerabend *et al.* 2004). That is to say, a government body holds the authority, responsibility and accountability for managing the protected area, determines its conservation objectives, subjects it to a management regime, and often also owns the protected area's land, water and related resources. RAE showcases the importance of integrated conservation and development programmes for protected areas (Wilshusen *et al.* 2003) and highlights the crucial role that can be played by local indigenous communities at the grassroots level. It demonstrates how a successful and innovative partnership, entailing a positive approach and attitude, can be forged among the government, NGOs and indigenous communities to ensure the best use of a natural resource without deviating from the underlying objective of conserving and protecting it. RAE can be seen as the vehicle that has the potential to ensure the long-term viability of the Lower Kinabatangan Wildlife Sanctuary.

Acknowledgement

The authors would particularly like to acknowledge the help and assistance of Sahdin Lias and other staff of Red Ape Encounters. Without their willingess to share information with us, this chapter would not have been possible.

References

Ancrenaz, M., Gimenez, O., Ambu, L., Ancrenaz, K., Andau, P., Goossens, B., Payne, J., Sawang, A., Tuuga, A. and Lackman-Ancrenaz, I. (2005) 'Aerial Surveys Give New Estimates for Orang-Utans in Sabah, Malaysia', *PLoS Biology*, 3: 1–8.
Boonratana, R. (1994) *The Ecology and Behaviour of the Proboscis Monkey (*Nasalis larvatus*) in the Lower Kinabatangan, Sabah*, unpublished doctoral thesis, Mahidol University, Bangkok.
Borrini-Feyerabend, G., Kothari, A. and Oviedo, G. (2004) *Indigenous and Local*

Communities and Protected Areas: Towards Equity and Enhanced Conservation, Gland, Switzerland: IUCN Publications.

Brechin, S.R., Wilshusen, P.R., Fortwangler, C.L. and West, P. (eds) (2003) *Contested Nature – Promoting International Biodiversity with Social Justice in the Twenty-first Century*, Albany, NY: SUNY.

Brown, J., Lyman, M.W. and Procter, A. (2006) 'Community Conserved Areas: Experience from North America', *Parks*, 16: 35–42.

Ceballos-Lascuráin, H.C. (1996) *Tourism, Ecotourism and Protected Areas*, Gland, Switzerland: IUCN Publications.

Dawson, S., Malim, P. and Taraan, J. (1993) *Density and Distribution of Elephants and Distribution of Other Large Mammals in the Lower Kinabatangan, Sabah*, Kota Kinabalu: Sabah Wildlife Department.

Dowie, M. (2005) 'Conservation Refugees: When Protecting Nature Means Kicking People Out', *Orion*, 24: 16–27.

Eagles, P.F.J., McCool, S.F. and Haynes, C.D.A. (2002) *Sustainable Tourism in Protected Areas: Guidelines for Planning and Management*, Gland, Switzerland: IUCN Publications.

Epler Wood, M. (1996) 'The Evolution of Ecotourism as a Sustainable Development Tool', paper presented at the Sixth International Symposium on Society and Natural Resource Management, Pennsylvania State University, 18–23 May.

Gadgil, M. (1990) 'India's Deforestation: Patterns and Processes', *Society and Natural Resources*, 3: 131–43.

Hockings, M., Stolton, S., Leverington, F., Dudley, N. and Courrau, J. (2006) *Evaluating Effectiveness: A Framework for Assessing Management Effectiveness of Protected Areas*, second edition, Gland, Switzerland: IUCN Publications.

Hutton, W. (2004) *Kinabatangan*, Kota Kinabalu: Natural History Publications (Borneo).

Kothari, A. (2006) 'Community Conserved Areas: Towards Ecological and Livelihood Security', *Parks*, 16: 3–13.

Kothari, A., Pathak, N. and Vania, F. (2000) *Where Communities Care: Community Based Wildlife and Ecosystem Management in South Asia*, London: Kalpavriksh, Pune and International Institute of Environment and Development.

Krishna, K.C., Basnet, K. and Poudel, K.C. (1999) *People's Empowerment amidst the Peaks: Community B at Annapurna Conservation Area, Nepal*, London: Kalpavriksh, Pune and International Institute of Environment and Development.

Lackman-Ancrenaz, I. and Ancrenaz, M. (1997) *The Kinabatangan Wildlife Sanctuary*, Sandakan: HUTAN.

Ministry of Culture, Arts and Tourism (1996) *National Ecotourism Plan*, Kuala Lumpur.

Nepal, S.K. and Weber, K.E. (1995) 'Prospects for Coexistence: Wildlife and Local People', *Ambio*, 24: 238–45.

Payne, J. (1988) *Orang-Utan Conservation in Sabah*, Kuala Lumpur: WWF Malaysia.

Payne, J. (1989) *A Tourism Feasibility Study for the Proposed Kinabatangan Wildlife Sanctuary*, Kuala Lumpur: WWF Malaysia.

Payne, J. (1990) *The Distribution and Status of the Asian Two-Horned Rhinoceros (*Dicerorhinus sumatrensis harrisoni*) in Sabah, Malaysia*, Kuala Lumpur: WWF Malaysia.

Payne, J., Cubbitt, G. and Lau, D. (1994) *This is Borneo*, London: New Holland.

Rajaratnam, R. and Ancrenaz, M. (2004) *The Second Faunal Survey of Sabah*, Kota Kinabalu: Sabah Wildlife Department and DANIDA.

Sabah Wildlife Department (SWD) (1994) *Wildlife Management Plan for the Lower Kina-batangan Basin, Sabah, Malaysia*, Kota Kinabalu: Sabah Wildlife Department.

Sebastian, T. (2000) *The Lower Kinabatangan Wildlife Sanctuary*, Petaling Jaya: WWF Malaysia.

Wang, S.W., Lassoie, J.P. and Curtis, P.D. (2006) 'Farmer Attitudes towards Conservation in Jigme Singye Wangchuk National Park, Bhutan', *Environmental Conservation*, 33: 148–56.

Wilshusen, P.R., Brechin, S.R., Fortwangler, C.L. and West, P.C. (2003) 'Contested Nature: Conservation and Development at the Turn of the Twenty-first Century', in S.R. Brechin, P.R. Wilshusen, C.L. Fortwangler and P.C. West (eds), *Contested Nature – Promoting International Biodiversity with Social Justice in the Twenty-first Century*, Albany, NY: SUNY.

Wilson, A. (2003) 'All Parks are People's Parks', *Policy Matters*, 12: 71–5.

Yoneda, M. (2003) 'Nature Tourism and Lodge Management in the Kinabatangan Area', in M. Mohamed, A. Takano, B. Goossens and R. Indran (eds), *Lower Kinabatangan: Scientific Expedition 2002*, Kota Kinabalu: Universiti Malaysia Sabah Press.

Young, M. (1992) 'People and Parks: Factors for the Success of Community-Based Eco-tourism in the Conservation of Tropical Rain Forest (Tavoro Forest Park and Reserve, Fiji)', paper presented at the IUCN World Congress on National Parks and Protected Areas, Caracas, Venezuela, February.

14 Adventures, picnics and nature conservation

Ecotourism in Malaysian national parks

Norman Backhaus

Tourism in protected areas

National parks are increasingly popular with tourists. Intensified discussions concerning their ecological integrity, and the link to sustainable and ecofriendly tourism, add to their attraction. Although most protected areas recognised by the United Nations explicitly allow the presence of tourists (IUCN 2003), tourism can destroy what it covets. The number and size of protected areas have increased during the last decades, but so has the tourist demand for 'nature' (Nyaupane *et al.* 2004). Today most protected areas are located in so-called developing or newly industrialising countries. This is remarkable since nature protection in its modern form – manifesting itself in 'national parks', 'wildlife reserves', 'World Heritage sites' or 'biosphere reserves' – is based on Western concepts that have globalised (Wright and Mattson 1996; Müller-Böker *et al.* 2001; Röper 2001; Soliva 2002). Consequently, the (local) implementation of conservation areas was guided by the notion of the dichotomy of 'nature *vs.* culture' (Ferry 1992; Schiemann 1996; Jagtenberg and McKie 1997), with consequences for the kind of tourism and associated style of management, which primarily cater for Western visitors and their needs. This contrasts with the (potentially different) needs of local communities (Müller-Böker *et al.* 2001) who, after the gazetting (or implementing) of a park, are prohibited from using its resources as they did before. Hence, more recent efforts – following the notions of political ecology (Robbins 2004) – seek to protect nature *and* culture with integrated management concepts, which postulate that nature protection can work only *with* the people concerned or optimally *through* them (Ellenberg 1993; IUCN 1993; Arbeitsgruppe Ökotourismus 1995). 'Protection by use' is the shorthand for the integration of the needs of different stakeholders, such as local people, government, tourists and NGOs (Ellenberg 1993; Müller 1994; Borrini-Feyerabend 1997; Pimbert and Pretty 1997; Müller 2001). As a result of protection, since governments lose many sources of income (for example, through logging or mining concessions) and local communities are restricted in their entitlements on natural resources, tourism seems to offer a potential alternative source of income (Müller 1994). This chapter focuses on the challenges park managements are confronted with as a result of increasing visitor numbers and their heterogeneity. It demonstrates that possibilities exist for sustainable tourist use of protected areas, which simultaneously consider the livelihoods of local people, through discussion of Malaysian national parks, with a primary focus on Gunung Mulu National Park (Mulu for short) in Sarawak.

Political ecology (Graner 1997; Krings 2000; Soliva *et al.* 2003; Robbins 2004) provides a theoretical backdrop to the chapter. Field methodologies on which this chapter are based include participatory and non-participatory observation, focused interviews with experts on tourism and nature conservation throughout Malaysia ($n=25$) and semi-standardised questionnaire surveys with tourists of Gunung Mulu National Park ($n=82$) as well as with residents of the greater Kuala Lumpur area ($n=500$).

Political ecology: highlighting the social construction of conservation areas

Today, nature conservation is regarded as a social practice that has consequences not only for the environment or ecosystems that should be protected, but also for people who are directly or indirectly involved or affected by nature conservation. A theory that grasps this human–environment relationship is 'political ecology'. Like many theories, political ecology is neither monolithic nor clearly defined, but consists of many branches with different roots across a broad epistemological spectrum (Robbins 2004). Here the focus is on key concepts and their application to nature conservation (Soliva 2002; Soliva *et al.* 2003). Nature, environment and environmental change are not only material manifestations but also phenomena of perception or objects of cognition, and therefore socially constructed. Social, cultural and economic characteristics of societies and of individuals are the framework through which nature and environment are perceived and utilised (Blaikie 1995 and Krings 1998, both cited in Soliva 2002: 19). The possibilities for the use and appropriation of the environment are being constantly redefined by society and thus ever-changing arenas for action emerge – as a consequence of human action and structuration – on different spatial levels. Since different actors or stakeholders have different notions about and claims on environmental resources, 'environment' is a contested field in which conflicts emerge. Hence, environmental issues are always political and have to be regarded within their sociohistoric context. Some social groups profit from a change in environmental politics, whereas for others the effects are negative. As a consequence national parks (and other conservation areas) are regarded as outcomes of different stakeholders' actions; for example, researchers who point out distinct and unique features of an area, members of national and international environmental NGOs who lobby for its protection, logging companies who lobby against it, tourist operators who want to develop the area, government officials who have to decide about the implementation of change and, last but not least, the local people who either think they can profit from conservation or fear its constraints to their livelihood. Once a park has been gazetted and is used as such, this setting or 'arena' changes and different stakeholders, the old and the new, such as tourists who want to visit a park, hold a range of views and make various claims. The following discussion concentrates on tourists as stakeholders in nature conservation.

Tourists as stakeholders in the 'arena' of conservation

According to an Australian study (Wight 2001), nearly half the entire range of tourists has an interest in nature and learning as part of their vacation, and ecotourism (Weaver 2001) is now the fastest-growing form of tourism (Wight 2001: 40–3). If this development continues – in industrialised as well as developing countries – many destinations that have previously catered to a small number of ecotourists (such as national parks in remote areas) could come under pressure because of an increasing overall demand, including from tourists with different needs and demands. Moreover, demand is not evenly distributed in protected areas; people would rather visit those parks with special attractions that are easily accessible. In addition, better accessibility enables people for whom a visit was too arduous before to visit a park. Thus, the heterogeneity of visitors – or rather of forms of tourism, because an individual tourist is not always a 'backpacker' or a package tourist, although he or she may have a disposition for a certain form – increasingly poses new demands for park management.

National parks and impacts of tourism in Malaysia

Tourism and revenues

Close to 5 per cent of Malaysian land is protected and most of the bigger protected areas are counted among IUCN-category II parks, which are open to visitors and have adequate tourist infrastructure. Although Malaysia's national parks are a major attraction for tourists – advertised with the slogan 'Malaysia nature-*a*lly' – almost no national park generates enough income to cover its costs. With an increase in entrance fees, that gap could be closed. However, it is government policy to keep access costs low, so that people with little financial means still can visit the parks. Moreover, the government is reluctant to introduce a differentiated fee system, for example one in which foreigners pay more than Malaysians, or 'Westerners' pay more than people from ASEAN countries. Therefore, to a considerable extent the taxpayer covers the costs of nature conservation, which also has an impact on the way Malaysians regard their parks. Tourist expenditures are important both for the maintenance of a park and for the local economy around a park (not *in* the park, for the Malaysian government generally does not allow people to live in parks, with the exception of a few indigenous, non-sedentary communities such as the Penan in Mulu). Local economies profit most from visitors who stay in locally owned and managed accommodation and from those who stay longer than the average one or two nights (Haigh 1995, cited in Scheyvens 2002: 151; Hampton 1998: 649; Gäth 1999). However, Malaysian tourism policy is aimed at 'up-market' tourists. Concentration primarily on affluent tourists can be a double-edged sword. For one thing, they tend to stay in hotels with international standards, which are mostly owned by foreign corporations or by people from urban centres, and not by local residents. This means that the benefit of more jobs per guest can be reduced by the leakage of a substantial part of the revenue.

Moreover, such a concentration ignores the fact that less affluent tourists may leave a substantial amount of money in the country, because they usually stay longer and a smaller amount of their spending leaks to other countries.

Environmental impacts of tourism

The major task of nature conservation is certainly the protection of the environment or ecosystem, without which there would no longer be any tourist attraction. Tourism in conservation areas has the advantage of generating revenues and bringing concepts of environmental protection to people; however, tourists can also have a detrimental impact on ecosystems. Tropical rainforests – covering almost all Malaysian parks – are very sensitive ecosystems, with low resilience to external impacts. Impacts can be attributed to the following tourist activities.

Transport

Flights in general have little local impact; however, they are a problem in terms of global warming, and almost all foreign tourists come to Malaysia, especially east Malaysia, by plane. Moreover, some parks, such as Mulu, are so remote that access by air is necessary and air services have been constantly improved. The take-off and landing of small planes disturbs animals in the parks, especially when the planes pass close to the canopy to allow passengers to see spectacular sites. In most parks the impact of cars is not an issue since car access is prohibited, yet roads and tracks at the fringes of conservation areas can have an impact, for without a sufficient buffer zone, they serve as means of access to parks by poachers. Nevertheless, boats are more often used than cars because the extensive river system serves as an excellent conduit. Although residues of fuel and paints can have an impact on water resources, Malaysian rivers suffer more from sediments washed down from areas where forests have been removed for development (Consumers Association of Penang 1996).

Moving around

The soils of tropical rainforests are sensitive to pressure. Footpaths tend to be washed out and broadened by constant use. Fortunately, in many parks the forest floor is covered by peat swamps, across which tourists are unable to traverse. In these areas plank walks or concrete paths keep visitor flows literally on track, although some people regard this as 'visual pollution'. Caves are also sensitive to visitor impacts, while noise made by tourists disturbs animals and drives them deeper into the forests. Among Malaysian experts, there is dispute about whether frequently visited parts of national parks are already disturbed. NGOs stress that, in some areas, the number of visitors is so high that parks are degraded beyond repair. Tourist operators and government officials deny this, emphasising that the natural resilience of these areas enables them to recover during the rainy season when fewer people visit. Park managers adopt an intermediate position, suggesting that they don't see evidence of major damage but cannot preclude this in the near future when, as is expected and to some extent desired, more visitors will come to the parks.

Hunting and collecting

Some people do not heed the ever-present appeals to 'take only photographs and leave only footprints' and not collect rare plants or leave litter. In many parks fishing in rivers is allowed, whereas hunting with guns or traps is clearly forbidden. The situation of local people in or around Malaysian national parks is highly dependent on their specific context. Moreover, it is necessary to differentiate between indigenous people who live or lived in the area of a national park and regard it as their living space, and others who have lived in the vicinity or moved to the area once tourism set in. The former have limited rights to continue their lifestyle, since the Malaysian government wants to resettle them in a sedentary life outside the parks. The latter have no privileged access to park resources and many try to make a living from tourism, either as guides, providers of lodging and/or food, or as employees or suppliers of hotels and resorts. In areas with many visitors, such as Taman Negara or Gunung Kinabalu, some people can make a living; in others it is quite difficult because of their limited skills (especially English speaking) and knowledge of tourist needs.

Forms of tourism in Malaysian national parks

According to surveys by the World Tourism Organisation, ecotourists from different Western countries have more or less the same expectations when visiting a national park. They place the highest priority on being in wilderness and viewing wildlife, followed by visiting indigenous people and archaeological sights. They seek all kinds of accommodation, but are more likely to spend their nights in cabins, lodges and inns than other tourists. Previously, most Malaysian national parks, owing to their remote locations, were visited only by well-organised groups of tourists. In effect they were not accessible by public transport, and private transport was too expensive for individual travellers. Hence park infrastructure was initially built for the needs of group tourists and did not meet the requirements of individual travellers. It was therefore interesting to examine how the needs of group tourists and individual travellers differ. Both groups here refer only to Western tourists, who, for a long time, were practically the only visitors to Malaysian national parks. However, on account of industrialisation, modernisation and urbanisation, a mostly urban Malaysian middle class has emerged, with the time, money and interest to visit their national parks. The survey of urban Malaysians revealed that the expectations of domestic tourists different from those of foreign tourists. The following sections discuss the general needs and expectations of Western group tourists and individual travellers as distinct from domestic tourists (who almost exclusively travel in groups, so there is no differentiation between domestic individual and group travellers). Other foreign visitors, consisting mainly of people from Singapore and other ASEAN nations, were not interviewed during this research, since their travel behaviour appeared similar to that of domestic visitors.

Adventure and safety

If we accept Urry's (2002) statement that tourism is demarcated from work and everyday life such that tourists draw a line between these 'ordinary life' activities and leisure, then they must often enter a realm of uncertainty as tourists. However, mass tourism, which is highly efficient, predictable and controlled, seems to do everything to remove uncertainty from tourists' itineraries (Meethan 2001: 75). There are great differences in levels of risk between lone or small group adventurers and mass package tourists who book all-inclusive arrangements. Tourism oscillates between 'safeness' and 'adventure', or alternatively between broader concepts of 'ontological security' and 'critical situations' (Giddens 1992). Since ontological security encompasses more than just feeling safe in a tourist context, and since an adventure is not necessarily a critical situation, the terms 'safeness' and 'adventure' are used here. From the stories people tell after coming back from their holidays, of little adventures they survived or unplanned walks on the 'wild side', things often do not go smoothly. These kinds of adventures can be seen as the 'spice' in a journey. Moreover, they can be perceived quite differently. Becoming lost in Mulu at dusk without adequate equipment and knowledge to find base camp, or not knowing whether sun bears are deterred or attracted by loud singing, can be an adventure. But an adventure can be much less dangerous and frightening, such as when it amounts to an argument with a Miri taxi driver about his refusal to use the meter, not knowing whether he wants to cheat or this is the local custom. For the European package tourist, bargaining for a souvenir might be enough of an adventure, whereas others are only satisfied when they can (coolly) tell how they almost died struggling with a twenty-foot python in the jungle of Borneo. The way in which a journey is perceived to be adventurous also depends on the physical condition of the tourist(s): exhausted from a strenuous walk, they may regard an adventure with less enthusiasm than if they had felt rested (Markwell 2001). As the spectrum of different types of visitors increases, notions about what is regarded as 'safe' and 'adventurous' become more heterogeneous and therefore the task of conservation area management to provide for the needs of visitors becomes more demanding.

Group tourists and individual travellers are two forms of tourist that are regarded quite differently from each other, although people have increasingly started to combine these forms when they travel. For example, people with experience 'backpacking' during the 1970s and 1980s have the know-how for individual travel but occasionally also like to book a package tour for certain destinations, or, as 'flash packers', stay in superior accommodation. Moreover, the number and quality of travel books has increased dramatically since the 1970s, which makes it easier to travel independently.

Picnics and contemplation: the opinion of urban Malaysians

Malaysian national parks have become an important destination for – mostly urban – domestic visitors. It is valuable to know what stakes Malaysian tourists have in 'their' national parks and whether they differ from those of foreigners. Not all Malaysian parks are visited by domestic and foreign tourists. Some, like Pulau Penyu in Sabah and Mulu in Sarawak, are visited more by foreign tourists. However, a larger cluster of parks, including Taman Negara, Bako and Gunung Gading, are visited equally by both, whereas others (Kubah and Gunung Kinabalu) are visited predominantly by domestic tourists.

Five hundred residents of Kuala Lumpur were surveyed about nature conservation and park visits. Whereas half the respondents had already visited a park, the overwhelming majority (94 per cent) would like to visit a national park or conservation area in the future, a clear sign of the acceptance of nature conservation by the Malaysian urban population. For them, activities such as camping and hiking (42 per cent) ranked higher than relaxing (22 per cent), contemplation (20 per cent), seeing fauna and flora (12 per cent) and being together with family and friends (3 per cent). Sighting plants and animals is not the main thing urban Malaysians expect from their visit; rather they are more interested in active, outdoor experiences, unlike the majority of Western tourists in Malaysian national parks.

The most visited parks are those easily accessible from urban centres in west Malaysia. Taman Negara is the most visited park. Reached from Kuala Lumpur within three hours, and the oldest and biggest park of Malaysia, its name simply means 'national park'. Of those who had visited a park, two-thirds had visited more than one park. A quarter went on day trips, another quarter stayed for two days and over half stayed three or more days. For many Malaysians it is not easy to stay overnight in a park, because many people work six days a week, so they had to either take a day off or drive to a park in the evening after work. However, the numbers show that most of the respondents could afford to stay two or more days. A typical group travelling to Malaysian parks and recreation areas consisted of four people (80 per cent). The other 20 per cent travelled in even larger groups and only one person went alone. This reflects the fact that Malaysians visit parks and recreation areas not primarily to see flora and fauna, but for activities such as hiking, camping or barbecuing. Most respondents (72 per cent) did not book a package tour and travelled to the parks independently.

When people were asked what they liked most about their most recent visit, natural attractions ranked highest (36 per cent) before fresh air and coolness (21 per cent). Sixteen per cent liked seeing special plants or animals, 7 per cent enjoyed outdoor activities (rafting, swimming, hiking), and 6 per cent liked non-natural attractions (canopy walkways, picnic areas and casinos). When asked what they did not like about their last visit, most (30 per cent) criticised insufficient infrastructure. Apart from that, people complained about rubbish, waste and vermin, or that the place was generally dirty (18 per cent). The majority (58 per cent) did not think that the parks and recreation areas were overcrowded, but a sizeable minority (42 per cent) thought there were still too many people there.

Nearly all respondents (96 per cent) would like to revisit the park they liked best. The overwhelming majority of the respondents (92 per cent) favoured the idea of Malaysia establishing more conservation areas in the future and thought that local people would benefit from tourism.

A World Heritage site: Gunung Mulu National Park

Although every park is different and has its specific context and problems, the case of Gunung Mulu National Park in Sarawak highlights issues common to park tourism. Mulu is referred to as 'the pearl of Sarawak's national parks', but has fewer visitors than other large parks such as Gunung Kinabalu Park in neighbouring Sabah or Taman Negara on the Malayan Peninsula. However, Mulu's World Heritage status, acquired in the year 2000, will further increase its attractiveness and possible visitor numbers. Its remoteness in the interior of Sarawak, on the border with Brunei Darussalam, as well as the fact that it is accessible only by air or water, are reasons for its comparatively low visitor numbers. The area is not densely populated and consists of minority ethnic groups; mainly Berawan and sedentary (and a few semi-sedentary) Penan live in its vicinity. The park comprises 528 square kilometres, was gazetted as a national park in 1974 and opened to the public in 1985. It has a humid, tropical climate, which is the precondition for tropical rainforest. February to July is regarded as the 'dry season' when rivers have less water. Higher visitor numbers occur in the drier months, but weather is probably not the reason for the peak (Map 14.1). Rather, those months correspond with the main holiday season in the northern hemisphere, which is between June and August. The increase in December is explained by a further 'holiday hypothesis', namely that Christmas and New Year is a holiday period common to many countries. The park ranges in altitude from 30 metres to 2376 metres and thus has a very high biodiversity. Despite Mulu having 16 vegetation zones and more than 60 mammal, 260 bird and 400 ant species, flora and fauna are not the main attractions (Thorsell 2000; World Conservation Monitoring Centre 2002). Instead, its main attraction is the cave system, whose major features are presented with superlatives as the greatest cave chamber (Sarawak Chamber), the biggest cave entrance (Deer Cave), and the biggest cave system of the world, with Gunung Api as the most cavernous mountain (Meredith and Woolridge 1992). Four caves are open to the public and can be entered without a guide, so the paths leading to the caves and the caves themselves are highly frequented. Others, such as the Sarawak Chamber, can be visited only with guides.

Tourist potential of Mulu

Although increasing numbers of foreign and domestic tourists are interested in visiting conservation areas, not all the areas have the same potential to attract and satisfy tourists. Not only are natural attractions such as unique species or spectacular landscapes important, but so are recreational activities. The World Tourism Organisation and the United Nations Environment Programme (WTO/UNEP

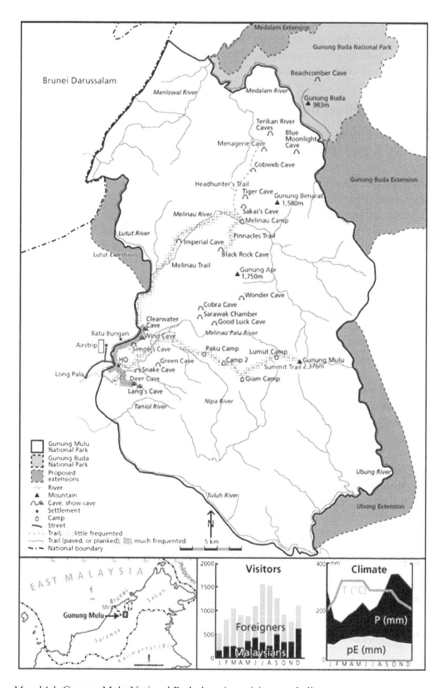

Map 14.1 Gunung Mulu National Park: location, visitors and climate.

1992: 17) has developed a checklist on the tourism potential of protected areas, which has been adapted and amended for Mulu. Questions from this checklist and the answers related to Mulu are summarised in Table 14.1, with ratings, where the sign (+) is used to indicate good, and the sign (−) to indicate bad.

Mulu is not well suited to meet many needs of domestic tourists. In particular, recreation facilities are almost non-existent, so domestic visitors hurry through the park and are likely to stay only one night. Consequently, the educational aim of nature conservation is not realised when the main attractions (visiting the caves and observing bats) are given priority by those on a short visit.

Group tourists and individual travellers

Although the majority of visitors tend to combine group and individual activities, these forms of tourism remain distinct regarding their specific needs within, and expectations of, national parks. In contrast to many tourist destinations, where backpackers and other individual tourists have been the pioneers who 'discover' a destination that subsequently becomes a mass tourism destination, conservation areas are for the most part established with group tourism in mind. Indeed conservation areas that are open for tourism generally provide the necessary infrastructure for group tourism, including planning for and managing the tourist influx. Since tourist authorities do not regard individual travellers (especially backpackers) as bringing in much revenue or even behaving appropriately (Scheyvens 2002), parks are generally not set up to attract them. The following findings from Mulu exemplify the differences between individual and group tourists.

As mentioned above, most people (75 per cent of the sample) reach Mulu in a small aircraft that flies low over the canopy and offers spectacular views of the landscape. There is also a boat connection, which provided the only access before the airstrip was built, besides some foot trails. However, it takes one day to get from Miri to Mulu, with frequent changes, and there are no fixed schedules. For groups it is easier to rent an entire boat (in the upper reaches of the river) and follow a planned itinerary. The boat option is a favoured form of access by groups, which often make the return trip by air. Both flight and boat ride are evaluated very positively, although neither can be regarded as comfortable or entirely safe. The journey thereby also acquires an adventurous quality. In the evaluation of transport, no differences emerged between group and individual travellers.

Although Mulu is advertised as a biodiversity hotspot with rare species, flora and fauna do not rank highly as a reason for visiting the park. Group tourists mentioned the following reasons (in order of frequency): the caves, nature in general, the bats, recreation, trekking, and wildlife viewing. Individual travellers mentioned trekking, the caves, nature in general, wildlife viewing, and recreation; they did not mention the bats. None mentioned adventure caving or visiting local communities such as the Penan or the Berawan.

A high percentage of the group tourists followed the 'usual' programme of visiting Wind Cave and Clearwater Cave in the morning and Deer Cave and Lang's Cave in the afternoon (Table 14.2). As the bats fly out of the mouth of Deer

Table 14.1 Tourism potential of Mulu

Question	Rating	Situation in Mulu
Is the protected area located near an international airport or major tourist centre?	+	The next international airport and main entry point to Mulu is coastal Miri. The flight takes 40 minutes. Most people come to Mulu by plane (75 per cent of the sample). There is a boat service too, but it takes roughly a day with frequent changes
Is successful wildlife viewing possible?	+/–	Wildlife viewing is difficult other than when millions of bats can be observed flying out of a cave mouth during good weather
Does the area offer more than one feature of interest?	++	The main attractions are the four show caves, which can be visited within one day. For those who want to stay longer, two trails (2–3 days) are another attraction. Tourists with only a few hours can take part in guided 'adventure caving'
What standards of food and accommodation are offered?	++ to – –	The available accommodation exceeds the number of visitors. In the park headquarters around 100 beds are available. The Royal Mulu Resort (RMR), a Japanese-owned four-star hotel outside the park, has 180 beds, and two small privately owned home-stays have around 20. The restaurants in Mulu – with the exception of the one in the RMR – provide basic 'pan-Asian' food and drinks
Does the area have additional cultural interests?	–	Not too many people live in the area and their cultural activities are not included in itineraries of travel groups, nor are they advertised. The Berawan and Penan peoples chose not to establish their community as a cultural attraction for tourists. However, the Penan village has nonetheless become a small tourist attraction where villagers sell some handicrafts
Does the area have recreation facilities?	–	Mulu does not feature many recreation facilities such as areas for swimming, for picnicking or for children to play. The attractions of Mulu are mostly, with the exception of watching the bats, relaxing and contemplative, if there are not too many people around. At the entrance to Clearwater Cave it is possible to bathe in a pond fed with cool water from the cave
Are the explanations in the information centre adequate?	+/–	Mulu has an information centre that is conveniently located beside the counter where visitors have to register. It consists of a large room with a sizeable, albeit eclectic, exhibition of materials on different topics related to Mulu

Table 14.2 Visits to and evaluation of Mulu's major attractions

	Group tourists		Individual travellers	
Attractions	Visited by	'Liked very much'	Visited by	'Liked very much'
Deer Cave	92%	68%	69%	52%
Lang's Cave	86%	64%	60%	40%
Wind Cave	92%	68%	91%	63%
Clearwater Cave	82%	73%	91%	48%
Watching the bats	82%	83%	75%	64%
Adventure caving	24%	84%	16%	66%
Trekking	10%	33%	68%	41%

Cave, tourists can watch them setting out for their nightly forage in the forest. Individual travellers visited the latter two caves much less, but group tourists, in contrast to their reasons for coming to Mulu, did spend time watching the bats fly from the cave. The biggest difference between the two groups is the extent to which they engaged in trekking. Only 10 per cent of group tourists made a trek, whereas two-thirds of individual travellers did. Although this is one of the main reasons individual travellers gave for visiting the park, it is remarkable, because trekking is possible only when a group of four or more people participate. It is not always easy to put together enough individual travellers willing to go on a trek on a certain date, but for group travellers this would be easier to organise.

In general the attractions rated very well, even in light of a possible bias towards features that attracted tourists in the first place. Individual travellers were more critical in their evaluation than group tourists. Some mentioned that the caves were too crowded or 'touristic' for them. The worst ratings related to treks, mostly resulting from the lack of cleanliness in the huts used for overnight stays but also to the limited knowledge (or capacity to share that knowledge) of the guides about the local environment.

The length of stay of visitors in conservation areas is a complex issue. On the one hand, a short stay enables a greater turnover of people who can enjoy the park and learn at least something about environmental issues. On the other hand, local businesses may profit more from people who stay longer. The air access makes shorter stays possible and therefore the average stay is a bit more than three days (two nights), though the mode is only two days (one night). Group tourists stay for a much shorter time than individual travellers, the latter mostly staying for three to four days. People who stay longer do not visit more attractions; rather it is the converse. The main reason for this is that most embark on a trek that takes two or three days and, especially with regard to individual travellers, treks cannot be planned well in advance. Hence, individual travellers leave more money in the local economy. Group tourists have less time to immerse themselves in the pristine environment of a rainforest, which, from an educational point of view, is less favourable. Consequently, some park managers question whether attempts to achieve outcomes related to education on ecologically sustainable development, including the protection of rainforests, have any beneficial impact on visitors who rush through their parks.

Mulu offers rather more beds suited to the upper end of the market and only a few in the lower or middle range. There is no firm distinction between group tourists exclusively staying in the more expensive rooms and individual travellers in budget accommodation; some individual travellers stay in expensive rooms simply because they are unable to find cheap accommodation. Expenditure on accommodation therefore did not vary greatly between types of tourists. The Royal Mulu Resort (RMR) was evaluated best by all, but then it is a four-star hotel, whereas the rooms in park headquarters have seen better times. Almost nobody stayed in the lodges, because they are neither well known nor easily found.

In Malaysia every national park has a so-called interpretation centre, where particularities of the specific park are explained. Most of these centres have a rather eclectic approach towards the content and form of the exhibitions. Nevertheless, visitors can find additional information in the centres. Usually they are located near the park's entrance and are thus easily accessible. However, Mulu's interpretation centre is visited by only a third of the respondents and very few group tourists went there, simply because their guides did not take them. Almost 60 per cent of readers of the Lonely Planet guide, the most popular travel book, visited the centre, because it was mentioned as worthwhile, thus demonstrating the powerful impact of guidebooks on certain destinations. The evaluation of the interpretation centre, however, was not very enthusiastic, with more than 60 per cent giving it average or bad ratings.

It is not easy to estimate what individual travellers as distinct from group tourists spend at Mulu; hence the calculations here are rough estimates. Sanggin *et al.* (2000) estimate that the roughly 150 inhabitants living near Mulu receive 400,000 Malaysian ringgit (RM) (US$100,000) per year from tourism. Of this amount, 35–40 per cent comes from employment at and services for the RMR, 50 per cent derives from general services, mostly from guiding, and the rest is from independent work. With roughly 10,000 visitors per year this means that an average of US$10 per visitor reaches the local people. According to the survey one-third of Mulu's visitors stay at the RMR. Consequently, up-market tourists do not actually contribute much to the revenue that goes to the local people, although they spend more per individual than other tourists. A typical individual traveller stays five days and four nights at the cheapest accommodation in the park headquarters at a cost of US$20, spends US$25 on food, takes part in a trek with three others and thus spends another US$30, totalling US$75 (US$15 per day). A group tourist books a package tour for two days (one night) in the RMR including food but excluding alcoholic beverages and the flight, and pays US$120 or US$60 per day. The amount spent by group tourists per day is roughly four times the amount spent by individual travellers, but a lower percentage of their expenditure is spent and retained locally.

This short insight into the behaviour and perceptions of group and individual tourists shows that, within the limited possibilities a park offers, there are substantial differences. The park management is mostly supportive of group tourists;

individual tourists are less welcome, even though they tend to spend more money locally. Considering the growing interest of tourists in nature and national parks, combined with improving access to parks, the parks have to be prepared to accommodate more individual travellers. Acknowledgement of their needs and a subtle adjustment in infrastructure (such as the provision of cheaper accommodation, a regular boat schedule, more places to just be and relax) and processes (such as more regular treks, empowerment of local guides, more local food) could prevent conflict, reduce pressure on sensitive areas (such as the caves, including the bats' habitat), and lead to further development of the local economy.

Conclusion

Malaysian national parks face considerable challenges that will become more prominent in the near future owing to the increasing number and also heterogeneity of visitors. Tourists are important stakeholders in the arena of nature conservation even though they are obviously not as strongly affected by environmental changes (and potential damage) as local people. Nevertheless, they contribute to the financing and image creation of specific parks and Malaysian parks in general. Traditionally, parks were set up for 'Western' package tourists who were mainly interested in fauna, flora and distinct landscape formations. This is intensified by the fact that Malaysian tourism authorities want to attract more affluent tourists who stay in prearranged up-market accommodation. However, surveys and interview results reveal that the range of tourist forms and tourists' wishes and needs are much more heterogeneous. Individual travellers are more likely to trek into the interior of a park and stay longer, whereas group tourists want to see the highlights of a park in as short a time as possible. Moreover, domestic tourists are mainly interested in leisure activities such as having a picnic, whitewater rafting, or swimming, for which many parks are not well equipped. Hence, a diversification of park activities would better accommodate different needs. Yet, since the distribution of these groups within Malaysian parks is also rather heterogeneous, this would not constitute a solution for all parks. However, decentralisation of activities, so that they do not inhibit each other, would help reduce their environmental impact and avoid conflict between different tourist needs. The same could be said of accommodation facilities and restaurants. In earlier stages of Malaysian nature conservation, accommodation was made available solely by park authorities. In recent years up-market resorts owned by people and companies from outside the region were established. Little room remained for local enterprises. Although there is a demand for cheaper accommodation from individual travellers and domestic tourists, local home-stay businesses experience difficulties filling their rooms. Some lack experience in meeting tourist accommodation requirements since their rooms can rarely be booked in advance.

As the management often does not have the means to invest in park infrastructure, let alone educate local enterprises about tourist needs and requirements, it remains focused on the core business of nature conservation and leaves the rest to market forces. With few exceptions, the park management does not work in tandem with local communities since most lack the skills and resources to assist in meeting tourist requirements. This applies particularly to those ethnic groups living from forest resources, who have a non-sedentary or semi-sedentary lifestyle and find it difficult to work according to pre-set units of time. This is the case with the Penan around Mulu and the Orang Asli on the Malayan Peninsula. Other ethnic groups in different socioeconomic contexts and with better access to formal education and the market economy, such as the Berawan around Mulu and the Kadazaan-Dusun near Kinabalu, are nevertheless rarely able to compete with people having long and extensive experience in tourist enterprises. Although effectively employed in less skilled positions in restaurants and hotels, some local people could nevertheless be trained to run their own canteens and home-stays, especially for the growing clientele of individual travellers and domestic visitors, and to work as guides or cultural interpreters. This could only come about, however, with the assistance and encouragement of park managements and government authorities, who ought to acknowledge first that individual travellers and domestic tourists contribute considerably to local economies even though they spend less per day than group tourists. Second, they should understand that tourists not only wish to have 'international' standards of accommodation and infrastructure (albeit with a local touch), but also appreciate local customs, foods and accommodation.

Rules and regulations regarding nature conservation and tourism in national parks always have a political dimension. It is difficult to achieve a balance that is at once beneficial for all stakeholders and not detrimental to the environment. In the case of Gunung Mulu National Park, the question remains over which stakeholder groups benefit from the current arrangements and which groups remain disadvantaged. Group tourists currently benefit most from the fact they can book suitable accommodation and activities in advance while having easy access to the park either by boat or plane. Individual travellers, on the other hand, often face difficulties in booking transportation, in organising treks, and in accessing cheap, functional and 'local' accommodation. Domestic visitors, for their part, almost all of whom come from urban centres on the Malayan Peninsula, do not have enough possibilities for the types of leisure activities they wish to pursue. Within the tourist industry the RMR has the greatest potential to benefit when, as is planned, more flights arrive, for they still have unused capacity in the up-market tourist segment. The same applies to the company that is to take over the accommodation managed by the park administration, if it is able to improve infrastructure and adapt it to different tourists' needs. Locally owned home-stay enterprises do not benefit particularly much, however, since they are neither marketed well nor able to take advance bookings from outside the local region. Some of the local Berawan people do benefit from visitors to the RMR since it offers them paid, albeit menial, jobs and provides a market for selling their products. Before the RMR was built, the Berawan opposed the project and claimed it was in fact on their land that the resort was to be erected. Obviously their claim did not succeed, but the RMR management adopted a policy of employing local Berawan people whenever possible. The Penan have been severely disadvantaged, however, because their rights to use the forest have been curtailed. They became instead one of the cultural objects of tourism and benefit only to a very minimal extent through selling a few handicraft items.

The park administration is challenged with new demands from tourists and a shortage of funds to improve the situation. The park, or rather its image, greatly benefited from the label 'World Heritage site'. The resources that led to the bestowal of that label, however, are starting to suffer, namely the show caves and their inhabitants, the bats, who are increasingly disturbed by the growing number of visitors. Who then is responsible for ensuring a more equitable and eco-friendly outcome for the various stakeholders in Gunung Mulu? Since final decisions regarding parks are made by the State of Sarawak, the Ministry of Environment and Public Health holds the decision-making power over park management policy, the distribution of funds, privatisation and concessions. At the local level the population would make their claims better heard if they could coordinate their forces to constitute a united platform for interaction with the state. Travel guidebooks, which have improved considerably during the last decade, can also play an important role. Their authors, besides providing information on what is permitted and not permitted within the park, could also act as advisors (along with NGOs) regarding the requirements and needs of tourists. Many NGOs are providing the park administration and government authorities with information on tourists while lobbying for an ecologically sustainable tourism agenda. All these organisations and groups play an important role. However, the greatest agency for triggering change lies with the critical intermediaries, the park managers, who see and hear what happens in and around their parks. Although they probably cannot fully control the processes of change, they can initiate and shape them.

Note

The fieldwork was supported by the Swiss Academy of Natural Sciences and the University of Zurich, Switzerland.

References

Arbeitsgruppe Ökotourismus (1995) *Ökotourismus als Instrument des Naturschutzes – Möglichkeiten zur Erhöhung der Attraktivität von Naturschutzvorhaben*, Munich: Weltforum Verlag.

Borrini-Feyerabend, G. (ed.) (1997) *Beyond Fences – Seeking Social Sustainability in Conservation*, volume 2, Gland: IUCN.

Consumers Association of Penang (1996) 'State of the Environment in Malaysia', conference paper, State of the Malaysian Environment Conference, Penang, April.

Ellenberg, L. (1993) 'Naturschutz und Technische Zusammenarbeit', *Geographische Rundschau*, 45: 290–300.

Ferry, L. (1992) *Le nouvel ordre écologique – L'arbre, l'animal et l'homme*, Paris: Bernard Grasset.

Gäth, P. (1999) 'Ökonomische Effekte des Trekkingtourismus im Langtang Nationalpark, Nepal', unpublished PhD thesis, University of Zürich.

Giddens, A. (1992) *The Consequences of Modernity*, Cambridge: Polity Press.

Graner, E. (1997) 'The Political Ecology of Community Forestry in Nepal', *Freiburger Studien zur Geographischen Entwicklungsforschung*, no. 14, Freiburg: Freiburg University.

Hampton, M.P. (1998) 'Backpacker Tourism and Economic Development', *Annals of Tourism Research*, 25: 639–60.

IUCN (1993) *Parks for Life – Report of the IVth World Congress on National Parks and Protected Areas*, Gland: IUCN.

—— (2003) *United Nations List of Protected Areas*, Gland: IUCN, UNEP, WCMC, WCPA.

Jagtenberg, T. and McKie, D. (1997) *EcoImpacts and the Greening of Postmodernity – New Maps for Communication Studies, and Sociology*, London: Sage.

Krings, T. (2000) 'Das politisch-ökologische Analysekonzept in der Umweltforschung – Beispiel der städtischen Brennstoffversorgung in Dakar (Senegal)', *Geographische Rundschau*, 52: 56–9.

Markwell, K.W. (2001) 'An Intimate Rendezvous with Nature? – Mediating the Tourist–Nature Experience at Three Tourist Sites in Borneo', *Tourist Studies*, 1: 39–57.

Meethan, K. (2001) *Tourism in Global Society – Place, Culture, Consumption*, Basingstoke: Palgrave.

Meredith, M. and Woolridge, J. (1992) *Giant Caves of Borneo*, Kuala Lumpur: Tropical Press.

Müller, B. (1994) 'Ökotourismus in Entwicklungsländern – umweltpolitische Leerformel oder wirksame Regionalentwicklungsstrategie', *Mainzer Geographische Studien*, 40: 361–74.

Müller, U. (2001) *Wie funktioniert 'Partizipation' bei Naturschutzvorhaben in der Schweiz? Beispiel: Erweiterung des Schweizer Nationalparks*, dissertation, University of Zürich.

Müller-Böker, U., Kollmair, M. and Soliva, R. (2001) 'Der Naturschutz in Nepal im Gesellschaftlichen Kontext', *Asiatische Studien*, 105: 725–75.

Nyaupane, G.P., Morais, D.B. and Graefe, A.R. (2004) 'Nature Tourism Constraints – a Cross Activity Comparison', *Annals of Tourism Research*, 31: 540–55.

Pimbert, M.L. and Pretty, J.N. (1997) 'Parks, People and Professionals – Putting "Participation" into Protected-Area Management', in K.B. Ghimire and M.L. Pimbert (eds), *Social Change and Conservation*, London: Earthscan.

Robbins, P. (2004) *Political Ecology*, Oxford: Blackwell.

Röper, M. (2001) *Planung und Einrichtung von Naturschutzgebieten aus sozialgeographischer Perspektive – Fallbeispiele aus der Pantanal-Region (Brasilien)*, Tübinger Beiträge zur Geographischen Lateinamerika-Forschung No. 22, Tubingen: Geographisches Institut.

Sanggin, S.E., Noweg, G.T., Abdul, R.A. and Mersat, N.I. (2000) 'Impact of Tourism on Longhouse Communities in Sarawak', paper presented at the 6th Biennial Borneo Research Conference, Kuching/Sarawak, July.

Scheyvens, R. (2002) 'Backpacker Tourism and Third World Development', *Annals of Tourism Research*, 29: 144–64.

Schiemann, G. (ed.) (1996) *Was ist Natur? – Klassische Texte zur Naturphilosophie*, Munich: Deutscher Taschenbuch Verlag.

Soliva, R. (2002) *Der Naturschutz in Nepal – eine akteurorientierte Untersuchung aus der Sicht der politischen Ökologie*, Munster: Lit.

Soliva, R., Kollmair, M. and Müller-Böker, U. (2003) 'Nature Conservation and Sustainable Development', in M. Domrös (ed.), *Translating Development – The Case of Nepal*, New Delhi: Social Science Press.

Thorsell, J. (2000) *Gunung Mulu National Park – World Heritage Nomination – IUCN Technical Evaluation*, Gland: IUCN.

Urry, J. (2002) *The Tourist Gaze*, London: Sage.

Weaver, D.B. (2001) 'Ecotourism in the Context of Other Tourism Types', in D.B. Weaver (ed.), *The Encyclopedia of Ecotourism*, New York: CABI.

Wight, P.A. (2001) 'Ecotourists – Not a Homogenous Market Segment', in D.B. Weaver (ed.), *The Encyclopedia of Ecotourism*, New York: CABI.

World Conservation Monitoring Centre (2002) Protected Areas of Malaysia. Online. Available <www.unep-wcmc.org> (accessed 30 July 2002).

Wright, G.R. and Mattson, D.J. (1996) 'The Origin and Purpose of National Parks and Protected Areas', in G.R. Wright (ed.), *National Parks and Protected Areas – Their Role in Environmental Protection*, Cambridge, MA: Blackwell.

WTO/UNEP (1992) 'Guidelines: Development of National Parks and Protected Areas for Tourism', Tourism and the Environment – UNEP-IE/PAC Technical Report, 13, Madrid: WTO.

15 Marginal people and marginal places?

Barbara Rugendyke and John Connell

Even in the most remote places tourism is becoming more important both as tourist numbers increase, as they have done on Easter Island, and because the challenges of developing other sectors of the economy have been too great – or perhaps less tempting. Countries like Brunei, once with petro-dollars and willing to stand aloof from tourism, now also seek tourists, and those where tourism has failed or faded, such as Papua New Guinea, are struggling anew to revitalise the industry. Nonetheless some of the most remote places in the region, like Niue, have literally and metaphorically failed to make the right connections, despite a quarter of a century of endeavour (Connell 2007). Other Pacific states have failed to attract significant tourist numbers mainly because of isolation: 'location, location, location' is even more appropriate than for real estate. Intervening opportunities, inadequate airline connections and ineffective marketing exclude certain places.

Small may be beautiful but it is also disadvantageous – whether as a small state (such as Niue) or as a remote village – in getting physical and virtual access to a remote and 'unknown' market, so that intermediaries are often the prime beneficiaries of tourism rather than the local people at the destination. Travel agents may be crucial, but most specialise in large markets and are linked to prominent airlines; many have simply never heard of remote locations (Connell 2007). Small states, and remote places, are particularly vulnerable to changes in access. Airlines are crucial. Between 2005 and 2006 the number of tourists to the Northern Marianas, in the northern Pacific, fell by 20 per cent after Japan Airlines abandoned its route to the main island of Saipan. Enabling equitable participation, letting local voices be heard, and even having marketing that does not distort the local scene arc critical issues for effective participation in grassroots tourism, here as elsewhere (Cater 1995; Stonich 1998).

Local people are never passive respondents to the incursions of tourism, but are creative, resilient, and often enthusiastic, even sometimes oppositional, in their attempts to shape and direct tourism into its most profitable and amenable form. Where tourist spaces have been constructed and commodified by those with power, there 'have been continuing processes of resistance and construction by the non-West' (van der Duim *et al.* 2005: 290). Several chapters have shown how, even in remote and seemingly disadvantaged locations, where economic alternatives are few, local people have not only been able to thrive from tourism, but have resisted forms of tourism that are anathema to their cultural concerns, or bring in little income, and have so developed a more locally beneficial form of tourism. Local involvement has never been easy; in most contexts it remains true that 'international tourism constructs as it commodifies, alienates as it appropriates, and dominates as it penetrates. Local authority is undermined, local empowerment is difficult to sustain, and local environments are changed for ever' (Conway 2002: 120). Consequently tourist–local relationships are often marked by ambivalence, tension and constantly fluctuating tripartite relationships as the balance of power shifts between insiders and outsiders, between 'tourists, locals and brokers' (van der Duim *et al.* 2005: 286). However active local people are in developing their 'own' tourism industry, invariably, as various chapters demonstrate, middlemen (and it is invariably men) and other intermediaries (such as transport companies) are often the key beneficiaries of tourist development, with the ability to shift tourism in particular directions and in favour of particular companies, places and ethnic groups. Rural and remote people and places are never well placed to control the industry.

Conflicts over the distribution of income from tourism are seemingly ubiquitous but tend to intensify as the significance of tourism grows and perceptions of exclusion mount (Walpole and Goodwin 2001), notably where people are denied access to resources they once used and/or owned. Establishment of national parks, and the simultaneous stimulus of tourism and control of park access, despite sometimes good intentions, has often failed to bring benefits for former or nearby residents of such parks. In and around Doi Inthanon National Park in northern Thailand tourism is organised and operated by outside tour companies, leaving local people with limited influence and economic benefits. Tour guides rarely facilitate cultural encounters, park management restricts villagers inside the national park from many traditional land uses, residents are fearful of tour guides reporting irregularities and anticipated jobs and incomes have failed to materialise (Kaae 2006). This situation is far from unique. In some parts of the world where local people have been excluded from national parks, success has followed long-term institutional support for those excluded, appropriate identification of the excluded community, transparency and accountability in income distribution and, ultimately, there being adequate resources to distribute (Archabald and Naughton-Treves 2001) – circumstances that are not always easy to achieve, as various chapters here indicate or imply. Goodwill on the part of governments, national or regional, is critically important.

In several contexts local people have been somewhat reluctant to depend on tourism, not because they are unaware of economic benefits, but because they need to maintain the 'old order' of agriculture to ensure diversity where tourism may be unreliable or inadequate. While tourism has boosted income this has not necessarily been adequate to ensure survival, especially where villagers have been displaced to new and unfamiliar locations, and lost local autonomy. Much contemporary literature on tourism emphasises the need for sustainability, yet it is rarely evident what this is or might be, other than that it is something that will provide livelihoods for future generations without destruction of environmental wellbeing. Ironically, sustainable tourism may actually be an outcome of unsuccessful tourism in the sense that, where numbers are relatively few, pressures on the environment are limited, so that tourism becomes just one limited means of achieving worthwhile economic diversity at village level, and social and cultural change are limited. (Yet such limitations may sometimes be unwelcome as villagers in remote places ponder more effective means of access to the 'golden hordes'.) Villagers the world over have opted for diversity where they have had a choice but, of course, it is a truism that 'people make their own livelihoods but not necessarily under conditions of their own choosing' (de Haan and Zoomers 2005: 43). So it is with tourism.

With the tourists

What these chapters have shown is that any notion that relationships between 'hosts' and 'guests' are straightforward, and even that such concepts have real meaning, can be discarded since, at the very least, social relations are much more complicated (see Aramberri 2001). Even in this volume, few studies examine that relationship from the primary viewpoint of the local people and their objectives. It is evident, however, that there is always some disappointment about expenditure being elsewhere, or simply not being enough to support local needs and goals, or about the manner in which the 'guests' treat the 'hosts'. The most basic multiplier effects that contribute to development of local transport, fisheries, agriculture and so on may be lacking, whereas tourist development can threaten all of these. Small restaurants are as likely to be part of global chains as they are local enterprises. Some resentment of unequal, one-sided relationships is natural and inevitable.

Exceptionally, local people have exerted greater control of the industry. Ironically this may follow external intervention. In northern Thailand a Thai entrepreneur, distressed over the imbalance between costs and benefits of tourism to hilltribes, initiated a project that educated tourists on the poverty of hilltribes and used 10 per cent of the profit of trips to establish schools, fund scholarships, build water supply schemes and develop sustainable agricultural practices (Dearden 1996: 224). Such initiatives are rare, may not be sustained over time or beyond the interest of a particular individual, and may be undercut by less ethical businesses.

Relationships are particularly complex on the edge of parks, reefs and similar protected areas, where local people may not benefit directly from tourist visits, can be excluded from areas that formerly provided resources, and may receive limited compensation. Various chapters indicate that frustrations mount when nearby tourism seemingly provides few benefits, especially if resources are denied to local people. Where protected wildlife damages local crops and disrupts animal husbandry, frustrations are naturally greater and there is resentment of parks and tourism, especially if compensation is perceived as inadequate, a situation again well demonstrated elsewhere (Archabald and Naughton-Treves 2001; Gadd 2005). Even ecotourism, much heralded as the saviour of small communities in remote places, though steadily growing, has yet to demonstrate long-term success and sustainability, and has impacts that are little different from, if much more slight than, those of mass tourism. This is well evident in Thailand (described in this volume), in Indonesia (Cochrane 1996), and in other global contexts (Carrier and McLeod 2005). Ecotourism is no panacea.

Local participation

Tourism is frequently perceived to be one possibility for poverty reduction in economically disadvantaged regions or nations, yet tourism as a development option has increasingly been criticised for its failure to include local people in decision-making, to conserve the environment, or even to ensure the distribution of some of its benefits to those who bear its social costs (Briedenhann and Wickens 2004). A future enabling more 'equitable distribution of power between Western and host cultures where interaction occurs in a "third tourist space", where decision-making responsibility involves the hosts and where they receive economic returns' (van der Duim *et al.* 2005: 291), would be invaluable, but is elusive. It is scarcely surprising that the idea of pro-poor tourism has acquired global significance, being developed as a response to issues of local participation, the leakage of incomes from small communities, and recognition that those who are most affected by tourism, in both social and environmental terms, may benefit least from it. Its proponents stress the necessity to improve incomes at the local level, ensure cultural, economic and environmental sustainability and enable incomes to be directed towards, and linked into, broader development goals (Mowforth and Munt 2003; Carbone 2005; Hall and Brown 2006; Harrison and Schipani 2007; Scheyvens 2007).

The growing rhetoric over pro-poor and ethical tourism has rarely been matched with actual practice, and many people living in the vicinity of significant tourist developments may be little better off than before such projects began. Some may be worse off. Even efforts to link tourism with other facets of the local economy have often been unsuccessful and more deliberate attempts may be required to ensure that local productive activities benefit from tourism (Torres and Momsen 2004). In southern Africa, what have been conceived as pro-poor tourism schemes have had mixed success, because of the particular difficulties faced by remote, rural people, including illiteracy, population pressure, HIV/AIDS and limited capital (Scheyvens 2002), but, as much as anything else, because of inadequate knowledge and management skills that prevent effective engagement with tourists and their intermediaries (Hill *et al.* 2006; Rogerson 2006; see also Bowden 2005). Similarly the notion of community participation is not readily applicable in many destinations in the region on account of local divisions and 'formidable operational, structural and cultural constraints to participatory principles in many developing countries' (Ying and Zhou 2007: 2). Such constraints vary within and between nations. Thus, in China, tourism has produced economic growth but little poverty reduction in Yunnan, but exactly the converse in Guizhou (Donaldson 2007). At the local scale, as at Xidi in China, through a villagers' committee, the community was able to take the initiative in determining the allocation of benefits from tourism, tourism investment and operations, and tourism planning. Yet, at a nearby, similar site for cultural tourism, the local community, Hongcun, was effectively excluded from decision-making by a national government decision to transfer rights to an external company (Ying and Zhou 2007: 8). Practice varies, even locally, but the sustainability of tourism relates to stakeholders receiving benefits, rather than mere participation in decision-making.

Issues of ethical practice remain. Despite much discussion during the past decade about ethics, participatory and sustainable development, the need for local community cooperation and involvement in planning and management (Timothy 2000; Singh *et al.* 2003), the relevance of fair trade tourism for community development (Evans and Cleverdon 2000), and the rise of various 'responsible tourism' organisations, the actual impact is extremely limited and invisible in most places. In large part this is because interventions in favour of more effective local participation must come from NGOs and governments, both of which are relatively powerless in the tourist industry. In the latter case, where governments are anxious to stimulate tourism, their interest is often in the more exclusive sector, rather than the much derided or ignored 'backpacker tourism' that is more likely to reach remote areas and involve local people. Rhetoric in favour of local participation tends to disguise a rather different reality – too often local people are passive beneficiaries rather than active participants (Sofield 2003). Government leaders, like local community leaders, are not always responsive to the needs of the poor.

In large part challenges to local participation exist because so much tourist area development is driven externally and only belatedly involves the local people, although there are numerous examples from the region of village people constructing tourist accommodation and somewhat forlornly waiting expectantly for the tourists to arrive. Major challenges remain for marketing small-scale, remote enterprises that do not enjoy significant external involvement and direction. Indeed it is readily evident that many of the conclusions reached in these chapters effectively, and unfortunately, demonstrate that various directions suggested in earlier studies for local community involvement have yet to have any practical outcome in the region (Timothy 1999, 2002; Tosun 2005, 2006). Local people largely continue to go it alone, as much in the face of government and private sector initiatives as alongside them. National marketing campaigns have tended to be precisely that, advertising the nation and supporting the larger national and international players, while making rhetorical statements about equity and regional interests. Enlightenment, enthusiasm and disinterest are as crucial as regulation and management.

Onwards and outwards

Tourism is a volatile industry, subject to fashions and whims, but especially subject to violent disruption, sometimes by natural hazards (as in the extreme case of the December 2004 tsunami for Phuket in Thailand) and more generally in various island states (Milne 1992: 199). Health crises, such as SARS (Zeng *et al.* 2005), political tension and even terrorism – as has occurred in Bali, where one of the most successful tourist destinations in the region has twice been thrown into chaos in the present century – provide an unpleasant reminder that tourist area life cycles can disintegrate. In some contexts local people can return to alternative livelihoods, as they have managed in some parts of Bali, but certainly not in others (Tom 2006). Both Fiji and New Caledonia have experienced similar but briefer long-term disruption (Connell 1987; Lea 1996), and there have been relatively concerted external attempts to discourage tourism in Burma (Myanmar) for ethical reasons (Henderson 2003). Whether from natural or political disasters, places if not people are slow to recover, as tourists (and especially travel agents) simply discover alternative destinations.

This volatility and the marginalisation of remote places may take on new guises as international tourism becomes increasingly challenged for its ozone-depleting, carbon dioxide-producing jet aircraft travel and hence its contribution to the accelerated greenhouse effect and global warming. As perceptions change, the much-vaunted enthusiasm for ecotourism has given way to the broader concept of 'responsible tourism', which emphasises not only the relationships between tourists and the environment but the crucial interactions between tourists, environment and local people (Russell and Wallace 2004). Notions of ecotourism as responsible tourism have been contested as 'an ideology that, while rhetorically people-centred, stressing "empowerment" and "community", involves tying the development prospects for these same people to severe localised natural limits' (Butcher 2007: viii). Thus, community participation invariably is considered as local, whereas the 'relationship between locally conceived development and national development is substantially unexamined' (ibid.: 163) and questions of social capital – of community commitment to mutually beneficial collective action (Jones 2005) – need further exploration. However, what is undoubtedly crucial is that the concept of responsible tourism be extended to include the intermediaries and 'middlemen' in processes of tourism expansion, and that this be more than merely a worthy and dull phrase.

The widespread absence of adequate longitudinal studies that enable reflection on Butler's (1980) model, and that can lead to abandonment of generalisations about economic and especially social change based on a single (often quite short) time period, constrains more pointed conclusions. This book, in presenting studies (such as those by Silverman and McCall) conducted in the same place for years, moves research a little nearer to providing a basis for more comprehensive analysis of long-term change.

Above all, what the various chapters demonstrate is that communities are never homogeneous and that however successful tourism may be in a particular place, and however that may be measured, benefits are always uneven. Tourism has rarely been about empowerment or equity, especially at the grassroots. Despite criticisms that the focus on the grassroots represents a 'retreat from development at the level of the nation' and the 'privileging of the local and small scale over the national' (Butcher 2007: 165), the experiences of the social and environmental impacts of tourism remain greatest at the local level. The focus on the local here is simply based on the premise that those who are most involved in the industry should be its principal beneficiaries. This notion was enshrined over two decades ago in Murphy's seminal work on a community approach to tourism, in the statement: 'The industry possesses real potential for social and economic benefits if planning can be redirected from a pure business and development approach to a more open and community-oriented approach which views tourism as a local resource' (1985: 37). Much has happened, but remarkably little has changed.

References

Aramberri, J. (2001) 'The Host Should Get Lost: Paradigms in the Tourism Theory', *Annals of Tourism Research*, 28: 738–761.

Archabald, K. and Naughton-Treves, L. (2001) 'Tourism Revenue-Sharing around National Parks in Western Uganda: Early Efforts to Identify and Reward Local Communities', *Environmental Conservation*, 28: 135–149.

Bowden, J. (2005) 'Pro-Poor Tourism and the Chinese Experience', *Asia Pacific Journal of Tourism Research*, 10: 379–398.

Briedenhann, J. and Wickens, E. (2004) 'Tourism Routes as a Tool for the Economic Development of Rural Areas – Vibrant Hope or Impossible Dream?', *Tourism Management* 25: 71–79.

Butcher, J. (2007) *Ecotourism, NGOs and Development*, London: Routledge.

Butler, R. (1980) 'The Concept of a Tourist Area Cycle of Evolution: Implications for Management of Resources', *Canadian Geographer,* 24: 5–12.

Carbone, B. (2005) 'Sustainable Tourism in Developing Countries: Poverty Alleviation, Participatory Planning and Ethical Issues', *European Journal of Development Research*, 17: 559–565.

Carrier, J. and McLeod, D. (2005) 'Bursting the Bubble: The Socio-Cultural Context of Ecotourism', *Journal of the Royal Anthropological Institute*, 11: 315–334.

Cater, E. (1995) 'Environmental Contradictions in Sustainable Tourism', *Geographical Journal*, 161: 21–28.

Cochrane, J. (1996) 'The Sustainability of Ecotourism in Indonesia: Fact and Fiction', in M. Parnwell and L. Bryant (eds), *Environmental Change in South-East Asia: People, Politics and Sustainable Development*, London: Routledge.

Connell, J. (1987) ' "Trouble in Paradise": The Perception of New Caledonia in the Australian Press', *Australian Geographical Studies*, 25: 54–65.

—— (2007) ' "The Best Island on the Globe": Constantly Constructing Tourism in Niue', *Australian Geographer*, 38: 1–13.

Conway, D. (2002) 'Tourism, Agriculture, and the Sustainability of Terrestrial Ecosystems in Small Islands', in Y. Apostolopoulos and D. Gayle (eds), *Island Tourism and Sustainable Development*, Westport, CT: Praeger.

Dearden, P. (1996) 'Trekking in Northern Thailand: Impact Distribution and Evolution over Time', in M. Parnwell (ed.), *Uneven Development in Thailand,* Aldershot: Avebury.

Donaldson, J. (2007) 'Tourism Development and Poverty Reduction in Guizhou and Yunnan', *The China Quarterly*, 190:333–351.

van der Duim, R., Peters, K. and Wearing, S. (2005) 'Planning Host and Guest Interactions: Moving beyond the Empty Melting Ground in African Encounters', *Current Issues in Tourism*, 8: 286–305.

Evans, G. and Cleverdon, R. (2000) 'Fair Trade in Tourism – Community Development or Marketing Tool', in D. Hill and G. Richards (eds), *Tourism and Sustainable Community Development*, London: Routledge.

Gadd, M. (2005) 'Conservation outside of Parks: Attitudes of Local People in Laikipia, Kenya', *Environmental Conservation*, 32: 50–63.

de Haan, L. and Zoomers, A. (2005) 'Exploring the Frontier of Livelihood Research', *Development and Change*, 36: 27–47.

Hall, D. and Brown, F. (2006) *Tourism and Welfare: Ethics, Responsibility and Sustained Well-being*, Wallingford: CABI.

Harrison, D. and Schipani, S. (2007) 'Lao Tourism and Poverty Alleviation: Community-Based Tourism and the Private Sector', *Current Issues in Tourism*, 10: 194–230.

Henderson, J. (2003) 'The Politics of Tourism in Myanmar', *Current Issues in Tourism*, 6: 97–118.

Hill, T., Nel, E. and Trotter, D. (2006) 'Small-Scale, Nature-Based Tourism as a Pro-Poor Development Intervention: Two Examples in Kwazulu-Natal, South Africa', *Singapore Journal of Tropical Geography*, 27: 163–175.

Jones, S. (2005) 'Community-Based Ecotourism: The Significance of Social Capital', *Annals of Tourism Research,* 32: 305–324.

Kaae, B. (2006) 'Perceptions of Tourism by National Park Residents in Thailand', *Tourism and Hospitality Planning and Development*, 3: 19–33.

Lea, J. (1996) 'Tourism, *Realpolitik* and Development in the South Pacific', in A. Pizam and Y. Mansfeld (eds), *Tourism, Crime and International Security Issues*, New York: Wiley.

Milne, S. (1992) 'Tourism and Development in South Pacific Microstates', *Annals of Tourism Research*, 19: 191–212.

Mowforth, M. and Munt, I. (2003) *Tourism and Sustainability: Development and New Tourism in the Third World*, second edition, London: Routledge.

Murphy, P. (1985) *Tourism: A Community Approach*, London: Routledge.

Rogerson, C. (2006) 'Pro-Poor Local Economic Development in South Africa: The Role of Pro-Poor Tourism', *Local Environment*, 11: 37–60.

Russell, A. and Wallace, G. (2004) 'Irresponsible Ecotourism', *Anthropology Today*, 20: 1–2.

Scheyvens, R. (2002) *Tourism for Development. Empowering Communities*, Harlow: Prentice Hall.

—— (2007) 'Exploring the Tourism–Poverty Nexus', *Current Issues in Tourism*, 10: 231–254.

Singh, S., Timothy, D. and Dowling, R. (eds) (2003) *Tourism in Destination Communities*, Wallingford: CABI Publishing.

Sofield, T. (2003) *Empowerment for Sustainable Tourism Development*, Amsterdam: Pergamon.

Stonich, S. (1998) 'Political Ecology of Tourism', *Annals of Tourism Research*, 25: 25–54.

Timothy, D. (1999) 'Participatory Planning: A View of Tourism in Indonesia', *Annals of Tourism Research*, 26: 371–391.

—— (2000) 'Building Community Awareness of Tourism in a Developing Country Destination', *Tourism Recreation Research*, 25: 111–116.

—— (2002) 'Tourism and Community Development Issues', in R. Sharpley and D. Telfer (eds), *Tourism and Development: Concepts and Issues*, Clevedon: Channel View.

Tom, E. (2006) *Bali: Paradise Lost?*, Melbourne: Pluto Press.

Torres, R. and Momsen, J. (2004) 'Challenges and Potential for Linking Tourism and Agriculture to Achieve Pro-Poor Tourism Objectives', *Progress in Development Studies*, 4: 294–318.

Tosun, C. (2005) 'Stages in the Emergence of a Participatory Tourism Development Approach in the Developing World', *Geoforum*, 36: 333–352.

—— (2006) 'Expected Nature of Community Participation in Tourism Development', *Tourism Management*, 27: 493–504.

Walpole, M. and Goodwin, H. (2001) 'Local Attitudes towards Conservation and Tourism around Komodo National Park, Indonesia', *Environmental Conservation*, 28: 160–166.

Ying, T. and Zhou, Y. (2007) 'Community, Governments and External Capitalism in China's Rural Cultural Tourism: A Comparative Study of Two Adjacent Villages', *Tourism Management*, 28: 1–11.

Zeng, B., Carter, R. and de Lacy, T. (2005) 'Short-Term Perturbations and Tourism Effects: The Case of SARS in China', *Current Issues in Tourism*, 8: 306–322.

Index

accessibility 2, 4–6, 31, 42, 46, 52, 54, 55, 61, 82, 99, 134, 138, 258, 273, 279
accommodation 13–15, 42, 45, 48, 51, 54, 169, 172, 183–4, 192, 202, 204, 207, 220, 246–7, 258, 260–1, 266, 268, 270; *see also* guesthouses, home-stays, resorts
administration 46, 110–11
adventure *see* tourism, adventure
advertising 80–2, 115, 120–3, 132, 137–8, 155, 184, 210, 230, 274, 278
Africa 89, 90, 91, 160, 277
agriculture 5, 8, 15–16, 21, 23–4, 30, 50, 54, 65, 103, 115, 139, 150, 157–8, 161, 164, 173, 176–9, 180, 185–7, 189, 190–2, 194–5, 205, 208, 223, 226–8, 231, 237, 239–40, 275, 276
aid 134, 140
Alor (Indonesia) 26
Anderson, Benedict 31
Angkor Wat (Cambodia) 22, 26
archaeology 45–7, 49, 53, 181, 260
art 3, 25, 60–73, 216
artifacts 41–4, 51, 61–73; *see also* handicrafts
Asian Development Bank (ADB) 143
Australia 48, 50, 62, 108, 167, 184
authenticity 8, 14, 24–8, 59, 63–4, 66–7, 70–1, 77–9, 100, 107, 108–9, 224, 234
autonomy 52–3, 55, 60, 100, 275

backpacking 3, 7, 8, 22–3, 33, 61, 108–9, 137, 261, 265, 278
Bali (Indonesia) 3–4, 5, 9, 13, 14, 22, 24, 25, 26, 28, 30, 70, 94, 148–63, 164–5, 167–70, 171, 173, 176–7, 170
Bangkok 198, 206, 208
barter *see* exchange

begging 8, 30
Beqa (Fiji) 9–10, 11, 17, 18
Bhutan 21
biodiversity 179, 198, 236, 242, 265
Borneo 236–73
Borobudur (Java) 22, 26
British Museum 45
brochures 30, 91, 92, 122–3, 137–8
Brunei 1, 274
bushwalking 182, 184
Burma (Myanmar) 14, 17, 28, 279
Butler, Richard 3, 279

Cambodia 1, 22, 26, 138
Cannibal Tours 26, 60, 66
capital 13, 66–7, 134, 160, 277; accumulation 46; cultural 91–2, 279
Caribbean 50, 119, 132
casinos 22, 53, 80
Chambri 28, 30, 63–4, 93; *see also* Papua New Guinea
Chile 41–55
China 1, 2, 17, 19, 25, 26, 29, 98–113, 165, 167, 277–8
Chuuk (Federated States of Micronesia) 24
civil society 209, 216
class 51, 65, 125
clothing 42, 45–6, 80, 86–8, 99, 103, 109, 152, 175
colonialism 41, 61, 72–3, 114, 116, 134, 165
commerce 45–6, 49, 54, 64–6, 139, 151–3
commodification 210, 223
commoditisation 26, 27, 59–72, 92, 93–5, 275
community xvii, 18, 19, 31–2, 109–10, 133, 139–41, 152, 154, 157–8, 165, 167, 169, 174–7, 179, 195, 198–204,

209–11, 214–18, 220, 222–5, 228–31,
234, 236, 238–45, 247, 252–3, 256,
266, 270, 277, 279–80
conflict 13–14, 15–16, 18, 165, 167–71,
173, 175–6, 191, 200–3, 209–11,
228–31, 236, 257; *see also* incomes
conservation 1, 20, 21, 140, 179–83, 185,
193–6, 198, 200–1, 209, 211, 214,
217–8, 222–3, 228, 231, 236, 238–9,
242, 244–6, 250, 252–3, 256–9, 261–3,
265, 267, 270, 277
consumption 152, 185, 227
Cook Islands 4, 10, 24, 132
corruption 65, 156–7
crime 22, 62, 65, 216
Cuc Phuong National Park 179–97
culture 10, 11, 22, 25–31, 77–95, 103,
115, 140–1, 152–4, 168, 171, 176, 182,
194, 200, 215–18, 220, 222–5, 227–9,
231, 234, 236, 239, 244, 246, 249–51,
256–7, 266; *see also* tourism, cultural
251

dance 30, 62–3, 88, 139, 222
decentralization 79, 229
deforestation 191, 200
development 5, 7, 17–18, 31–2, 51, 65,
79–80, 100–5, 139–42, 145, 164, 167,
169, 198–9, 201–2, 204, 209, 214–18,
220–1, 223–4, 228–9, 234, 239, 244–5,
253, 274–8
disasters *see* natural hazards
diversification 139, 145, 153, 155, 159–61,
276
drought 164–5

Easter Island *see* Rapanui
East Timor (Timor Leste) 4, 165
ecology 179, 257
ecotourism xvii, 1, 2, 4, 7, 9, 17–18, 20,
60, 108, 198–9, 201, 214–16, 218,
220–4, 227–31, 234, 236, 238–40,
243–7, 249–50, 252–3, 256, 258–9,
276–7, 279; community-based
ecotourism (CBET) 214, 219, 229, 234
education 5, 8, 29, 30, 54–5, 87, 101–2,
104, 112, 151, 164, 168, 171–2, 174,
176, 187, 198, 226, 228, 245, 250, 252,
265, 267, 270
egalitarianism 55, 158
electricity 53, 106, 192, 240
employment 4–6, 7, 9–10, 12, 47, 50, 55,
102–3, 105, 117, 139, 150, 151–2, 156,
157, 159, 164, 169–72, 174–7, 180,

187, 189–92, 195, 205–6, 208, 214,
216, 232, 236, 241–2, 252; informal
sector 10–12, 151, 159, 168, 170,
173–4, 176, 211
empowerment 6, 10–12, 86–9, 110, 124,
231, 275, 279
energy 132, 142
entrepreneurs 8, 13, 14–15, 20, 28, 77,
82–6, 92, 168, 173, 230, 232–3, 237,
244, 276
environmental degradation 2–3, 20–2,
43, 53, 108, 132, 142, 165, 179–80,
182–3, 185–6, 191–2, 195–6, 201, 210,
215–17, 220, 222, 228, 242, 259, 262,
279
equity 9, 19, 32, 109–10, 139, 144, 154,
158, 168, 227
ethics 9, 117, 276, 277–8, 279
ethnicity 8, 10, 14, 30–1, 32, 53, 103,
115–7, 165, 168, 170–2, 175, 185, 187,
209, 221–2, 238, 240, 263, 265, 270
Europe 48, 49
exchange 43, 45, 50–1, 63–4
exploitation 6, 8, 10, 11, 65, 237, 239

Fanon, Frantz 127
festivals 48–9
Fiji xvii, 2, 4–6, 7, 8, 9, 10, 16, 17, 18, 19,
21, 22, 24, 25, 27, 32, 88, 114–30, 279
financial crisis, Asian 1, 155
firewalking 18, 19, 27
fishing 6, 7, 21, 45, 50, 54, 139, 164, 175,
240, 242, 247, 252, 260
Flores (Indonesia) 13, 19, 29, 32, 170
folk villages 88
food 23, 49, 53–4, 88, 152, 165, 169, 172,
174–5, 182, 184, 186, 190, 192, 194–5,
201, 204, 206, 211, 220, 233, 236, 247,
260, 266, 270
forest 179–80, 189, 200–3, 205–6, 208,
210, 221–3, 225, 234, 237, 239, 245–6
four-wheel drives 46, 85, 92
France 45, 49, 184
French Polynesia *see* Tahiti

Galapagos 52
gambling 11
garbage 53
gender 10–12, 17, 25, 140, 152–3, 168–9,
171–2
Germany 44, 173, 184
globalisation 2, 23, 25, 31, 112, 256
Goa (India) 132
Goffman, Erving 120, 124, 126

golf courses 14, 22, 53, 202, 204
governance 46, 169, 200–3, 229, 277–8
graffiti 53, 185
Gramsci, Antonio 116
guidebooks 28, 80–2, 91; *see also* Lonely
    Planet
guides 6, 8, 14, 42, 45, 47, 108, 150, 169,
    174–5, 183, 185, 189, 226, 232–3, 236,
    247, 249, 251, 260–1, 270, 275
guesthouses 31, 49, 50, 54, 65, 92, 99, 102,
    109, 131–45, 158, 170–2, 184; *see also*
    home-stays

handicrafts 3, 5, 6, 8, 11, 12, 24–5, 42–4,
    46, 50–1, 54, 64–71, 150, 160, 174,
    176, 186, 194, 223, 225, 232, 236, 247,
    266, 271
Hawai'i 52, 55, 85, 88, 105
health 17, 44, 101, 164, 167, 174, 178,
    186, 277
hedonism 22
heritage 26, 29, 215, 217, 256, 271
Heyerdahl, Thor 45
hilltribes 8, 24, 27 103, 180, 222, 276; *see
    also* Hmong and Muong
Hmong 8
home-stays 7, 11, 12, 13–14, 19, 23, 182,
    194, 223–4, 226, 230, 232–3, 246–7,
    250, 252. 270; *see also* guesthouses
Hong Kong 20, 32
hotels 47–50, 52, 54, 116–17, 123, 131,
    135, 151, 164–5, 167–74, 176, 182,
    184, 260, 270
houses 5, 29, 103, 151
hunting 7, 21, 105, 179–80, 185–6, 260

Iban 27
identity 26–30, 62–4, 66, 68–73, 86–95,
    134, 168
images 12, 25, 26, 27–8, 58–60, 80–2,
    92–3, 104, 112, 118–24, 138, 224
incomes 4–5, 6, 8, 11–12, 17–20, 21, 49,
    50–1, 54, 65, 84, 86, 87, 95, 101, 102,
    109, 139, 144, 151–3, 182, 189–94,
    199, 207, 216, 218, 220, 222–3, 230–3,
    236, 240–1, 244, 247, 249–50, 252,
    275; disputes xvii, 7, 87, 275
indigenous people 6, 8, 41–57, 114–30,
    135–45, 164–5, 168, 180–1, 185,
    221, 236, 238, 240–2, 253, 258, 260,
    265, 270; *see also* hilltribes, Hmong,
    Muong, Tibetans
individualism 66–7
Indonesia 5, 15, 16, 22, 24, 26, 90, 148,

    164–5, 173, 178; *see also* Bali, Flores,
    Komodo, Lombok, Sulawesi, Sumatra
inequality xvii, 12, 18–19, 63, 124–6
informal sector 9, 11, 12, 13, 168; *see also*
    employment
infrastructure 5, 17–18, 21, 79, 109, 195,
    199, 204, 207, 240, 258, 260, 262, 265,
    270
Islam 10, 14, 148, 156

Japan 173, 184
Java 11, 14, 22, 26, 165, 167, 170, 173
John Frum cult 80, 82, 86
Jiuzhaigou 7, 98–113; *see also* China

Kalimantan (Indonesia) 165
kava 82
Koh Samui (Thailand) 10, 12, 13, 14, 15,
    21–2, 31
Komodo (Indonesia) 6–7, 19, 32
Korea 184
Kuta 3–4, 12, 16, 21, 22, 31, 148; *see also*
    Bali

labour *see* employment
LAN-Chile (airline) 48–9, 52–4
land xvii, 5, 14–17, 134–5, 158, 186,
    239, 253; alienation 3, 10, 14, 15–16,
    177; commoditisation 3, 6, 15, 16,
    117; degradation *see* environmental
    degradation; disputes 7, 15–16, 18,
    169; ownership 48, 52, 134–5, 141,
    155, 208
language 6, 7, 8, 9, 11, 14, 23, 46–55, 104,
    125, 164, 172–3, 175–6, 224, 245, 277
Laos 1, 138
leadership 141
livelihood 187, 192, 20–1, 203–6,
    210–1, 236, 257; *see also* sustainable
    livelihoods
loans 153–4
localisation 70–3
logging 180, 200, 222, 239, 250, 256–7
Lombok (Indonesia) 10, 12, 15, 23, 133–4,
    164–178
Lonely Planet 82, 107, 137

MacCannell, Dean 60, 100
Malaysia 8, 10, 13, 19, 23, 165, 236–73;
    see also Sabah, Sarawak
management 173, 183, 198–204, 207,
    210–11, 215, 220–4, 227–31, 234, 245,
    253, 256, 258, 261, 265, 267
manufacturing 159–60

marginalisation 6–7, 59, 61
marketing 247; *see also* advertising
markets 4, 29, 63, 151, 174, 204, 210, 223, 234
marriage 48, 55, 71
mass tourism *see* tourism
media 82, 112, 167, 182, 204, 209–10
Melanesia 14, 58–97
Micronesia, Federated States of 18, 21, 24, 26
middlemen 275, 279; *see also* entrepreneurs
migration 4, 10, 12, 13, 14, 16, 19, 25, 44, 46, 48, 87, 140, 150, 152, 154, 172, 205–6
mining 198
moai 45, 51, 53
modernity 32–3, 64–7, 79, 87, 91–2, 153–4, 161, 206–7
multinational corporations 116, 132, 141, 159
Muong 181, 185–7, 189–94
music 3, 25, 222, 240, 249
Mustang (Nepal) 6–7

National Geographic 80
national parks 1, 6, 21, 98–100, 105–6, 179–80, 185–6, 194–5, 198–204, 207, 209–10, 222–3, 225, 229, 256–60, 262–3, 270, 275, 276
natural hazards 135, 278
nature-based tourism 1, 6, 107–8, 179–97, 199, 214, 218, 225, 236–42
neo-colonialism 59–60, 215
Nepal 6–7, 9, 18–19, 22
New Caledonia 14, 279
New Zealand 134
Niue 274
non-governmental organisations (NGOs) 199, 209, 214, 218, 222, 227, 230, 236–9, 242, 253, 256–7, 259, 271, 278
Northern Marianas 274
Nusa Tenggara Barat (NTB) (Indonesia) 164, 168, 171

oil palm 237, 239–40
orientalism 30, 117

Pacific 1, 11, 19, 30, 52–4
Papeete 44, 49, 52; *see also* Tahiti
Papua New Guinea 4, 7, 12, 26, 28, 29, 30, 58–76, 90, 94, 274
participation 169, 175, 195, 199–200, 215–18, 220, 224, 230–1, 245, 252, 257

Pattaya (Thailand) 22
Penang (Malaysia) 13, 263
Philippines 10, 11, 21, 22, 172
pollution 185, 216
Polynesia 114, 118
population 48–9, 55, 163, 164, 180, 186; decline 44
post-colonialism 59
post-tourism 92
poverty 11, 159, 195, 201, 218, 276, 278
pro-poor tourism *see* tourism
prostitution 11, 14, 33, 44, 167, 172

racism 120, 125
Rapanui 41–55
religion 14, 25, 26, 82, 86–7, 102, 103, 114, 139, 152, 165, 167–8, 171–2, 175, 227; *see also* Islam
remittances 4, 14, 25, 87, 150, 223
resettlement 179, 186–7, 193, 195, 202
resistance 6, 18, 53–4, 71–2, 118–20, 125–6, 127–8
resorts xvii, 2, 5–6, 15, 123, 131, 143, 165, 174, 204, 260
resources 20, 202, 205–6, 209–10, 276
risk 145, 158, 160, 274, 278–9
Routledge, Katharine 44

Sabah (Malaysia) 236
Said, Edward 117
Saipan (Northern Marianas) 10, 274
Samoa 7, 11, 15, 27, 131–47
Sapa 7, 12, 32, 33; *see also* Vietnam
Sarawak (Malaysia) 27, 28, 29, 257
SARS 167, 278
security 164, 169, 202, 261
Sepik *see* Papua New Guinea
service industries 1, 23, 164, 182, 185, 192, 195, 218, 220, 224, 226, 230, 265
sewage 53, 141
sexism 125
sexuality 72
socialism 103
Solomon Islands 4, 8, 10, 15–16, 18, 26, 27
souvenirs 24–5, 42–5, 50, 59, 64–5, 218, 261
Spain 42
status 9, 91, 103, 132, 144
Sulawesi (Indonesia) 10, 14–15, 165, 170
Sumatra (Indonesia) 5, 6–7, 10, 66
surfing 3
sustainability 6, 32, 105–6, 11–2, 135, 143, 164, 167–8, 171, 174, 195–6, 201, 214,

220, 227, 238–9, 242, 250, 253, 256–7, 267, 275–6
sustainable livelihoods 5, 139, 143, 144, 155, 276; *see also* livelihood

Tahiti 14, 28, 52, 55, 118
Tanna *see* Vanuatu
Tapati festival 48–9
technology
terrorism 167
Thailand xvii, 7, 9–10, 11, 12, 13, 14, 20, 21, 22, 23, 24, 25, 138, 198–235, 275, 276, 278
Tibetans 98–111
Toraja (Sulawesi, Indonesia) 14–15, 25–6, 27–8, 29, 30, 73
tourism; adventure 45, 226, 261, 265–7; cultural 46, 52; cycle 3–4, 13, 33, 133, 278–9; domestic 1, 99–100, 137, 143, 260; elite 6, 51, 58–9, 61, 132, 143, 278; mass tourism 3, 33, 54, 98–113, 132–3, 144–5, 215–6, 261, 265; plans 48, 135, 145, 149–50, 165, 169, 193, 195, 217; policy 2, 7, 9, 32, 110–11, 135, 143, 144–5; pro-poor tourism 2, 32, 276, 277; *see also* ecotourism, nature-based tourism
tourist: expenditure 139, 258; gaze 120; *see also* income
trade 41–4
tradition 25–31, 65–6, 77–95, 118, 144, 167, 216, 223, 225, 234, 246; invented 25, 26, 27, 88, 93–5, 216, 234
training 48, 54, 140, 143, 171–3, 175
transmigration 165
transport 11, 12, 23, 42, 43, 45, 48–9, 50–1, 59, 106, 134, 161, 192, 220, 226, 232, 234, 247, 259, 270, 274, 279; *see also* accessibility

travellers *see* backpacking
trekking 222; *see also* backpacking
tsunami 167

Ulithi (Federated States of Micronesia) 21
unemployment *see* employment
United Kingdom 44
United Nations 256
urbanization 98

vandalism 182–3, 185
Vanuatu 5, 7, 8, 15, 17, 18, 19, 29–30, 77–97
Vietnam 1, 2, 7, 12, 16, 20, 22, 32, 33, 179–97
violence 4, 14, 42, 62, 148, 155–6, 161–7, 278–9; *see also* conflict
visitor needs 46; numbers 46, 48–51, 54–5, 61, 79, 99–100, 135, 161, 165, 169, 174, 181, 198–9, 206, 210–11, 216, 224, 230, 246–7, 256, 263
vulnerability *see* risk

water xvii, 5, 22, 30, 53, 87, 132, 142, 185, 190, 208, 216, 240, 242, 253, 260, 276
websites 137
West Papua (Indonesia) 1, 24, 165
wildlife 179, 185, 200–2, 222, 237–42, 244–5, 249, 250, 260, 265–6
women 44, 48, 51, 70, 132, 133, 159, 168, 171–2, 176, 189–90; *see also* gender
World Tourism Organization 260, 263
World Travel and Tourism Council 168

Yap (Federated States of Micronesia) 18, 19
Yasur volcano (Tanna) 80, 89, 91
Yunnan (China) 1, 14, 17, 27, 28, 277